To Jo Booth from Karina

Carol Jahme is a science writer, film maker and broadcaster. She lives in London and is married with two children.

Hope u find this book just as enjoyable to ponder as I did.
April 2005

D0273746

Beauty
and
the Beasts

Woman, Ape and Evolution

CAROLE JAHME

Dedication

To my mother, who gave me the need to understand the complexities of human nature.

To my father, who subtly guided much of my understanding.

A *Virago* Book

Published by Virago Press 2001
First published by Virago Press 2000

Copyright © Carole Jahme 2000

The moral right of the author has been asserted.

A CIP catalogue record for this book
is available from the British Library.

ISBN 1 86049 775 6

Typeset in Minion by M Rules
Printed and bound in Great Britain
by Clays Ltd, St Ives plc

Virago
A Division of
Little, Brown and Company (UK)
Brettenham House
Lancaster Place
London WC2E 7EN

Contents

Acknowledgements

The following experts have, over the last couple of years, been especially helpful by sharing their wisdom and personal stories with me, either directly, face-to-face, or by e-mails, faxes and lengthy phone calls.

Leslie Aiello, Jeanne Altmann, Karl Ammann and Ape Alliance, Pam Asquith, Molly Badham and Twycross Zoo, Robin Baker, Mark Bellis, Ken Binmore, Christophe and Hedwige Boesch, Hilary Box, Sarah (Sally) Boysen, Stephen Brend, Stella Brewer Marsden, British Natural History Museum & Library, Bob Campbell, Mick Carmen and London Zoo's primate keepers, Janis Carter, Julia Casperd, David Chivers, Noam Chomsky, Leda Cosmides, Richard Dawkins, Frans de Waal, Alan Dixon, Robin Dunbar, Nick Elerton, Linda Fedigan, Alison Fletcher, Robert Foley, The Dian Fossey Gorilla Fund, Roger Fouts, Evelyn Fox Keller, Liza Gadsby, Biruté Galdikas, Jane Goodall, Vanne Goodall, The Jane Goodall Institute, Patty Gowaty, Rebecca Ham, Francesca Happé, Donna Haraway, Alexander (Sandy) Harcourt, Kristin Hawkes, Robert Hinde, Mariko Hiraiwa-Hasegawa, Sarah Hrdy, The Human Genome

Project, Bill James, Alison Jolly, Richard Leakey, Mary Leakey, Phyllis Lee, Ashley Leiman, LSB Leakey Foundation, John Manning, Debbie Martyr, Lisa Mather, Tetsuro Matsuzawa, Bill McGrew, Shirley McGrill and The International Primate Protection League, Virginia Morell, Umeyo Mori, Desmond Morris, National Geographic, The Orang-utan Foundation, Craig Packer, Amy Parish, Francine (Penny) Patterson, Steven Pinker, The Primate Society of Great Britain, Ian Redmond, Vernon Reynolds, Matt Ridley, Cyril Rosen, Thelma Rowell, Sue Savage-Rumbaugh, Bill Sellers, Devendra Singh, Vernon Smith, Barbara Smuts, Kelly Stewart, Paul Stewart, Hiroyuki Takasaki, Jo Thompson, Sherwood Washburn, Wenner Gren Foundation, Andy Whiten, Liz Williamson, Margo Wilson, Richard Wrangham, Adrienne Zihlman.

Special thanks also to Peter Blake, Ailsa Berk, Peter Elliott, Dan Westfall, Dan Topolski, and Nick Gordon, who all spoke to me about their particular affinity for primates and primatologists.

Thanks to Heather Osborn and The National Film Theatre. Lino Mannocci and Emma Dooley also offered useful suggestions.

I am also very much indebted to Shaheen Hafeez, who not only gave me a year's free credit as she cared for my baby while I tried to write, she also cooked me and mine hot dinners into her generous bargain. Thank you, Shaheen.

Thanks must also go to my boys Noah and Oscar, who did try to be quiet, and to Nicholas Dunbar, whose support was vital.

My editors made the book a reality. A big thank you to Sally Abbey, Jill Foulston and Andrew Wille, and all at Virago; they worked hard keeping me in line, and thanks to my agent Sara Fisher for always being positive.

Beauty
and
the Beasts

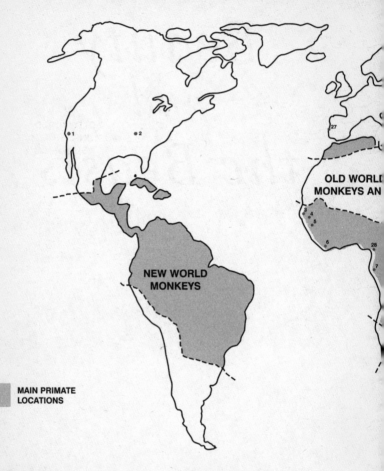

OLD WORL[D]
MONKEYS AN[D]

**NEW WORLD
MONKEYS**

**MAIN PRIMATE
LOCATIONS**

1. Penny Paterson and Koko at Gorilla Foundation in California
2. Sue Savage-Rumbaugh and bonobo Kanzi at Georgia State University
3. Stella Brewer and Janis Carter's chimpanzee rehabilitation project on the Baboon Islands in the River Gambia National Park
4. Mount Assirik, Niololo Koba National Park, Senegal
5. Janis Carter's 'Projet de Conservation des Chimpanzés en Guinée'
6. Tai Forest, chimps
7. La Lopé Reserve, chimps, gorillas and numerous monkey species

8. Wamba, the Japanese bonobo field site
9. Lukuru Wildlife Research Project, site of Jo Thompson's camp, Yas[ ]
10. Gombe Stream Reserve, site of Jane Goodall's camp
11. Mahali Mountains, site of Mariko Hiraiwa-Hasegawa's chimp study
12. Sterkfontein caves where paleontologist Ron Clarke discovered 3.5 million-year-old *Australopithecus* 'Little Foot', oldest human ancestor ever found
13. Madagascar, where lemurs live. They have been studied by Alison Jolly for 35 years
14. Laetoli, site of sixty-nine fossilised hominid footprints dated 3.75 million years old. Discovered by Mary Leakey
15. Olduvai Gorge, where the Leakeys discovered *Homo habilis* and *Australopithecus boisei*

ZOANTHROPOLOGY RESEARCH SITES OF NON-HUMAN PRIMATES

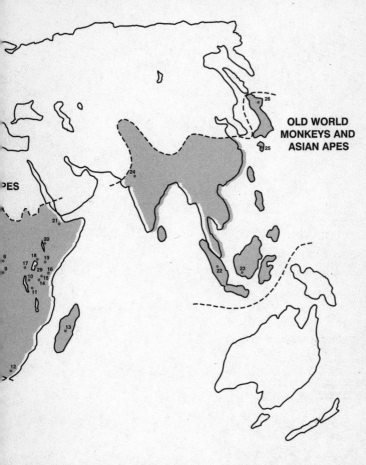

OLD WORLD MONKEYS AND ASIAN APES

PES

6. Amboseli National Park, site of Jeanne Altmann's study of baboons

7. Virunga Conservation Area, site of Dian Fossey's camp, Karisoke, now run by Liz Williamson

8. Queen Elizabeth Park, site of Thelma Rowell's baboon study

9. Gilgil, site of Barbara Smut's baboon study

10. Koobi Fora, site of Richard Leakey's discovery of Turkana Boy, 1.5 million-year-old *Homo erectus*

11. Hadar, site of Don Johanson's discovery of Lucy, 3.2 million-year-old *Australopithecus afarensis*

12. Sumatra, location of Debbie Martyr's search for the bipedal orang-pendek

13. Tanjung Puting National Park, site of Biruté Galdikas's Camp Leakey

24. Mount Abu, site of Sarah Hrdy's research into langurs

25. Koshima Island, site of Umeyo Mori's macaque study

26. Shiga Heights, the most northern location of any wild primate

27. Lapedo Valley, Portugal, site of 24,000-year-old hybrid skeleton of young boy, showing anatomical traits of both Neanderthal and anatomically modern humans which indicates interbreeding of the two subspecies

28. Afi Mountains, Nigeria, site of Liza Gadsby's drill research

29. Rusinga Island, Lake Victoria, site of Mary Leakey's discovery of 17.5 million-year-old *Proconsul*, the oldest fossilised primate skull ever found

1
Danger and Obsession

'I love my work so much I feel like the luckiest person to ever have walked the face of this planet.' Jo Thompson is an ordinary thirty-eight-year-old American woman who feels compelled to live an extraordinary African life. She studies a very rare and only recently discovered ape, the bonobo, sometimes called the pygmy chimpanzee, though in fact they are not chimpanzees. In August 1997, Thompson flew from Ohio to Kinshasa, the capital city of the politically unstable Democratic Republic of the Congo. After hours of permit checking and bureaucracy she met up with white American missionaries, whose jeep collected her and supplies of food and medicine and then drove to the one-horse town of Bandundu, where she boarded a small light aircraft. From her aeroplane seat Thompson looked down at the vast canopy of tropical foliage and the winding Sankuru river below. The forests of the Congo basin are the least explored in Africa. As the forest thinned out a little patches of open grassland could be seen in between the wooded areas; this is bonobo country.

Jo Thompson was happy to leave her husband behind in

Georgetown, Ohio, and did not expect to see him again for at least a year. Life in the Congo is hard for the indigenous people, many of whom would love to swap roles with Thompson and live in her comfortable home in Georgetown, where clean drinking water pours out of taps, refrigerators keep food fresh and death from infectious diseases has been almost totally eradicated. But Thompson has fallen in love with a group of wild bonobos, and nothing can keep her away from them. The hardship of the primitive lifestyle will not dissuade her from her calling; in fact, Thompson finds the simple and very basic way of life an added attraction. Her overwhelming commitment to the animals is typical of women primatologists.

Thompson works alone, the only Westerner involved with the Lukuru Wildlife Research Project that she established at Yasa. But, she says, 'I have not experienced loneliness yet. I like being by myself. I embrace the local Kindengese people and they have accepted me. They call me "Mama Tofuku", which means, "the bonobos' mother". My marriage and my choice to work completely alone for long periods of time is quite unique compared to other researchers. My husband has never been to Africa and doesn't intend to go, yet he is a critical component in my ability to do this work. He manages all my personal and professional affairs in my long absences. He loves me very much and believes in what I do. He also sees that this vocation is my life.' Thompson adds that she has decided not to have children as there are too many people in the world.

Jo Thompson's love of apes can be traced back to when she was five. She requested that her parents take her and her friends to the Ohio Zoo for her birthday party, and she still has a treasured photograph of herself taken that day, standing outside the chimpanzee cage. Today the Ohio Zoo has the finest exhibit and collection of captive bonobos anywhere in the world and the zoo's staff are committed to the species. Thompson has retained a close relationship with the zoo and today it helps to fund her research.

Working with primates can be hazardous and field work in itself offers manifold dangers. Primatologists have caught malaria carried by mosquitoes, been killed and eaten by lions, chased by elephants, and gored by buffalo. In addition, women researchers have been raped not only by local tribesmen but also by orang-utans.

Yet many women are reluctant to leave their study sites and are able to endure much hardship in exchange for the autonomy of the life of a field researcher. Canadian Biruté Galdikas, who famously studies the only Asian great ape, the orang-utan, has certainly experienced more than her fair share of physical hardship. Much of her study area in Borneo is swamp, and daily Galdikas would wade, up to her neck, through cold, black water infested with blood-sucking leeches. She contracted a fungal infection because of being wet so much of the time, and she also permanently damaged her neck as she strained her head backwards to look up into the forest canopy for elusive, solitary, tree-living orang-utans. One primatologist told me that men belong in cities and women belong in jungles. Most of the long-term field studies have, in fact, been undertaken by women who form tangible bonds with their primates. By comparison, men, generally, are suspiciously hasty to finish their theses and return to the safe, man-made environments of university tenure. Men do not seem to develop the same emotional attachment to the individual animals they study and it is easier for them to leave particular animals or species behind.

A number of different women scientists have told me that studying the social behaviour of primates is like watching a TV soap opera. They become emotionally involved with the animals' daily lives and fascinated by the animals' families – the quarrels, the reconciliations, the shifts in individual status, the births and the deaths – though because evolution writes this timeless script happy endings are not obligatory.

Within the mammalian order of primates there are just over three hundred different species and one hundred and thirty of those species are critically endangered. Primates are separated into two suborders, the *prosimii* and the *anthropoidea*. The *prosimii* include the lower primates such as the lemurs, lorises, aye-aye, angwantibos, pottos, galagos and the tarsier. The *anthropoidea* contain the higher primates, the monkeys, apes, and humans. Non-human primates inhabit tropical and sub-tropical climates. Monkeys are geographically classified; species found living in Africa and Asia are known as Old World monkeys, including macaque, baboon, colobus and vervet species and monkeys living in South and Central America are known as New World monkeys, including howlers, capuchins, marmosets and tamarins. There are no native contemporary species of primate in Australia, North America or Europe.

There are three sub-groups within the *hominoidea* super-family: the lesser (smaller) apes, the gibbons and the siamangs; the great apes, the chimpanzee, bonobo, gorilla and orang-utan; and the hominid family in which humans are the only living member, all other hominids being now extinct. At a glance one can see apes are bigger than monkeys, and unlike monkeys apes have no tail. There are nine species of lesser apes, found in India, through South-East Asia and down as far as the Indonesian island of Java. The gibbons and siamangs have hairless bottoms and walk bipedally if they come down from the trees. There are two distinct species of gorilla that divide into five sub-species, they are isolated from each other and all reside in Central West Africa. The four sub-species of chimpanzee are geographically more widely spread than gorillas. The chimpanzee sub-species are the pale-faced chimpanzee of West Africa, the black-faced chimpanzee of Central Africa, the newly recognised *Pan troglodytes vellerosus* from Nigeria and the smaller, long-haired chimpanzee of East Africa. As far as we know there are no sub-species of bonobo and the species is found in only one isolated area of Central Africa. There are two distinct species of orang-utan, Asia's only great ape, the Sumatran and the Bornean.

Genetic information is suggesting the two may be as different from each other as chimps are from gorillas. There are three sub-species of the Bornean orang-utan. The Sumatran orang-utan is slightly smaller and has longer and redder hair than its Bornean counterpart.

Primates inhabit varied environments, from the cold, lush, dense forests of the mountain gorilla to the arid, open spaces inhabited by the savannah baboon. They have many behavioural and physical attributes in common, such as a basic dental pattern, but there is also much diversity in morphology and ethology. One way to unite all primates is to acknowledge the evolutionary fact that all living primates originate from a common ancestor.

Sixty-five million years ago, during the Palaeocene epoch of the Tertiary period, a small, probably terrestrial, shrew-like insectivore lived in the forests of the Atlas Mountains of Morocco. Discovered in 1990, *Altiatlasius* was an adaptable survivor of the Cretaceous climatic catastrophe that claimed the dinosaurs and many other mammals. Ancestral placental mammals had to compete for food with other insectivores, such as reptiles and birds. As these species were predominately diurnal it is likely our first ancestors were nocturnal. As the first primates were insectivores and nocturnal, smell would have been a more important sense than vision. These creatures were probably polygynous (single dominant male controls fertile females); foraging for food on the forest floor would have been a solitary affair; infants were probably hidden in a nest and not permanently carried. These were not gregarious, social creatures, as we think of primates today.

Today's prosimians are a living link to this earliest primate. *Altiatlasius* probably looked a little like the mouse lemur, *Microcebus*, of Madagascar. Some 35 million years ago, during the Oligocene epoch, most diurnal prosimians were replaced by the fast developing higher primate species. Only a few nocturnal prosimians have managed to hang on to their ecological niche in isolated areas of Africa and Asia. But prosimians do dominate the island of Madagascar. The Malagasy prosimians were saved from extinction or radical adaptation when continental drift caused the 120-million-year-old Mozambique Channel. Allowed to evolve without much competition from other more aggressive primates, the Malagasy lemurs have not undergone the same ecological pressures that forced mainland African primates to adapt and change.

The new, more intelligent monkeys were mostly diurnal, arboreal and frugivorous. The change in diet from insects to fruit meant that monkeys were no longer solitary hunters but instead highly social and hierarchical animals that lived in trees. To help them survive predation from birds they grew in size and became organised as a cohesive group. Apes started to develop some 25 million years ago during the Miocene epoch. Fifteen million years ago there was an abundance of ape species and monkeys were in a minority. The origin of hominids can be traced to this time.

Officially, we are the only bipedal ape; bipedalism evolved in our ancestors between 10 and 5 million years ago. Two million years ago there were a number of bipedal hominid species; some were robust and some were gracile, some belonged to the genus *Australopithecus* and some to the genus *Homo*. Classification of hominid fossils often causes controversy amongst palaeontologists. From the fossil record the earliest human relative is *Ramapithecus* (an ancestor of the orang-utan). From the discovery of 8- to 14-million-year-old fossilised fragments of jaw-bone, it is evident that *Ramapithecus*'s diet consisted of tough food morsels, probably gathered by hand at ground level. This was the period when some primates left the trees, never to return. During the Pliocene, about 5 million to 1.5 million years ago, ape species were declining in numbers, and species of Old World monkeys increased. The Pleistocene, about 1.6 million to 10,000 years ago, witnessed the development of even more monkey species, the extinction of all other hominids and the arrival of anatomically modern humans.

These early people migrated out of Africa and managed to successfully colonise the world.

Many of the physical (morphological) and behavioural (ethological) aspects that we take for granted in primates evolved from a life in the trees. Claws are not needed to swing through the trees, so they were replaced by nails (intermediate claw-like structures are still found amongst some lower primate species such as marmosets, the lorises). To assist swinging through trees the clavicle bone developed to support the muscles in the arm; a primate's shoulder socket is designed to allow swinging by one arm at a time. Large, forward-looking eyes are an adaptation of arboreality: the skull developed a bony socket to protect the improved eye, and good vision allows for agility along branches. To aid this coordinated dexterity an opposable thumb evolved, allowing primates to hold on. The development of opposability allowed primates to pick up objects for study and for defence, and to carry objects over a distance and make and use primitive tools.

Primates vary immensely in size, from the mouse lemur, which weighs as little as 60 g (2 oz), to the gorilla, exceeding 180 kg (400 lb) in the wild. Gestation periods also vary from 2 months in the mouse lemur to approximately 9 months in the gorilla. Longevity is difficult to estimate in wild animals owing to predation and disease, and captivity does not suit primates as they are sensitive to human disease and physically designed to range widely; however, some captive apes and monkeys have lived more than 45 years.

A large number of primate species exhibit sexual dimorphism. The males are usually physically larger than the females. Humans are sexually dimorphic, as are gorillas: the male is almost twice as big as the female. Styles of locomotion vary in primates. The brachiation (alternating arm swing) from branch to branch of gibbons is different from the tarsiers' leaping and clinging to tree trunks, which is different again from the terrestrial bipedalism that humans have adapted to. Primates are behaviourally and physically flexible and can use a mixture of the above, but for the most part they are dominated by quadrupedal (four-legged) locomotion. Most monkeys walk on the flat of their hands. Orang-utans walk on the side of their hands, and gorillas and chimps 'knuckle-walk', using the fists or knuckles. No primate is capable of flight or burrowing and few primates like water.

A much larger brain was needed to accommodate the new social lifestyle, improved vision, motor skills and primitive technology. Proportionally, higher primates have the largest brains to body size of all mammals. The optic nerve from each eye runs to both halves of the brain, resulting in a superimposed image that allows depth perception. The sense of smell became less important to higher primates as their eyesight adapted. The emphasis on vision led to the growth of the brain's occipital lobe and the olfactory (smell) area of the brain's decrease in size. These new gregarious, coordinated animals with good, binocular vision and large brains were a highly successful adaptation.

Most primates are herbivorous (plant-eating) or frugivorous (fruit-eating). Some have specialised in a few plant species and other primates dine on a wide selection. Opportunistically primates will feed on insects, eggs, birds, reptiles and small mammals. Chimpanzees and baboons have been observed to hunt, kill and eat larger mammals, such as monkeys.

The sex life of primates affects much of their social behaviour. Primates (except for lemurs and lorises) do not have a breeding season, but instead the individual female has a regular menstrual cycle and a sex drive which peaks just before ovulation. Because female primates are frequently sexually receptive, male primates have to be constantly present and fertile females are rarely found alone. Male primates are potentially sexually active at all times. To be assured of reproductive success a monkey or ape mother can spend years caring for one infant (possibly twins) at a time. The male primate has a different strategy for reproductive success: he will attempt to mate with as many fertile females as possible.

Today all primates are social animals. Even the loris, which leads a predominantly solitary life, is social when given the opportunity. Generally, nocturnal primates are less gregarious than diurnal ones. All higher primates (monkeys, apes, and humans) are diurnal except for the owl monkey from South America. Social units of the higher, more gregarious primates vary in size, composition, and organisation, starting with the family unit consisting of an adult pair and their immature offspring, such as among gibbons, ranging up to large multimale aggregations numbering in the hundreds, such as among baboons.

Different communities within one species of primate will have their own culture and their own unique way of problem-solving. Separate communities within a species of primates, such as chimpanzees, will exhibit unique cultural behaviours akin to the cultural differences found between groups of people. For example, chimpanzees in East Africa have differences in patterns of grooming, opening and eating fruit and making and using tools from West African chimpanzees. Within a social group of primates there will be much variation of behaviour, with the youngest having a different strategy from older, high-ranking individuals. Throughout an individual primate's lifetime she will adapt to the changes in her social rank and exhibit various strategies. Much primate behaviour is learned and is acquired by the individual with age. As primates are long-lived animals they have extended, playful childhoods, the time when crucial information on survival within their group and within their environment is gathered. A mother teaches her infant most of what it needs to know and later peer-group pressure will dictate various behavioural strategies. Primatology has shown us that relatedness, or kin-bonding, is a very important part of primate life. Primates remain physically and emotionally close to their mothers for years.

The sophisticated ability of primates to learn, adapt and solve problems is an obvious indication of their high intelligence. Primates are highly vocal animals with calls that signify certain things. South American howler monkeys are the loudest animals on earth. When spoken language evolved in hominids is contentiously debated. Communication between non-human primates consists of a complex exchange of vocalisations, physical gestures and scent marking. Compared to other mammals, primates have a greater communicative repertoire. To express affection they hug, kiss, groom, mount, and lip-smack; if aggressive, they flash their eyes, yawn, hold eye contact, display and charge. Dominance and reciprocity have a place in primate society. For example, a high-status individual will be groomed by a low-status individual in exchange for protection from an aggressor. But the time invested in the reciprocity may not be even; two weeks of grooming may equal only two minutes of protection.

In terms of genetic similarities and biochemical similarities based on blood research, chimpanzees and bonobos are man's closest living relatives. Chimpanzees and bonobos are actually closer to humans than they are to gorillas, although all four ape species are within 1 per cent of each other. Since they share almost 99 per cent of their DNA with us, blood transfusions between chimpanzees, bonobos and humans are possible.

Primatology is the study of primates. Primatologists come from the worlds of biology, zoology, anthropology, sociology and psychology. A primatologist may work in a laboratory and design specific man-made experiments to test the intelligence of captive primates, or may choose field work and study the natural behaviour of wild primates. Over the last thirty years a great number of long-term field studies on wild primates have been set up. These long-term observational projects have revealed fascinating data on the importance of long-term social relationships between individual animals.

Observation has shown that kin support each other in fights and share food.

Many primatologists today specialise in behavioural ecology, the study of wild primates in their natural habitat, a focus necessary for the conservation of a species. The ecologist will determine whether the species is diurnal or nocturnal, locate the food and water resources, quantity and quality of foods eaten. The scientist will ascertain numbers of individuals and species of predators, location of sleeping sites and interactions with other species including humans. Other primatologists might come from the school of sociobiology, concerning themselves with the genetically programmed behaviour of individual animals. Sociobiologists study either captive or wild populations of primates, and their initial assumption is that all behaviour is inherited. This is to say that all behaviour has evolved and adapted to help increase reproductive success. Sociobiologists argue passionately about how much behaviour is genetically innate and how much is learned and imitated. They also tend to assume that the primates they study are well adapted to their environment; but they may not be. The schools of evolutionary psychology and evolutionary biology have been born out of sociobiology.

These two sub-disciplines examine the development of behavioural and physical adaptations. Charles Darwin's books – *On the Origin of Species by Means of Natural Selection* and *The Descent of Man* – explain his theories of natural and sexual selection, and his theories form the theoretical basis a primatologist uses to understand evolution. The theory of natural selection states that certain advantageous traits in a species have come to dominate the species because the descendants of the first individual to exhibit that trait had a better chance of survival. Some traits are spread widely through a population because they are successful and have been selected for through countless generations. The earliest ancestor of the cheetah that could run as fast as a modern-day cheetah had a better chance of escaping predators and catching prey. Therefore it survived and its gene for speed was passed on to its young, who in turn passed on the gene for speed to their young; the cheetahs that could not run fast died out. Eventually, the most advantageous genes dominate a species. Genes are often linked to each other; it is possible that the gene for speed is also linked to the cheetah's gene for camouflage and its spotted coat evolved at the same time. If we look at primates, we say that our binocular vision and our opposable thumb have been naturally selected. Darwin believed that natural selection was a random process but that sexual selection is motivated by the subconscious desires of individual creatures.

Darwin's theory of sexual selection has gained momentum in primatology, evolutionary biology and psychology (and the study of animal behaviour in general) since the 1970s. Science and culture (particularly Freudian interpretations of sex) resisted Darwin's theories for years. The theory states that the sexes have two opposing strategies towards reproduction. Males actively and sometimes aggressively court females and females selectively choose males. Males look for certain traits in females, such as intelligence, high status and maternal abilities, while females look for traits such as intelligence, strength, stamina and high status in males. When individuals search out genes they consider good, they pass those good genes *and* the desire for those good genes on to their young. Over time, through natural selection these genes start to dominate. But they can also mutate, drift and link with other genes over generations and so the quest for a good gene is not failsafe.

In 'runaway' sexual selection, the desire for a trait has meant that the trait has grown larger, brighter, heavier or even louder and is now costing the animal much personal energy to sustain it. Runaway sexual selection explains the peacock's tail. A large, brightly coloured peacock's tail is calorifically expensive to keep in good working order; its existence tells the hen that the male is healthy and a good mating partner, so these males have more breeding opportunities than males with smaller tail feathers. But a large, cumbersome tail makes the bird more vulnerable to predatory attack. If the peacock's tail got much bigger, it could hinder the bird's basic survival rate and sexual selection would fall prey to natural selection. Sexual selection affects the social, behavioural, and cultural life of a species as well as its physical appearance. For years primatologists underestimated how courtship, sexual competition and mate choice profoundly affects the evolution of primates, including humans. The scientist uses non-human primates as models from which to extrapolate much about the evolution of human behaviour and the methods employed to study non-human primates will also be used to study humans.

Before field primatologists can start their research they must habituate a group of primates, which means locating and daily following a group of wild animals that are not used to humans. Over time the animals become accustomed to the scientist's presence. If a primatologist visits an established field site, the resident primates will be used to human observers and the scientist can get straight to work; if not, habituation may take months. Once habituation is achieved the observer must be able to recognise individual animals, which again takes time. Scientists carry photographs of the animals and sketches of individual's noses to help identify which primate they are watching.

Field primatologists have three main ways of collecting data. They use instantaneous sampling when studying certain behavioural patterns in a group of primates. In this method the scientist will have a strict time-scale for observing the members of a group. For example, if the scientist wants to test how many naps are taken and for how long, during the day the scientist will perhaps observe each group member for one second every twenty minutes over a ten-hour period for three months. At the end of that period the scientist will have statistical data to show, for example, that infants and high-status individuals have had more and longer daytime naps than any other group member. Instantaneous sampling also allows a field primatologist to observe a number of different behaviours at one time.

A scientist can also use the method of 'focal animal sampling'. During a focal animal sample the scientist may decide to follow one or two particular animals for perhaps three hours or even the whole day, for a finite period such as a month. During the day the scientist will note down exactly what the animal does and in what order. Behavioural patterns soon emerge. The problem with this method is that some individual animals are more tolerant of follows than others. Today, *ad libitum* sampling is a method employed if the terrain is hard or the behaviour rare. Sometimes a constant follow is impossible and strict timings pointless. In the early days of field work, ad lib sampling was the most common method used. It allows scientists to jot down anything they are lucky enough to witness, but it does not allow scientific comparisons between groups of primates to be made. It can be used to supplement other methods of data collecting. For instance, while using an instantaneous sampling method, one might be surprised to see three low-status primates ally themselves against a high-status primate and take its food away. This unusual and significant behaviour could be sampled *ad libitum*.

It is possible to use experimentation in the field. A stuffed leopard was once placed near a group of chimps and the chimps' aggressive behaviour towards the predatory leopard was filmed and analysed. Primate calls have been recorded and played back to the same group of primates and their behaviour observed; again this experiment was filmed. Food resources have been reduced or highly desired foods introduced to test the behaviour of dominant individuals. More recently, DNA taken from hair follicles and from blood samples has been tested to show paternity. Blood samples are also used to test hormone levels and for parasitic infection. Plastic sheets are placed under night nests to catch the primates' first urine sample of the day, so that the scientist can evaluate oestrogen and testosterone levels. Once data have been collected and hypotheses tested, the scientists can begin to theorise over their findings.

In Britain, the main centres for primate study are St Andrew's, Stirling, Cambridge, Liverpool and London Universities. In America, the main centres are Wisconsin, Michigan, Oklahoma, and Atlanta Universities, UC Davis, UC Santa Cruz, and the Yerkes Primate Research Center. Pisa University in Italy, the Max Planck Institute in Germany, and Senshu & Kyoto Universities in Japan also offer courses linked to primatology.

Primatology is the only area of science where women dominate men in sheer numbers. Women have spent years studying these extraordinary and frequently long-lived animals. Women run the primate conservation charities around the world and women undertake most of the scientific primate field research. Sixty-two per cent of members of the World Directory of Primatologists are women and 90 per cent of primate sanctuaries are run by women. At a recent meeting of the PSGB (Primate Society of Great Britain), of seventy-two members present, forty-five were women. Of five lectures given that day, only one was given by a man. It is not just the high-profile apes and monkeys that receive female attention; prosimians are just as popular. Alison Jolly is the world expert on the lemur. She first went to Madagascar in 1962 and to this day she remains committed to studying and conserving the matriarchal lemur.

A few women primatologists who are exponents of 1970s Californian feminism are very much opposed to women's dominance being acknowledged. They fear that if their science becomes known as a 'female vocation' their work will be diminished within the world of science, which is still male dominated and inherently chauvinistic. Some say that primatology is already regarded as a soft option simply because of the numbers of women the subject attracts.

But a young woman entering the 21st century who loves the great outdoors, wants to study apes, monkeys or prosimians and cares about the conservation of the animals does not see her vocation as a soft option. The fact that most of her fellow researchers are also women is not something to be ashamed of. One female Ph.D. student at Liverpool University told me that there was only one man in her tutorial group of ten. He was considered an honorary woman because he wanted to study the traditionally female domain of primate mother and infant bonds, instead of a typically male area of research such as the animal's diet or the male's territorial range. In thirty years sexism has gone full circle.

As the numbers of women attending university primate courses increase and more women volunteer for primate conservation, a critical mass of women is growing, developing a gravitational pull, sucking in more and more women to the world of primate study. On average eight out of ten Ph.D. students in primatology are women and all places are vastly oversubscribed. The women offer each other support as role models and motivation as competitors. There is a great deal of competitiveness and professional rivalry between the intellectual camps. Jealousies between female primatologists are common. The belief that you observed certain behaviours first but that your discovery has been ignored in favour of another reseacher's work is a common gripe.

Primatology is one of the few areas where women have become professional scientists purely through their amateur enthusiasm without having to go to university. A male equivalent of this is astronomy. Eager non-professional astronomers can find permanent places in the history books, such as the amateurs Hale and Bopp, who recently identified a new comet.

As the years pass by, women have remained bonded and committed to their animals, almost to the point of obsession. Could it be that the study of primates allows researchers to revisit their own lost innocence? The women can revert to a time in their own early development before they learned to talk, when they were just learning to walk. Primates are so like us and feel so many of the same emotions as us, and they have a dignity that can elude human beings. Could it be that women are drawn to a world of basic simplicity that is now lost to them? Is it, perhaps, a search for *complicité*, as the French say, that motivates these women? Finding *complicité* with another being allows one to communicate telepathically; language is superfluous to this type of special friendship. It is an elusive form of love, truth and intimate understanding that only mothers and babies and the very best of friends can enjoy. Humans who share *complicité* need only give a look or a nod for their partner to understand their meaning. Relationships between apes and women work on this level too.

When women choose to study and communicate with non-human primates they are effectively choosing to return to a time before language evolved. Sixty-five per cent of human language is non-verbal. Most of the time we instinctively communicate with one another by means other than the spoken word, using eye contact, facial expression and gesture. Much non-verbal instinctual communication used by humans is also used by non-human primates. All mothers go back to a time before language when they have their first baby, drawing on their innate and sophisticated emotional intelligence to 'read' their baby's desire. This is necessary because for the first two years of that child's life it will be unable to express itself coherently. Women are at ease 'reading' these other types of communicative signals, whether they come from babies or adults or other primates. Because non-human primates are very similar to us, women intuitively understand much of what the animals are feeling and are contented to watch over them for years.

The International Primate Protection League, IPPL, is a charity with a global network of offices. It was established in 1973 by Shirley McGreal. While living in Thailand, McGreal saw the pitiful sight of baby monkeys and gibbons piled high in tiny crates on the baking hot tarmac of Bangkok airport, awaiting shipment to American laboratories. McGreal then visited the US Army Gibbon Laboratory in Bangkok, which was rumoured to use gibbons in biological war games; she became convinced that an international primate welfare organisation was necessary to protect the lives of apes, monkeys and prosimians.

Although there is still a thriving black market in apes for private collections the IPPL's record in wildlife-smuggling investigations and convictions of animal dealers is unsurpassed. Of the IPPL's 12,000 members, 85 per cent are women. Throughout the world, it has women working undercover, spying on pharmaceutical companies or local ape traders to help bring about convictions of cruelty. Often these women knowingly put their lives at risk, such as Heather McGiffin and Rosalind Alp, who in 1989 worked for the IPPL in Sierra Leone, providing information on poachers for the international ape trade. A number of convictions were made after their covert operation.

Daily the IPPL's office in London receives letters from all types and ages of women, begging to undertake field research and do their part to save primates from extinction. This is a typical letter:

> It is my life's goal to be able to work with monkeys or apes on a reservation anywhere in the world. I realise this may be an impossible dream but none the less I will pursue it. I was hoping you could put me in touch with reservations that care for orang-utans or chimpanzees. I know I could achieve something decent with my life if given the chance.

These women range from housewives to barristers, from students to retired women; many have families and mortgages that they are more than willing to leave behind. But in their idealism to become a field researcher they are often disappointed. Funds are always low and only those who can demonstrate an established interest in primates, such as a relevant degree or years of voluntary work, are in with a chance. For all the women undertaking primate field studies, there are many hundreds more who can only ever dream of such an opportunity.

Today the IPPL has field representatives all over the world. Iqbal Malik is their contact in India, and is the first woman to have received a doctorate in the behaviour of wild monkeys in India. Whenever there is a confrontation between people and monkeys, Malik is called. She sometimes translocates monkeys to new, safe areas if it is necessary. Malik established her monkey charity, 'Vatavaran' to help her pay the costs of primate conservation. She has also organised educational programmes to show people how to co-exist with monkeys, and is now known as 'The Monkey Mother of India'.

Young women may feel innocently drawn to primates through their concern for conservation and their love of animals. They may then take up a scientific career and undertake new research on wild primates. But after years of field research and making evolutionary discoveries these women may find themselves pulled back in the direction of their first love, the animals themselves and the conservation of the species. Working in the field many women are confronted by poachers and illegal loggers, and the individual animals they study die one by one. Faced with this, their scientific research has to take second place to conservation. As orang-utan expert Biruté Galdikas says, 'What is the point in studying the behaviour of animals that are about to become extinct? One's first duty is to protect them.'

Primatology became a recognised area of science in the 1960s, but decades before, women amateurs were laying down its foundations. During the Victorian and Edwardian eras, long before British women had the right to vote, intrepid women pioneers would set off with missionary zeal to explore the British Empire, and this period of history marks the beginning of women's obsession with our primate cousins. Watercolour paintings depicting exotic animals, thrilling adventure stories, dead and live primates would all be brought back from African jungles to entertain and educate the public. Ordinary women became intrigued by these human-like creatures.

Alyse Cunningham was a childless British woman who, in 1918, found herself a maternal role caring for John Daniel, an infant lowland gorilla. Cunningham bought John Daniel when he was two years old from the London store Derry and Toms for the enormous sum of £300. Cunningham lived in London's still fashionable Sloane Street, and John Daniel was given a bedroom of his own, with an electric radiator for heating. He was dressed like a child and learned to use light switches, sinks and toilets correctly and to sit politely at the table to eat. He became very fond of Cunningham's three-year-old niece and the two of them would play harmoniously together. It is thought that on average gorillas have a mental age equivalent to that of a three- to four-year-old child, so when young, gorillas and children are mentally on an equal footing. Civilised tea parties would take place where John Daniel and a small group of specially invited children would happily sit together, napkins on their laps, eating high tea. Later they would all play together. Many years later, as adults, when the former playmates of John Daniel were interviewed, they all said they'd had great fun. Playing with a gorilla of their own age had seemed the most natural thing in the world. As adults they felt privileged to look back on some extraordinary and precious memories. John Daniel was also exhibited for a few hours a day at London Zoo, and by 1920 he was a very famous ape. He was an intelligent and sensitive animal, who became devoted to his adoptive family and hated to be alone. He would happily sit at their feet and if he were ever told off he would cry pathetically and hold on to the ankles of the person scolding him.

By the time John Daniel reached five years of age, Alyse Cunningham was finding him too difficult to keep. In just three years he had blossomed from a 32-lb. infant to a 110-lb. juvenile who stood 3 feet 4 inches tall. The tea parties had to be stopped and a solution had to be found. Depressed, Cunningham felt compelled to sell him, and the New York Zoo bought him. If he had gone to London Zoo Alyse could have visited him regularly and the trauma of a new environment would have been eased, but John Daniel was packed into a small crate and dispatched by ship to New York. On arrival he was found to be losing weight and apparently lacking the will to live. Cunningham heard of her gorilla's plight and quickly made plans to sail over and nurse him back to health. Refusing food or water he pined for her and eventually died of a broken heart before Cunningham could reach his side.

Alyse Cunningham became a mother to two more gorillas, but both died in infancy. Today the stuffed and mounted John Daniel can be seen in the primate hall of New York's Natural History Museum, specimen no. 54084.

Since the 1930s Hollywood has exploited the movie-going public's passion for apes, from *King Kong* to *Mighty Joe Young*. Among the long list of ape movies, a recent addition is *Buddy*, 1997 starring Rene Russo, which tells the life story of Gertrude Lintz, another passionate Edwardian woman. Gertrude Lintz's earliest memories are described in her autobiography, *Animals Are My Hobby*, where she talks of visiting London Zoo and crying for the caged animals to be released. Born Gertrude Davies in London in 1889, she was one of five children whose half-French mother died in childbirth when Gertrude was eight. By then a Presbyterian minister in Tennessee, Gertrude's father had a nervous breakdown and the children were all fostered into separate homes.

Luckily for Gertrude, when she was sixteen years old she was taken in by 'Uncle David', a philanthropic New York millionaire who had shown an interest in Gertrude when the family had lived in the city. After a failed singing career, aged eighteen she remembered her family had had a St Bernard when they lived in London, so she decided to breed St Bernards. Uncle David, like a fairy godfather, set her up financially, and she started her breeding farm in New Jersey with two male dogs. Gertrude was a proper young lady of the time and knew nothing of sex. Her mistake was corrected and her farm quickly filled with 250 St Bernards, her dogs winning many prizes. Gertrude married William Lintz, a doctor, and soon afterwards she discovered that her dogs were dead or dying. It is thought a rival dog breeder tormented with professional jealousy poisoned most of Gertrude's dogs, though this was never proven. Distraught, Gertrude had a nervous breakdown.

With William's support Lintz recovered and they moved to a large estate on the Hudson river, where she could continue to indulge her love of animals. Lintz would never become a mother in the real sense, but she did have a menagerie to dote on, including a Komodo dragon, a leopard, guinea pigs and owls. But the pride and joy of Gertrude's collection was Maggie, a baby West African chimpanzee.

There was no restriction on the trade in exotic animals at that time and Gertrude Lintz's collection of apes quickly grew. By the early 1930s she had the largest private collection of primates in the world. While young the animals behaved themselves and were dressed in the finest of children's clothes and bootees. Accompanied by her odd-looking family, Lintz would dance into fashionable New York stores where she would purchase the latest velvet two-piece suits for the infant apes to wear. Lintz may have adopted so many chimps and lowland gorillas because of a frustrated maternal instinct, but she conducted herself like a scientist. She started one of the very first language experiments with her apes; the animals were treated like children and encouraged to 'talk'. She spoke to the press of 'looking for clues to the evolution of man' through her observations of her apes. With the animals in tow – the gorillas, at 300 -lb., were now three-quarters fully grown – Gertrude Lintz visited the 1933 World Fair and toured Canada, where her strange family was mobbed by fans.

Many of the infant apes had arrived on her doorstep in desperate need of medical attention. Entirely self-taught, she nursed the animals back to health. Her two gorillas Massa and Buddy arrived in a terrible state, Massa suffering with pneumonia and Buddy cruelly treated on the voyage from Africa, when he had been sprayed with acid that left him both physically and emotionally scarred. As the apes, especially her beloved gorillas, reached maturity, Lintz found it harder to care for them and keep them under control.

When Massa was fully mature Lintz accidentally dropped a pan of water on him. It startled the animal and in a defensive attack he bit her. Lintz's life was saved by the timely intervention of a friend who hit Massa on the head with the wrought-iron saucepan, which momentarily stunned Massa and gave enough time for Lintz to get out of his cage. She was left with gaping wounds that required nearly 100 stitches. Massa immediately became aware of his physical power over a mere woman and Lintz realised this near fatal shift in status meant the gorillas would have to find new homes.

The year was 1939 and the RKO blockbuster *King Kong* was on general release. Thousands of American people were lined up in queues outside picture palaces to see Fay Wray scream as the beast, Kong, tightened his grip around his love object and gently removed her garments layer by layer in a scene censored and cut from the British version of the movie. The last line of dialogue in *King Kong* is uttered by the cynical movie director, now moved by the beast's sacrifice: 'It was beauty killed the beast.' Thus immortalised on the silver screen, mature silverback lowland gorillas achieved great commercial value overnight. Philadelphia Zoo wanted Massa as a crowd puller and the Ringling Brothers Circus wanted Buddy for the same reasons. Kept in a confined space and on public view these two animals became great stars.

After their departure Gertrude Lintz said of her gorillas, 'If a gorilla has a moral sense what else can I call him but man?' She never forgave herself for selling the animals and in the 1950s found herself mother to another lowland gorilla, which died in infancy. The Lintzes eventually sold their estate and moved to Florida, where they both died in a nursing home in the early 1960s, just as Jane Goodall was being crowned queen of the apes. Buddy remained with the circus and was renamed Gargantua the Great, dying in 1949. Massa lived at the zoo until he was fifty-four, and died in 1984, toothless, almost hairless and alone. His is the longest lifespan for any captive gorilla.

The tale of Gertrude Lintz and her apes is much more than an anecdote about a misguided and eccentric, childless woman. In her own way Lintz showed the long-term commitment, love and passion for primates echoed in every other woman conservationist and scientific researcher mentioned in the subsequent chapters of this book. When a woman commits herself to primate, she does so for life. This bonding process is sometimes described as primate proselytising. Today we know that gorillas shouldn't be imported into the West for a life of captivity. Gorilla families protect their infants from poachers with their lives. For every infant poached its mother, father and other adult relatives die. The trauma of this experience kills most infants within a few days. But Lintz was a product of her age and would not have understood how baby gorillas are captured. In wanting to study the ape's behaviour, Lintz was a woman a head of her time.

Today, thanks to charities like the IPPL, it would be almost impossible for either Alyse Cunningham or Gertrude Lintz to acquire apes in the United Kingdom or the United States with such ease. Women with passions like those of Gertrude and Alice have to go to where the apes live, and many women pack their bags and do just that. Today, when university-trained primatologists plan to visit research centres in Asia, South America and Africa, they are usually required to read documents to prepare them for life at the research centre, then sign on the dotted line in declaration that they have read the warnings; should an accident happen when they are not following the recommendations, it is considered the researcher's fault. Reading field researcher briefs forms a salutary awakening for an inexperienced primatologist. The following extract has been taken from three different sources. St Andrew's University, the Kibale Forest Reserve in Uganda and the Tanjung Puting Park, Borneo all have official guides for field primatologists:

It is quite possible that the subjects you are studying could bite you and they might be rabid. Dying from rabies is very nasty. If you are bitten by a rabid primate or any other animal you must immediately scrub the wound for at least five minutes under running water. You must then thoroughly douse the area with iodine or alcohol and you must quickly get to a doctor for a booster injection. If you have never been immunised against rabies you will need a course of injections. Do not delay. If the bite is bad the incubation period for rabies is one week; if it is more superficial and as far from your brain as possible, i.e. on your ankle, you may have more time. Incubation can take some weeks. If you cannot get to a hospital immediately it is still advisable to have the injections when you can rather than hoping that you've had a lucky escape. Do not have the wound stitched as the suturing of animal bites can increase the possibility of infection spreading. Tetanus is also a hazard. Scrub the wound and apply alcohol as before.

Some hospitals may be as good as any in Europe or America. It pays to know which in advance of need. Because of AIDS accept a blood transfusion or stitches only as a last resort. It is essential to take the University First Aid Course before you go. Water is collected in tanks from the roof. During the dry season this runs out, then we haul it in jerrycans from a small well. Bring water sterilisation tablets with you. Most people who stay for more than a few weeks will suffer gut infection; we have outbreaks of Giardia, Entamoeba histolytica and other unidentified sources of diarrhoea. Personal hygiene is critical. Rubber boots are recommended in swampy regions and also to protect against army ants, which can climb easily over any other type of footwear and will have you stripping off once they start biting. It is also worth ironing all your clothes to kill any eggs that parasitic flies may have laid in the fabric. It is painful and itchy to have parasitic larvae burrowing into your skin. One recommended medical 'bible' is called *Where There Is No Doctor*. Bring your own copy.

Some of the animals co-existing with primates are dangerous. You should give all major predators and very large game, such as elephant, rhino and buffalo, a wide berth. You will be working in areas inhabited by lethal snakes. Snakes don't eat people, like some of the big game can, but you may surprise the snake and provoke it into attacking in self-defence. Most snakes detect footsteps and get out of the way, so clomp about! The most dangerous snakes are those which do not get out of your way such as the African puff adder or in Central America the fer-de-lance. Don't ever go out in sandals; even wearing shorts is a calculated risk as snakes strike low. Monkeys and apes will see predators before you do, including big snakes, so when your primate avoids an area, do likewise. Anti-venom treatment is a complicated affair and can often do more harm than good. Field workers tend to take their chance, however some research centres do have the best treatment available. Find out which ones do and then speed is of the essence.

Even small game and insects, especially bees, are surprisingly dangerous. All African and most Central American bees are so-called 'killer bees'; they are very aggressive if they think their nest is threatened. A female student was stung to death when she accidentally fell into a bees' nest. Spiders and scorpions are also dangerous, and a bite will cause intense and prolonged pain. Check your boots just like they do in the movies! But the biggest danger is that you could become immobilised in the bush with no one back at camp aware of how to find you. A broken leg or sprained ankle could kill you if it means forty-eight hours in dry bush without water. Dehydration is a killer.

It is not abnormal or shameful to have negative feelings about being essentially alone in a strange and at times hostile environment. Even tears are quite permissible, but feelings of malaise can become the excuse for inertia. Living by oneself for a prolonged period may make it difficult to adapt when other researchers arrive. Do not become possessive about the site or unsympathetic to the needs of a newcomer and conversely the newcomer must respect the needs of the resident. There is nothing like getting thoroughly immersed in the research for easing the loneliness and giving oneself a sense of purpose. Indeed, most people discover a positive side to solitude – the peace, the beauty of the surroundings and the primates themselves all contribute to a deep sense of satisfaction. You must remember that you are doing something that will make you very proud. It will change your view of the world. Few are so lucky to have such an experience.

When working with chimpanzees never forget they are potentially very dangerous animals. We must protect the chimps from human disease and humans from chimp aggression. So never approach a chimpanzee closer than 16 feet and never offer food to a chimpanzee. Chimps are much stronger than people and can prey on children; never take small children into the bush to see wild chimpanzees. Remember chimps eat meat. If a chimp looks directly at you, look away: your stare may be threatening. Grunt softly, be self-involved and groom yourself. Avoid sudden movement including arm-pointing; a raised and pointed arm is a threat to a chimp. If a chimpanzee glances at you several times, move away.

If working with orang-utans, menstruating women must take care. Adult and sub-adult male orang-utans have been known to forcibly grab, disrobe and examine menstruating females. If a menstruating woman has a heavy flow it is advisable to stay away from the orang-utans altogether. Orang-utans have a reproductive strategy of rape and have been known to rape women. Orang-utans are more afraid of men than women. If as a woman you are concerned about safety do not work alone.

In 1990 Alison Fletcher was embarking on field research for her

Ph.D. Like the late Dian Fossey, Fletcher wanted to study the mountain gorilla. In Fletcher's case it was specifically the social development of juvenile gorillas that interested her. She planned to work for three years at Karisoke, Dian Fossey's original research site in the Virunga Mountains in Rwanda. Since Fossey's death in 1985 the Dian Fossey Gorilla Fund has run the site.

Normally a group of gorillas would consist of one silverback and one or two blackbacks, sons of the silverback who are maturing and must leave their natal group to start their own family groups elsewhere at a later date. Quite often blackbacks will emigrate and live alone or team up with a brother, male cousin or friend and roam about together for years waiting for their chance to have their own family group. Or, alternatively, if the silverback is old, a blackback may challenge him for control of the group. Maturing young females will also leave their natal groups and transfer to another established group; females do not live alone, but move from one family set-up straight into another family, directly from the role of being a daughter to the one of being a wife. Gorillas are strictly patriarchal.

The mountain gorilla lives in the Virunga Park, which consists of 400 square kilometres of rainforest and straddles the borders of Uganda, Rwanda and the Democratic Republic of the Congo. The small mountain gorilla population of 600 has been living on the Virunga Mountains for the last couple of thousand years. It is suspected that the Virunga volcanoes erupted and separated a small group of lowland gorillas from the rest of their community as they fled east into the mountains. Over the next few generations the gorillas would have adapted to the change in atmosphere, growing thicker hair and their diet changing from mostly fruit to mostly leaves. This initial group of individuals would have bred with each other and their descendants have formed the foundation of the genetic make-up of the mountain gorilla of today.

In 1990 group 5, Alison Fletcher's study group, had four adult males, which meant the group was unstable. In addition poaching and shooting were causing the gorillas anxiety. Fletcher found the nature of field work lonely and exhausting. The gorillas live 10,000 to 14,000 feet up where it is wet and cold and the air is thin. Fletcher had become accustomed to moving on group 5's periphery and things seemed to be going as well as could be expected until one day one of the group's infants became caught up in a poacher's bamboo and wire snare. The infant screamed as it tried to free itself, but only succeeded in tightening the wire around its body. This type of gorilla entanglement in a snare designed to kill a duiker or bushbuck is very common, especially amongst youngsters, and it usually causes the slow death of the gorilla from gangrene. Fletcher's group became very agitated, and the atmosphere quickly changed from the peaceful munching of leaves and general grooming to one of panic and aggression.

In the chaos the infant was hit and pulled by members of the group. At this point Fletcher decided to wade into this potentially very dangerous scene and rescue the infant. She was successful in releasing the animal and the group made a hasty exit. A few days later Fletcher located group 5 and settled down as usual to observe them, but her relationship with them – or, more precisely, with one particular blackback – had changed without her realising. Over time the blackback became increasingly hostile.

On reflection Fletcher thinks the animal must have connected her to the poacher's trap and the ensuing turmoil. Whatever the reason, he was now her enemy. He left the group and sought her out. Sensing his aggression, Fletcher backed away. Menacingly the gorilla followed her for forty minutes at close quarters as Fletcher slid down the side of the mountain desperately trying to distance herself from him. When he was satisfied he'd run her off his group's immediate territory he watched her go, and then made his way back to the peace and quiet of gorilla life.

Although Fletcher was not injured she was shaken by this sinister and uncharacteristic behaviour from a male gorilla, but she remained undeterred. The next time she approached her study group, clipboard and stopwatch in hand, he was ready. He ambushed her before she could back away, grabbed her and bit her arm. A gorilla is a herbivore, but it possesses enormous canine teeth. Fletcher sought medical attention back at camp. At this point she was urged to leave Karisoke, go back to her university and arrange for another Ph.D. But she didn't want to leave. She felt committed to her thesis and to the animals.

Fletcher was bitten twice more on subsequent occasions. The last attack resulted in the gorilla charging her, knocking her down and sitting on top of her for forty-five minutes. Fletcher could not move under his weight and this time he bit through her knee. Finally he let her go; he didn't want to kill her, which he could have done easily, but just wanted her to leave them in peace. As far as the gorilla was concerned, he was defending his group from a potential poacher.

Fletcher needed hospital treatment and this time she was ordered to leave. It was under duress that she left Karisoke. She said to me that throughout the blackback's uncharacteristic vendetta against her she always felt that she, and not the gorilla, was in control of the situation.

Rebecca Ham must be one of the most enthusiastic primatologists I've met. She punctuates her conversation with a generous laugh and adjectives such as 'amazing!'. Ham, Canadian and now in her early thirties, has based herself in Washington and is working for the World Wide Fund For Nature (WWFN) on a project entitled the Biodiversity Support Programme. It has been devised to research how wars affect wildlife, and Ham will shortly begin researching the aftermath of war on the Rwandan ecosystem. She says her mother has been an inspiration to her. 'My mom's amazing. She encourages me to do field work. She's in her sixties now and she's going to China with VSO, she sold the house, all her belongings, I must be like her.' As a child Ham felt inspired by the famous National Geographic primatologists, Jane Goodall, Dian Fossey and Biruté Galdikas. She also felt 'the need to be a good person and to do something decent with my life'.

For her Ph.D. Ham attended Stirling University, where Bill McGrew, a great supporter of women primatologists who has helped a number of women amateurs to become professional, became her supervisor. McGrew has commented, 'When it comes to big projects, particularly involving the return of captive apes and monkeys to the wild, I cannot think of a single man that has played a leading role.'

For her thesis Rebecca studied a monkey called the grey-cheeked mangabey, which lives in the forests of Gabon. In 1991 she flew to the dense La Lopé Reserve, a lush Gabonese jungle, where it took her three years to complete her research. Fellow primatologist Caroline Tutin studied the lowland gorilla in La Lopé and invited Ham to rent a room from her while she accumulated observational data on the mangabey.

This part of the world is rich in primates. Gorillas and chimpanzees live in the forests of the La Lopé Reserve and unique to this jungle is the very rare and little known sun-tailed guenon, a monkey only discovered in 1985.

One day Ham was walking alone, along a sandy bank, upstream into dense tropical forest searching for her mangabeys when she found herself face to face with an injured and thirsty black forest buffalo. A young adult male, its guts were hanging down from a gash in its stomach. The bull had probably been in a territorial fight with another male, and had lost.

A disoriented buffalo is the most dangerous animal in Africa. Though quite often one can smell buffalo before seeing them (they smell similar to the concentrated odour of domestic cattle), unfortunately on this occasion it smelled her before she smelled it. She backed into the river but the bull charged her, its horn piercing her left breast as it lifted her out of the water. Face to face with the bull, putting her hands on its forehead, she managed to push against the animal's head to lift herself off its horn, but the horn exited from another part of her breast.

Ham collapsed on to the ground and the animal charged her again. This time its horn passed through her ankle and it tossed her into the bushes. She kept very still and watched the bull. It was unsure what to do next; eventually it went back to the water's edge and limped off downstream.

Soaking wet Rebecca Ham went into shock through blood loss, but her catatonic state may have helped to prevent her from panicking. If the adrenaline in her system had not subsided and her heartbeat had not come down she would have lost much more blood. She hauled herself on to her feet. It was a 2-kilometre walk back to the village, and her life depended on it. She was unsure exactly what her injuries were, but she realised that if she fainted she might never wake up again. She had to keep moving. Limping and pulling herself along through the trees, Ham finally reached the village and alerted people to her injuries. The attack happened at 2.30 in the afternoon, and she was finally admitted to the National Hospital in Gabon at 7 a.m. the following morning. In the hospital, before her ankle was bandaged, Rebecca remembers seeing a live army ant crawl out of the open wound. She spent ten days in this hospital before she was strong enough to fly home for more hospital treatment.

A hunting party went out in search of the bull, which needed to be shot before it killed someone. They found it near to the spot where it had attacked Ham. It had been killed and partially eaten by a leopard.

Ham told me, 'I suffered with PTS (post-traumatic stress) disorder afterwards, and had nightmares and flashbacks. I was home and in physio from 13 June '93 to August, then I flew back out to finish my research. But when I returned to Africa I was scared to venture into the forest again. A hunter gave me valuable advice; he said, "Let it go." And I did, I pushed it from my mind. I don't relive it. At the time everything went into slow motion, just the way people describe car crashes. Like watching a video, I was watching myself being gored by the bull from the outside. I know I'm lucky to be alive.'

Even if her emotional scars have not healed, Rebecca Ham's physical ones have. 'The scar just above my left nipple has healed well. It looks like a bullet hole and is not nearly as impressive as the story. Five years after the attack, scar tissue from my ankle came to the surface as a hard lump under the skin. One day, when I was at home, it was right under the surface of my skin. I cut into it and sand came out, just like stories of shrapnel moving around the body and emerging again – all this sand came out like a time capsule from that moment when the bull and I met on the banks of the stream. The human body is amazing.'

After completing a two-year chimpanzee census in Guinea in July 1997 Rebecca Ham decided to return to the US to try to integrate herself back into society. 'But I felt like a visitor when I returned. I had a very bad culture shock coupled with my second near-death experience. It was a terrible time.'

Malaria is a standard occupational hazard for primatologists and for anyone living or working in the tropics. Ham had suffered a bout once before and recognised the pattern of symptoms, but at this time she was under a good deal of stress, trying to wind down the project. The fever and headaches grew worse, and she took a single massive dose of her chloroquine tablets, anti-malaria pills that she should have taken regularly but had forgotten.

Rebecca Ham boarded her flight home. She felt hot and sick as the air hostess served lunch. She unwrapped the vacuum-packed food but she couldn't face eating anything. The fever escalated and while she sat watching the in-flight movie she lapsed into unconsciousness. Ham told me, 'At this time I felt very, very bad.' Back in the United Kingdom she made it, somehow, to her university, who sent her straight to hospital. Although medical staff could see she was ill, they doubted her explanation that she had malaria and tested her for other tropical diseases without treating her malaria for thirty-six hours. During that day and a half, lying delirious in a hospital bed, she nearly died. But Ham is lucky; her guardian angel keeps protecting her. The anti-malaria treatment the hospital finally gave her managed to pull her back from the brink.

When Rebecca Ham was recovered she finished her academic business at St Andrew's and left Scotland for Washington, DC. Back in the United States, the culture shock at being home made home anything but sweet. Ham had travelled through Guinea, a very poor Third World country and had been deeply affected by the experience. People's worries and woes in Washington and presidential scandals seemed unreal and unimportant by comparison. In Africa Rebecca Ham had dealt with an epic narrative on a daily basis; life and death were all around. The clinical environment of Washington's middle classes seemed alien to her for a long time.

Many women primatologists find after years in the field that they feel more at home in the jungle than where they originally came from. This seems to be the case for Jane Goodall and Biruté Galdikas and was certain true for Dian Fossey. These three women are the most famous female primatologists and they have all risked danger in pursuit of their obsessions. Goodall is devoted to chimpanzee welfare and conservation, Galdikas to orang-utans and Fossey was utterly devoted to gorilla conservation. In their obsession with their chosen species they have become heroic figures and much of the truth of their stories has been masked by the romantic notion of a life at one with nature. These three women represent the original archetypal primatologists; they form points to a triangle found at the centre of primatology. Goodall, Fossey and Galdikas have unique images which encourage a following from three different types of primatologist and three different types of fan. Jane Goodall's supporters tend to be middle-class women who admire chimpanzees greatly and believe chimps to be the most intelligent and important of apes because of their closeness to man. Dian Fossey's supporters are often women who feel disappointed with mankind in general, who seem to idealise the natural world and relate vividly to Fossey's own isolation and love for gorillas. Biruté Galdikas's fans tend to be spiritual conservationists, often women who want to save rainforests and all the forest's inhabitants, in which orang-utans are seen as a royalty species.

It was the palaeoanthropologist Louis Leakey who guided Goodall, Fossey and Galdikas in their obsessions and through certain dangers. The published images of the vulnerable women 'alone' with their apes was so powerful that it captured the imagination of the general public completely. Today their media-constructed images precede them wherever they go and Goodall and Galdikas have become very good at using the media for conservation purposes. Goodall, in particular, has become a living icon. She is so respected and influential now that heads of state turn and listen attentively when she speaks.

gorilla

2
Leakey's Ladies

If we want to search for clues to the evolution of human behaviour, the study of the great apes and other gregarious wild primates such as baboons is a good place to start. Today evolutionary biologists find comparative studies of behaviour, anatomy and psychology essential to answering such questions, but this approach is relatively new. Primatology is a young science, conceived in the 1950s out of its parent disciplines, palaeoanthropology and psychology, developing into a fully fledged field in the 1960s. When Louis S.B. Leakey originally sent a young woman, his protégée Jane Goodall, into the wild forests of Tanzania to study apes and metaphysically look through a window on to our primal past, many people thought Leakey was irresponsible.

The human brain is relatively three times the size of that of our closest living relatives, the chimpanzee and bonobo. Consequently, for many years scientists believed there was an insurmountable gulf between the evolution of *Homo sapiens* and that of other primates. How can we be closely related to these apes when our brains are so much bigger? The study of apes within an evolutionary context was therefore considered to be a waste of time. Before the 1950s it was generally assumed that we needed to dig hard for the fossilised remains of early hominids to piece the human jigsaw together.[1] In the 1920s and 1930s just a handful of men such as Robert Yerkes and Solly Zuckerman were studying apes and monkeys, but it did not register en masse that these dumb animals might be even more interesting. Only in the late 1940s did one man, Louis Leakey, become convinced that a comparative study of ape behaviour was essential for an understanding of the big picture.

Louis Leakey was known as the Darwin of Pre-history. He was a palaeoanthropologist, a maverick and a pioneering scientist who existed outside the scientific establishment and he worked this to his advantage. A white Kenyan but a citizen of Great Britain, Leakey was born to missionary parents in Kenya in 1903. Leakey spoke Kikuyu fluently and throughout his life he considered himself to be Kikuyu rather than English. He played daily with the local Kikuyu children and learned their skills of tracking and trapping animals. Many years later he would pass these ancient skills on to Jane Goodall.

When he was a child the Kikuyu would speak to Louis of earlier ground-dwelling pygmy people who once inhabited the land. Fact or fiction, Louis's imagination was inspired. Finding prehistoric artefacts as he played in the woods, he was convinced these were arrow heads made by Stone Age man. Leakey collected the pieces together and took them to the Nairobi Natural History Museum to confirm his identification. One day he would run the same museum. Field work, whether it involves excavating fossils or observing chimps, is an outdoor way of life demanding obsessive commitment from those involved. Archaeologists and primatologists rarely fall into their professions accidentally when adults, and usually have tended their interest since childhood.

In 1919, after the end of the First World War, Louis's family was able to return to Britain and Leakey was sent to Weymouth College for Boys. He was humiliated on arrival because he was behind with his academic progress, but in spite of this he remained determined to follow in his father's footsteps and go to Cambridge University. Leakey worked hard at his studies and ignored the suggestion that he should give up and get a job in a bank. Encouraged by his English teacher, Leakey attended Cambridge for a series of interviews. He took part of his entrance exam in Kikuyu and in 1922 convinced the university dons into accepting him. Talking people into things was one of Louis Leakey's greatest skills.

Leakey graduated with a double first in French and anthropology. He told his university supervisor he was going to return to Kenya to prove that Darwin was right – that man did originate in Africa. His supervisor told him not to waste his time – man came from Asia – but this put-down only fed the fire of resolve in Leakey's belly. He managed to talk the university into awarding him a grant enabling him to do just that. It would actually take another thirty years before his dream would be realised and he would bring *Australopithecus (Zinjanthropus) boisei* and *Homo habilis* back to life, but during those thirty years Leakey's pioneering research opened up Africa to many fossil hunters who followed.

Leakey was a romantic and an instinctive scientist. By the time Jane Goodall introduced herself to him in 1957 Louis had suffered many setbacks, but he had also made many successful headlines in the world of palaeoanthropology. He was also in his second marriage.

Leakey met his first wife, Frida Avern, in Kenya in 1927. She was a teacher from Cambridge, England and was on a tour of Africa with a girlfriend. Both young women were seeking adventure and when introduced to Leakey they both found his stories of fossil-hunting fascinating. Twenty-four hours after meeting Frida, Leakey had proposed; they married a year later back in England. Frida supported Leakey in his work, travelled with him to Africa and helped him excavate sites. As she was a good artist, she also illustrated his first fossil books. But when their first baby, Priscilla, arrived, Frida had little time to support Leakey. During Frida's second pregnancy she suffered a threatened miscarriage and had to spend months in bed. With the arrival of his first children, Leakey turned to other women for attention. By the time he fell in love with his second wife, Mary Nicol, he had been unfaithful to Frida many times over.

Louis and Mary met at a dinner given by the Royal Anthropological Institute in 1933. At the time, Mary Nicol was an assistant to the acclaimed archaeologist Gertrude Caton-Thompson. Mary did not have any academic degrees but she was a talented amateur archaeologist and had spent some of her childhood excavating Neanderthal caves in the Dordogne, France. She was also a gifted artist and had just finished illustrating Caton-Thompson's latest book.

Born in February 1913, Mary was the only and much loved child of Erskin and Cecilia Nicol, both painters. Her father was also an adventurer, who had travelled through Europe and Egypt, where he had lived for years with the Bedouins. Between the ages of five and thirteen Mary travelled with her parents through Europe. Her parents were attracted to the beautiful landscape of the south of France and the prehistoric cave paintings found there. The Nicols spent time with French Palaeolithic archaeologists. Mary, sometimes assisted by her father, would sift through the earth excavated from the caves. She found many prehistoric artefacts and, just as Louis had as a child in Kenya, Mary carefully collected, cleaned and tried to classify her finds. Mary's great-great-great grandfather, John Frere, was the first Englishman to acknowledge that certain peculiarly shaped stones were actually Stone Age flint tools made by early man.

Mary adored her father and thought him 'the best person in the world'; she was shattered when he died of cancer in 1926. Mary and her mother returned to London and Mary was sent to a girls' convent school. But she had a wild streak and was expelled from the first convent for refusing to recite poetry, and from the second convent for purposely making and exploding a primitive bomb during a chemistry class. Years later Mary remembered the bomb: 'the explosion was quite loud and quite a lot of nuns came running, which will have been good for some of them'. That was the end of Mary's schooling. After her years in France Mary only wanted to excavate prehistoric sites and when she was seventeen she wrote to respected amateur archaeologist Dorothy Liddell to ask if she could assist Liddell on her Neolithic excavation site, Hembury Hill, in Devon, England. Liddell inspired Mary and made it 'quite clear that females could go to the top of the tree'. Mary illustrated some of Liddell's publications on the tools excavated. Caton-Thompson saw Liddell's books, admired Mary's skill and invited Mary to help her.

Louis and Mary sat next to each other and while chatting over dinner they found they had much in common. Mary was quiet and shy, but Louis, who was ten years older, had a way of making people open up, especially women. Their first encounter resulted in Mary agreeing to work as an artist for Louis. Mary may have felt some similarities in spirit between Louis and her late father. While Mary sketched his fossils, Louis taught her more about palaeontology; excavating fossils was different from excavating human settlements. Mary fell in love with Louis, and he with her. In January 1934, with baby Colin only a month old, Louis told Frida he wanted a divorce. In October 1934 Louis left for Olduvai Gorge in Tanzania. Six months later Mary followed. She could 'hardly wait to see Olduvai'. Frida filed for divorce in 1936, but had hoped Louis would come back to her. On 24 December 1936, having returned to Britain for more funds, Louis and Mary had a modest marriage in a registry office. Mary's mother was present but she was very unhappy about the marriage as were many of Louis's friends and relatives. Three weeks later the newly-weds returned to Africa. Louis didn't see Frida or their children again for twenty years.

A great self-publicist, Louis was always pleased to take the limelight for his and Mary's discoveries. Not as overtly ambitious as

Louis, Mary was shy of cameras and press conferences, and professionally cautious when naming fossils; she did not like to announce a new discovery until she was sure she could correctly identify it. Louis was prepared to take risks in his announcements because he wanted to be the first palaeoanthropologist to comment on an extinct species. After Louis's death, on occasions Mary's reluctance to name a fossil meant other palaeontologists would presumptuously and publicly announce, comment on and name her fossils for her, angering her greatly. But for many years, with Lousis's public panache and Mary's eye for detail, they were an unbeatable team.

In the 1930's Mary and Louis entered a race with other fossil hunters to find Africa's oldest human remains. They excavated primarily at Olduvai Gorge. Leakey chose to excavate in the Gorge because it was lower than 4,000 feet above sea level and therefore would not have been too cold for early hominids; additionally it was the site of an ancient lake that would have supplied the necessary water and grazing animals to prey upon. Both Mary and Louis Leakey dearly wanted to breathe life back into these early people. Often Louis would endow them with a life history drawn from his imagination rather than supported with fossilised remains. Mostly his hunches were right, as with his intuitive faith that human lineage was bushy and ancient, but without factual evidence to back up a theory scientists undermine themselves, and from time to time Leakey did exactly that.

Factual evidence is sometimes hard to supply. Most primate field research relies on first-hand anecdotal observation, but science demands proof through replication. If the primatologist is researching behavioural patterns in a wild animal's life, rather than anatomical adaptations seen under a laboratory microscope, observations will remain anecdotal. Additionally, they may never be observed a second time, not so much because the animal never repeats its behaviour, but because the primatologist is never again in the right place at the right time.

Thorough field study necessitates a great deal of luck as well as long hours of surveillance. Backing up observations with factual evidence is very difficult, and the best way to prove an animal did what you said is to film the creature carrying out that behaviour. But like some humans, non-human wild primates are camera-shy. They do not like the arrival of new equipment and a cameraman. A further factor a researcher has to deal with is the cyclical behaviour of long-lived animals; some behavioural patterns may only be repeated after a number of years have passed. Many primatologists have risked their scientific reputations in reporting discoveries that they cannot replicate. Scientists themselves, being very competitive, are notoriously quick to discredit one another's research, presenting a further major hurdle for the psychological and intelligence studies done on captive apes, such as the language work of primatologist Sue Savage-Rumbaugh.

Palaeontology was a family affair for the Leakeys. Mary and Louis had three sons. Jonathon, Richard and Philip and the boys' Dalmatian dogs would go together on digs to Olduvai Gorge. Their sons also became masters in the field, discovering the bones of a number of prehistoric animals such as giant pigs and hippopotamus. Louis always kept a rifle close by his side during digs. Guests at the camp were advised to stay in their tents and to sleep through the nightly visits from lions and hyenas on the prowl for water and food (apocryphally it was believed lions would not tear canvas, though Jane Goodall would later experience otherwise). Leopards would come at night and try to prey on the Dalmatians. Irritated by being woken by the howls of her petrified dogs, Mary would set off firecrackers to frighten the predators away.

Hand axes and arrowheads littered Olduvai, and Louis Leakey wanted to meet the primitive men and women who had made these tools. He believed that 2–3 million years ago the hominid line was bushy and that at least two, probably three or even four species of hominids had evolved simultaneously and lived alongside each other. Leakey was later proved right, but at the time he was out-numbered by those who believed in the more popular theory that human lineage developed with only one single species of hominid existing at any one time, stage by stage evolving into later species. Today different species of great apes live successfully side by side: gorillas and chimps in Africa and orang-utans and gibbons in Asia. It seems logical that a number of early species of people similarly lived in close proximity but specialised in different sources of food.

With each prehistoric ape fossil he dug up, Leakey became more determined to make a comparative study of contemporary ape behaviour. He wanted to understand more about the lives of these prehistoric primates, but behaviour cannot be preserved in fos-silised remains. Leakey's simple conclusion was that the study of our nearest relatives, the great apes, would allow us to deduce much about how these ancient ape-people lived and how our primal ori-gins evolved.

The Leakeys excavated not only at Olduvai but also further north on Rusinga Island in Lake Victoria, where in 1948 Mary found a very well-preserved skull of *Proconsul*. Resembling a gibbon, this 25-million-year-old creature is a forerunner of all apes, including man. Mary wrote about her find: 'This was a wildly excit-ing find which would delight human palaeontologists all over the world. For the size and shape of a hominid skull of this age, so vital to evolutionary studies, could hitherto only be guessed at. Ours were the first eyes ever to see a *Proconsul* face.' At this time the Leakeys loved each other very much. The *Proconsul* skull thrilled them both so much that they decided to celebrate that same night by making another baby. Nine months later Philip, their youngest son, was born. The Leakeys also found ape bones that were 18–20 million years old. All modern apes, including humans, are descen-dants of these species of apes. What intrigued Louis Leakey further was that he and Mary also found the fossilised remains of plants and seeds from the same period. Leakey was now able to visualise clearly the ecosystem of these creatures. Chimpanzees have not evolved as radically as humans have over the last 15 million years. The human and chimp common ancestor lived in the Miocene epoch. Leakey imagined their lives were not dissimilar to that of a group of chimpanzees that lived in the Gombe Reserve of forested hills on the edge of Lake Tanganyika, further south in Tanzania. Leakey had wanted someone to study the behaviour of these chimpanzees since he'd first heard of them in 1945, and felt sure that a study of the territorial, feeding and social lives of the Gombe chimps would help us to start to understand ourselves.

At that time it was not fashionable to think of chimpanzees as a model for early humans because the chimpanzee brain is so much smaller than the human brain, although we now know the brain of our 3.5-million-year-old ancestor, *Australopithecus*, was similar in size to that of a modern chimp. But Leakey chose to ignore the paradigm of the day and trusted his instincts instead. From 1935 to 1972 the Leakeys collected hundreds of excavated hominid fossils at Olduvai. By 1961, they had found fragments of skulls and teeth from a prehistoric man that Louis instinctively knew came from the *Homo* lineage, even though the fossils did not match the already identified species *Homo erectus*. The brain cavity in these skulls was as large as that of *Homo erectus* and considerably larger than that of the contemporary hominid *Australopithecus*.

In 1962 one skull that they pieced together was that of a small, adult female that they named Cinderella. By 1964 the Leakeys had fossils of eight individuals. The remains were always found alongside tools, at this time another indication of the *Homo* lineage. The fossilised teeth and skull of this species were smaller and thinner than *Homo erectus*. Finally in 1964 the world of palaeontology agreed that the Leakeys had indeed found a new hominid species. Cinderella was recognised as being 2 million years old, and for a time she became our oldest known ancestor. She and her family were given the name *Homo habilis*, meaning 'handy man', as a result of the stone tools found beside the bones. It was presumed that tool use was the main development that had made men out of these beasts. From Leakey's research it was proved that our ancestors did in fact share the African landscape with other hominids, distant cousins who had become estranged through millions of years of evolution and whose descendants did not survive nearly as long as we have.

Even though *Homo habilis* and *Australopithecus* were contemporaries and would have gazed out across Olduvai Gorge together, it is believed that they would not have done so hand in hand. They were different species, with different brain capacities. They probably couldn't interbreed and were most likely enemies. As the *Australopithecus* line died out it is presumed *Australopithecus* was not tooled, and defenceless against the *Homo* line, which exterminated them with their arrows and axes.

The theory of man as tool maker lasted for a couple of decades until Jane Goodall entered Louis Leakey's world. With Leakey's support, Jane Goodall has single-handedly changed the way we regard human evolution and revolutionised how wild animals, especially apes, are studied.

When twenty-three-year-old Jane telephoned Leakey for the first time in 1957 he already had several young, eager women working for him, as well as a series of extra-marital affairs behind him. Jane Goodall, Dian Fossey and Biruté Galdikas, who respectively studied the chimpanzee, gorilla and orang-utan, are known as the trimates, an obvious pun on the word primate. These three are the most famous women associated with his name, but they are certainly not the only women shown favour by Leakey. At least another fifteen women found careers working with apes and monkeys thanks to Leakey's patronage. Most of them came to Africa initially to work in Louis's Tigoni Primate Research Centre on conservation projects in Nairobi while Louis discussed other options with them.

By 1965 Louis and Mary's marriage was over. Although they continued to work together, the love had gone. Louis had started treating Mary the way he had treated Frida. Just as he had invited Mary to his and Frida's home back in 1933 and flagrantly betrayed Frida, so he treated Mary. Louis invited all the women he had crushes on back to his and Mary's home. Mary was confronted by a series of extramarital affairs, some of them carried out in her bed, which left her feeling bitter and resentful towards any woman Louis took an interest in.

Leakey loved to be surrounded by the opposite sex. He had very few male friends and frequently expressed his preference for the company of women. Throughout his life Louis Leakey watched male and female human behaviour, arriving at a conclusion that has set a precedent in this area of science and has entered the public consciousness at a cultural level: women do field work better than men. For twelve years he had been looking for someone to devote themselves to the study of the Gombe chimps. He did not have the time to undertake the chimp study himself, as he was so busy juggling various research projects. He needed an inspired, young, enthusiastic person to take up the challenge. Often Leakey would mention the 'nameless man', a young man who in 1946 had badly disappointed him and 'failed utterly' in the job of studying the Gombe chimps. He had found the ten-hour days of following wild chimps through the Gombe woods excessively tough, and gave up for the same reasons that Leakey didn't want to do the job himself. Days can pass when you do not see any animals, and it takes a particular mentality to be happy to be alone and to feel contented with a slow pace of life. You have to want to do that job and nothing else. Leakey was intellectually greedy; there was much he wanted to accomplish in his lifetime and a long-term field study would prevent him from achieving those other things.

Because Leakey was not a university professor, he did not have students in the traditional sense, but many young people gravitated towards him in their quest for adventure. Leakey loved young people and welcomed their interest in his work. Although he did encourage young men, it was young women he particularly liked to inspire. Ten years after the failure of the nameless man, Leakey sent British Rosalie Osborn into the bush to study gorillas in Uganda. But her study lasted only a few months because Leakey could not raise enough funds to keep her there. Raising money for field research is as hard today as it was then. But if you happen to see your animals doing something unexpected, something human-like, financial support is easier to find. This is what happened to Jane Goodall.

As far back as Goodall can remember she has been enthralled by stories of Africa. Born in London in 1934, as a baby she had been given a toy chimpanzee by her mother, and she still owns this favourite comforter. As a child she loved the story of Dr Doolittle, the Edgar Rice Burroughs Tarzan Stories and the Tarzan and Jane movies. She read natural history books and taught herself about African wildlife. One of her earliest memories is of creeping into the hen-house and sitting quietly for a few hours to discover where eggs come from, only to make her poor mother frantic by her sudden and prolonged absence.

While still a child Jane had decided she would live in Africa and write books about the wildlife. Goodall's father, Mortimer Morris-Goodall, had joined the British Army on the first day of the Second World War and after it was over remained in service for another ten years. During that time Jane and her younger sister Judy grew up mostly without the support of their father, and their mother Vanne became estranged from her husband. When Jane was sixteen years old her parents divorced and her father remarried.

Although Jane's father was absent for much of her formative years he nonetheless had a great influence over her. Morris-Goodall was himself an adventurer. He drove an Aston Martin sports car, competed in motor racing and founded the Aston Martin Club. In the late 1990s Morris-Goodall is still very active, attending British sports-car racing dinner dances and receiving awards for his lifetime of competitive racing. No doubt Jane's father's exciting reputation as a risk-taker in the pursuit of self-fulfilment influenced Jane to seek out her own interests. Goodall's father has always been very proud of his daughter's pioneering work in Africa.

Living in Bournemouth with Vanne, Judy and two aunts and working as a secretary was not how Goodall envisaged her future. Jane Goodall had ambition. In 1957 she visited Kenya for a holiday, carrying the hope that she might find work studying animals. Goodall was a committed amateur natural historian, and was no evolutionary biologist. It was not an overwhelming desire to understand the origins of mankind that motivated her, but a love of animals. She stayed with an old school friend in Nairobi and found herself a job as a secretary. During a conversation with her friend's parents she explained her love of wildlife, and they immediately responded that Jane should telephone Dr Leakey at the Coryndon National Natural History Museum (later renamed the National Museum of Kenya).

By this time Leakey was a very well-known character in Kenya, and was often in the papers as a result of the discovery of some new fossil at Olduvai. In 1944 as curator of the Coryndon Museum in Nairobi, his first act had been to open its door to all races in a blaze of publicity. The Coryndon Museum was the first public place in Kenya to allow admission irrespective of race. Black and Asian visitors came, though for a while some white members of the public stayed away in protest, but soon all races walked freely together in the museum.

Irven and Nancy DeVore first met Louis Leakey in 1959, when Irven was twenty-five and Louis was in his late fifties. Irven DeVore is today a professor of anthropology at Harvard University; back in 1959 he was still a Ph.D. student. DeVore remembers Louis as overweight, white-haired and almost toothless, but what really surprised him was that women, including Nancy, found Louis Leakey sexy. A great raconteur, Leakey would tell personal anecdotes, such as the time he had charged yelling, unarmed and naked at a pack of hyenas to frighten them away from their kill because he wanted to share an experience with early man. Louis Leakey was extremely charismatic and curious about everything. Even though his ego was vast, he showed an interest in everyone he spoke to and made them feel special. Women, especially, loved to talk to him.

Leakey was peripatetic in this work ethic, continually rushing from project to project. When Mary Leakey found what Louis thought were fossilised faeces of early man, containing the bones and skin of small mammals, Louis wanted to test his theory and begged DeVore, as an experiment, to eat a rat raw, so that a comparison could be made. Louis enthused that the experiment would make DeVore famous; he explained that a young man with a full set of teeth was needed and it would be a poor experiment if he tried it himself. DeVore vociferously refused this offer of celebrity, as did Louis's three sons. Louis never did find a guinea pig to eat his rat.

Leakey admired and respected women as intellectual equals; married to Mary he couldn't very well do otherwise. Leakey was pleased to employ women, and not just in traditional roles as secretaries. He wanted to use women as explorers. Jane Goodall says that when she first rang the museum and hesitantly asked to speak to Dr Leakey, Leakey barked back at her, 'I'm Leakey what do you want?'

To begin with Goodall would have been innocent to Louis's eccentricities. She rang Leakey the prehistorian, naively wanting advice on animal-watching, and didn't realise she would be speaking to Louis the philandering adventurer. When Leakey answered his phone he would have heard yet another young, quietly spoken English female voice ask for an appointment. He invited Goodall to come in to see him the next day. Rather than staying in his office

Louis decided to take her out for a drive in Nairobi Game Reserve. As they drove through the park he listened as she talked about her love of animals. Goodall was slim, blonde and beautiful and passionately enthusiastic about wildlife. Leakey tested her, asking difficult questions, and was very pleased when Goodall could describe an antbear[2] and understood the meaning of ichthyology.[3] Enchanted with Goodall, Leakey was about to fall in love again. To her delight that day he offered her a full-time job as his personal assistant, and she accepted. She insists the job was offered to her because of her knowledge of the natural world and that Leakey's visceral attraction to pretty women had nothing to do with it.

Leakey invited Goodall to his home and introduced his latest personal assistant to Mary. Although Mary and Louis's relationship by this time had evolved from a passionate marriage to that of business partners nonetheless Mary still succumbed to feelings of sexual jealousy whenever her husband introduced yet another young female assistant. Mary and the boys had just returned from a ride. Jane admired their horses, and intentionally Mary suggested Jane mount one particular horse and take a short ride. But as soon as Jane sat in the saddle the horse started to buck and walk backwards. The horse was known by the Leakeys to be a difficult animal, and this was its typical behaviour. To Mary's satisfaction Goodall was distressed. She insisted the animal must have saddle sores, and when the saddle was removed Jane was proved right. Mary Leakey respected Goodall's empathy with the horse; both women shared a love of animals.

The Leakeys took Jane to Olduvai. The skull *Zinjanthropus*, Nut Cracker Man, was about to be discovered and Olduvai was still very inaccessible by road. There Louis trained Goodall in palaeontology techniques and Goodall helped Mary and Louis to excavate. Louis passed on to her the Kikuyu art of animal observation and tracking he had learned as a boy. At this time she formed a friendship with Richard Leakey and they made plans to capture and train a zebra, although their friendship was short-lived because of their rivalry for Louis's affection.

At this time Louis Leakey was negotiating with anthropology graduate Cathryn Hosea Hilker for her to set up camp at Gombe and study its chimpanzees. Hosea Hilker wanted the job but at that time in her life was unable to commit the required five years to the quest. But Leakey didn't want any more false starts after the disappointment of the nameless man's short-lived study, so Hosea Hilker returned home to the United States. Many years later, while working in big cat conservation at Cincinnati Zoo, Cathryn Hosea Hilker met Jane Goodall, who was on a lecture tour, telling her that she hated her but confessing that she hated herself more for not taking up Leakey's original offer.

In 1958 a few months after their first meeting Goodall hinted to Leakey that she wanted to study the chimps and after Hosea Hilker turned down the offer Louis gave Goodall the opportunity to go to Gombe in her place. She did not hesitate in accepting, but privately she was terrified she would let Leakey down. The mantle he was handing over to her felt like an immense responsibility. It took two years for Leakey to raise the funding for Goodall's study. Institutions were doubtful that an untrained, unqualified woman could cut the mustard in the same way as a chap. But Leakey was not deterred. He never doubted the money would come eventually. He used the delay as an opportunity to dispatch Goodall to London for exposure to chimpanzee observation at London Zoo and anatomy training from the British primatologist John Napier.

The night before Goodall left on her 1958 summer sabbatical to London Leakey confessed he was in love with her. Goodall was appalled. She admired him greatly and loved him filially, but not romantically. Goodall told him so and could see that she had hurt Leakey's feelings. She was afraid he would take back the offer to work at Gombe, but he didn't. The research project came first. Leakey never mentioned his physical attraction to her again.

In 1960 Leakey raised the money for Goodall's study from a successful American businessman, Leighton Wilkie. Wilkie was a tool manufacturer and felt passionately about the evolution of tool use, and would later also give support to Dian Fossey and Biruté Galdikas. The three women would remain eternally grateful to Leighton Wilkie. Back in Nairobi and desperate to get to Gombe, Jane experienced another delay. She was told by Gombe Park officials that a lone woman was not allowed to camp in the woods. Everyone feared the worst, including the bad publicity the death of an Englishwoman would cause.

Goodall's mother, Vanne, a bubbly woman who had never remarried, was worried about her daughter's safety and was persuaded to leave Judy in Bournemouth with her aunts and accompany Jane to Gombe. Vanne was worried but she did not want to deter Jane so she agreed to come along. Leakey found an African cook and a handyman called Dominic to accompany the women. Together this unlikely team of English mother and daughter and African assistants reached the shores of Lake Tanganyika in July 1960 and set up camp. By November Jane had proved a white woman could survive without injury in the Gombe forest and on a limited diet of tinned sardines. Eventually the authorities agreed that Vanne could fly home, leaving Jane with Dominic and her new African helper, Hassan, though for much of the time when Jane was following chimps through the woods she was entirely alone within the 50 square miles of chimpanzee territory containing, perhaps, 150 chimps.

To begin with, things were very hard for Goodall. She was lucky to even hear chimp hoots of alarm and branches cracking as the animals fled from her advance. Sometimes the powerful animals would charge her and threaten her. Ferociously strong, an adult chimpanzee could kill an unarmed man with ease. It would have taken great courage to keep pursuing the animals when confronted with threatening displays of hostility. Primatologists are advised to stand their ground, without flinching, if a fully grown wild chimpanzee charges, and Goodall has had to hold her breath and keep still many times. She has never been bitten by a wild chimpanzee, although a disturbed laboratory chimp did remove part of her finger once.

Leakey and Goodall wrote to each other regularly. Goodall's letters would be wretched notes, telling Leakey she was failing at habituating the chimps and he would write back urging her not to give up. Encouraged, Jane persisted. She has shown herself to be a very committed woman who knows what she wants from life. Over a period of a few months the Gombe chimps began to accept her presence and Goodall was delighted to recognise individual chimps. This is when Goodall first met the now famous chimps, Flo and David Greybeard, the two animals that accepted Goodall before any of the other chimps.

Leakey had asked Goodall to collect sample specimens of chimp dung and sample species of leaves and fruits that they ate, for identification, and she did so. Sitting some 40 metres away and looking through binoculars Leakey had specially bought for her, Goodall watched the animals' social interaction, noticing much disparity in behaviour between individuals. Some chimps were more extrovert or more aggressive than others. The competitive nature of chimpanzee hierarchies was becoming evident.

Four months after Goodall's arrival, she witnessed something no one had ever seen before. David Greybeard killed and ate a wild piglet, sharing the meat with his chimp friends. And only a few days later Goodall saw something far more revolutionary, with David Greybeard once again as the protagonist. He settled down next to a termite mound and for an hour feasted on termites by fishing them out of the mound with the use of a tool. Carefully selecting long stems of grass and pulling the chosen blade out of the ground, the chimp then made sure to remove any irregular pieces of vegetation, leaving behind a long, straight and strong implement with which to poke into a termite hole. Once the modified stem of grass had entered the mound, termites would bite it and hang on to it with their mandibles. When David Greybeard felt the termites tugging on his 'fishing rod' he would gently remove the piece of grass and lick off the termites, then reinsert his tool to catch further termites.

After sixty minutes and a decent protein feed David Greybeard ambled off and Jane took the opportunity to study what was left of the termite nest. Goodall copied the chimp's actions and, sure enough, after feeling the tug of the termites biting into the stem of the grass she removed it to see several juicy-looking termites hanging on to her tool. Here was a revelation: the wild Gombe chimps were tool-using meat-eaters, like us. This activity, seen practised by many of the Gombe chimps, is commonly known as 'termiting'. Over the years Goodall has also seen her chimps dipping bundles of moss and grass into narrow hollows in trees. Like a sponge, the moss soaks up the collected rainwater. Retrieving the saturated mass, the thirsty chimps suck it dry.

Since Jane Goodall proved that a long-term field study of wild chimps was possible, other biologists have set up research stations in Africa and some of them have noticed a different manipulation of tool use. Hedwige Boesch and her husband Christophe have studied chimps living in the Taï Forest, Ivory Coast for the last twenty years. Their chimps have a taste for the local coula and panda nut, but the shells are hard to crack open. The chimps have solved this problem by inventing an anvil and hammer. The animal collects a pile of nuts and settles down next to a fallen branch or a root with a depression in which the chimp can firmly secure a nut. A good anvil such as this is precious, and chimps will fight over it. The chimp carefully places a nut into the anvil and then hammers it with a stone, the shell neatly splitting apart. When the nuts are in season the chimps will remain at the site cracking nuts all day long. Habituating the chimps was difficult for the Boesches. Hedwige told me the first time she was able to get close to a 'spectacular' wild male chimp she cried with joy. She said she will always treasure the memory of that moment and when the Ebola virus infected her study group, leaving only two males alive, she felt utter despair.

The chimps' approaches to ecological problems represent cultural differences between communities of chimpanzees and differences between individual family lineages. Infants learn these skills by watching their mothers, and the techniques are passed on to subsequent generations. When young females depart from their natal group and join other communities, they take knowledge with them that they will pass on to their babies, and eventually the new groups assimilate the skills. (This is comparable to the spread of ideas as man migrated around the globe, such as the Arabic number system that was adopted by various cultures.) The study of 'Chimpanzee Culture' is a new, and fast growing, area of primate research.

After nearly forty years of ongoing study, Goodall has discovered many fascinating things about chimpanzees that have shocked and amazed the public, but bringing to our attention the fact that these animals adroitly make and use tools is her most astounding achievement. Before her discovery, there had been anecdotal reports from a hunter who had seen a chimp use a stick in much the same way to fish out honey from a bees' nest, but Goodall's field report was the first well-documented account of tool use in a wild ape. The theory of man the tool-maker was about to explode. Now, according to Jane Goodall, a young, untrained woman, not only were tools used by man's *Homo* ancestors, but the chimpanzee, a more distant cousin than even *Australopithecus*, also had the technology to make and use them.

This report threw prehistorians into a state of flux. Leakey couldn't have been more proud of his protégée when Goodall brought this fact to light. At the time of Goodall's discovery, he said, 'We now have to redefine man, redefine tool or accept chimps as humans.' In interviews Leakey would pronounce that, obviously, before man's ancestors used and made stone tools they modified sticks and grasses just as Goodall's chimps did. Amid all the excitement, Goodall continued to study the chimps for their own sake; that Leakey and others could extrapolate information and theorise on man's origins from her observations was a fortunate by-product for her.

Leakey had taken a chance with Jane Goodall. One of his many hunches had proved right and Goodall's success after the unnamed man's failure gave credence to Leakey's belief in women's special talents for this type of work. Women, Leakey reasoned, mothered babies, and were therefore better suited to non-verbal communication because they could understand the needs of a creature who couldn't talk. He also believed that a woman's presence would not agitate patriarchal apes in the way a man would. Leakey thought that women were inherently patient and silent observers, whereas men were ambitious hierarchical careerists. A life of near poverty watching animals live out long lives for ten hours a day, decade by decade, was not going to turn men on at all.

Irven DeVore believes Leakey was right. DeVore noticed that when he, his wife and his child tried to walk with wild baboons in Nairobi National Park, the animals accepted his wife and child's presence but not his. His wife Nancy was able to get close to the animals and watch their behaviour, but the baboons wouldn't tolerate him nearly so easily. DeVore had to throw nuts at the baboons to encourage them to approach him.

In some ways, Louis Leakey's theories have been supported by the evidence. On average men do not stay longer than two years in the field with any one group of primates. They return to university, publish papers, push their careers forwards and move on to ask other questions of other animals. On the other hand, time and again there are examples of women sticking out the gruelling existence with their particular group of animals for decades. Leakey's other ape women substantiate this claim. Dian Fossey had spent

eighteen years watching mountain gorillas before she was murdered. Louis's third in line, Biruté Galdikas, has studied the
orang-utan for twenty-nine years. Long-term field studies are a
largely female preserve. Women become emotionally connected to
the animals they study and do not want to leave them.

As well as bonding especially closely with certain individual animals, the women also love the freedom of the life of a field
researcher. Another, though unintentional, outcome from Leakey's
legacy is that from the sea of female researchers more and more
women have chosen to study the life of the female primate. Because
of this we now know much more about female (and male) primates than when Jane Goodall first went to Gombe. Through
evolutionary theorising these studies help us to understand better
the behaviour of modern man and woman.

Leakey declared he did not want the young women working for
him to be scientists prejudiced with theories. He wanted women to
be intellectual blank slates, lacking preconceptions on animal
behaviour, and to go naively into the field with an open mind and
write down everything they saw. If Goodall had come to Leakey
with a biology degree already in place, she would have been biased
towards the existing paradigm of the day. Leakey did not want
people starting on his field projects who already thought they knew
the 'truth'. He actively sought out intelligent but uneducated
women who would question everything they saw in a fresh, new
way. He also knew that if the women lacked the experience of
employing rigorous scientific methods, he would remain their
much needed mentor and they would seek his help at every turn.

Neither Goodall nor Fossey had any scientific qualifications in
the area of ethology or zoology; Goodall was a secretary and Fossey
was an occupational therapist. Biruté Galdikas had an undergraduate degree in anthropology, but was also a long way from working
for a doctorate when first noticed by Leakey. Was Louis Leakey a
man of objective and sound judgement, or was he persuaded by his

own primal urges? He may have been a Svengali figure and a well-known womaniser, but his trimates became world experts on their particular chosen species of ape and have trodden a path into the jungle for other women to follow.

Louis Leakey's obvious love for the Goodall family was noticed and rumours and jealousies spread. Richard and Mary both resented the time and effort Louis dedicated to Jane's career. It seemed to both of them that other women had always come between them as a family. Louis Leakey had become romantically involved with Vanne Goodall, and when visiting Britain he always stayed with her in her flat.

Today, Vanne, now in her nineties, says, 'Louis was a genius, a brilliant man who was in a rush to discover as much as he could, as he knew the span of his own life was like a wink of an eye compared to the millions of years in man's evolution.' On 1 October 1972 Louis was staying with Vanne when he collapsed and was rushed to hospital where it was discovered he was dying. At the time Vanne Goodall had been helping him finish the second volume of his autobiography.

The year before Jane Goodall was born Louis Leakey had been in Britain and rumours spread that Louis was in fact her father and that his romance with Vanne had begun then, his paternity of Jane explaining his commitment to the Goodalls. The Goodalls say the rumour is absolute nonsense and Jane certainly physically resembles Mortimer Morris-Goodall.

I asked Richard Leakey about the rumour. 'There are a hundred rumours ranging from Louis being Jane's father to Louis being her lover. I have no idea if either or if both stories are true; I don't think they can be. Jane is emphatic that she knows who her father is and it's not Louis and Jane is emphatic that she was never Louis's lover. Now Jane is alive and Louis is dead and no one has come up with evidence to prove otherwise so we have to accept that.' I went on to ask Richard about Mary's resentment towards the primate studies. 'My father had a very sharp, prescient mind. He was interested in the broad picture of the origin of the human species. He realised the common denominator between ourselves and apes today is equal to the common denominator between ourselves and our ancestors. My mother was a trained archaeologist, interested in detail and hard evidence, the bones themselves. When her husband spent his time and money supporting the primate studies she lost her fundraiser and public spokesperson; Louis was publicising the primate studies and not her own research. My father had his theories on women's abilities. Louis was gender-sensitive: he preferred women. I do too. The resentment my mother felt was within the marriage rather than a resentment of primatology itself.'

Every day at Gombe, rain or shine, Goodall would rise at dawn, dress in her plimsolls, khaki shorts and shirt, tie her hair back, leave her tent, have a cup of tea and walk up into the wooded hills searching for tracks. At first, months passed and Goodall had only seen fleeting glimpses of the wild chimps. Often depressed, Jane would write to Louis, telling him she was failing at the task, starting her letters 'Dear FFF' – Fairy Foster Father – and signing off 'Love FC' – Fairy Child. Louis had opened doors for Jane and had shown a great deal of affection to her mother and also to Judy, for whom he had arranged a job as a laboratory technician at London's Natural History Museum and the Goodall women loved him in return. Louis had pet names and acronyms for all the women in his life, and his secret name for Jane was Mwendwa, Kikuyu for 'My Most Beloved'.

Goodall's discovery that chimps were meat-eating predators that made tools came, luckily, just as the funding was running dry. Had she not seen such an amazing sight the financial support for Goodall's study might have dried up and she might have been forced to return to her secretarial training, but now the anthropological Wenner-Gren Foundation took an interest in her research and provided a grant. A year later Leakey had persuaded the

National Geographic to support the project financially in return for photographs. At first Goodall refused, as she didn't want a camera crew around disrupting her routine, and she was also very camera shy. But Louis loved to court publicity for his projects, and knew it was the only way to be sure the money would keep rolling in.

Urged by Leakey, Goodall eventually relented and the National Geographic photographer Baron Hugo Van Lawick arrived at camp. Van Lawick's romantic films of Jane Goodall at Gombe effectively set the tone for all subsequent films made on the life and work of women primatologists and the 'National Geographic Effect' and the 'Jane Goodall Phenomenon' were born. In August 1963, the first article, written by Goodall, 'My Life Among the Wild Chimpanzees', was published in the *National Geographic* magazine. Van Lawick's photographs helped enormously to popularise primate studies.

Van Lawick was as handsome as Goodall was beautiful. They shared a love of African wildlife and eventually, with Leakey's encouragement, fell in love and married. Leakey knew it would be easier for Goodall to stay at Gombe if she could share her private life with a kindred spirit. In 1967, five years after they met, their son, Hugo Eric Louis Van Lawick, known to all as Grub, was born.

When Goodall first went to Gombe she and Leakey both expected the study to last just a couple of years and then she intended to go on to study the bonobo, the gorilla and the orangutan. But this was not to be. Goodall bonded with the Gombe chimps, forming lifelong friendships with the animals, especially with David Greybeard and Flo. She couldn't leave them; Gombe had become her family home. Just before Grub was born Goodall studied hyena behaviour with Van Lawick in the Ngorongoro Crater, but after Grub's arrival she returned to Gombe.

Goodall picked up tips on mothering her baby from watching the mother and infant relationships of her chimps. She decided to breastfeed on demand for a year. Chimps suckle for two years on demand and for the next two years they continue to supplement their infant's chimp diet with breast milk. Like a chimpanzee mother, Goodall never left her baby to cry. Chimpanzee mothers carry their babies wherever they go. Possessing a similar strong prehensile grip to that of a newborn human infant, the chimpanzee baby will hang on under its mother's torso, in the ventral position, as she moves and when a little older it will sit up and ride on her back. Constant physical contact is important for the mental well-being of the infant chimp, as it is for human babies. Like the chimp mothers, Jane carried baby Grub constantly, though protecting a toddler from baboons and chimpanzees is no easy task.

Goodall heard of two African babies that had been taken by chimps locally and eaten. When Grub began to walk a large enclosure was built for him, and during his early years he was caged within this space during the day so that neither visiting baboons nor chimps could hurt him. Even when the apes climbed on to the roof they could not get in. They could only peer through the bars at Grub as he looked out at them.

The chimpanzees of Gombe have brought Goodall fame and fortune, although she never sought that. She has written countless books and is a wealthy woman, living modestly and travelling the world giving lectures on chimpanzee welfare. She hasn't left the world of science, but conservation and the care of captive chimps are her motivating forces now. She still heads the Gombe site and has also formed the Jane Goodall Institute, which runs four chimpanzee sancturaries in the Congo, Tanzania, Kenya and Uganda, which are homes to orphaned chimpanzees rescued from the ape trade. She does not attempt to rehabilitate these animals back into the wild, as other women have done, because she controversially believes it is impossible to do so.

Because the National Geographic nature films of Goodall are so well known, Jane has to travel in disguise. Wearing sunglasses and abandoning her ponytail help her to retain her anonymity. Her celebrity has encouraged countless numbers of women to follow in her footsteps. The image of an elfin white woman up against the elements continues to inspire and the wheels within the Jane Goodall Phenomenon keep on turning. Over the last thirty years many other women students have chosen to study the Gombe chimpanzees for their Ph.D. theses. Collectively we now know more about those particular chimpanzees than any other group of wild chimps.

Eleven years after its foundation the Gombe site had been visited by many students who had not only been alerted to Goodall's research through her well-publicised field study but had also been taught about her discoveries at university. Goodall and her collection of students studied in detail the chimps' diet, ecology, social relationships, mother–infant bonds and group hunting for meat. They were helping her to amass a great deal of data on the complex lives of these long-lived animals. Barbara Smuts arrived wanting to do something new, wishing to specialise in the study of the lives of female chimps and particularly the friendships and bonds between them. Arriving at Gombe and working alongside Goodall was, for Smuts, like walking right into the National Geographic TV specials she had always loved and starring alongside her heroine.

There have been a number of threats to Goodall's safety over the years. On 19 May 1975, Zairian guerrillas came to Gombe in the middle of the night to kidnap white hostages as Barbara Smuts and other students slept. Working for the late Laurent Kabila and his Marxist Partie de la Révolution Populaire, or PRP, the kidnappers sailed 34 miles across Lake Tanganyika, Africa's deepest lake, from their lakeside base in eastern Zaire.

There was no time to wake all the students, the unlucky four were left behind, sleeping in their beds. Barbara Smuts awoke to hear fellow student Steve Smith screaming and thought he must have been bitten by a snake. In pitch blackness she ran to his hut, as did another student, Carrie Hunter. The terrorists grabbed the women when they arrived; Smuts and Hunter were confronted with an image of fellow students Emilie Bergman and Steve Smith forced to their knees with their hands tied behind the backs. Smuts and Hunter were treated likewise while other terrorists searched the trees for more white hostages, though they didn't find them.

Barbara Smuts was at the right place, the place of her dreams, but at the wrong, nightmarish time; she had been there barely a month. The terrorists took their hostages back on the seven-hour boat journey across the lake. Guns were pointed at their heads all the way and they were beaten if they cried.

It was the rainy season, and the four students were held hostage in cold, wet mud huts and were fed sparingly on 'ugali', a vegetarian paste. The guerrillas spoke little English and the students spoke next to no Swahili; Smuts spoke some French, as did the guerrilla leader, so French became the language of their liberty. The students were ordered to write ransom letters demanding $500,000, weapons and Mobutu's release of PRP rebels. If the terms were not met the students would be killed. After a week Barbara was freed on the condition that she put her captor's demands to both the American and Tanzanian governments. Six weeks later Hunter and Bergman were released and a month after that Smith was freed. The students' families paid the ransom, but Smuts is not sure if the other demands were ever met.

All the students suffered with parasitic infections and stress disorders. Back in the United States the four abductees remained in touch. They wanted their government to publicise what had happened to them, but in those days the American government did not want to embarrass their anti-communist dictator ally Mobutu. Nobody seemed interested in their ordeal, so the four students decided to block it out of their minds and to continue with their separate lives. Barbara Smuts is now the only one out of the four who still studies wild primates.

Twenty-three years on the four abductees are together again, and have been in touch with the Senate. Now that Kabila's men run the Democratic Republic of the Congo, the four want the terrorist history acknowledged, but again the American government is indifferent, claiming it wants to remain on friendly terms with the new 'democratic' country because Kabila didn't want foreign aid, he wanted financial independence. The Congo has massive untouched reserves of oil and minerals deep in the jungle, including uranium. If Kabila's men could tap into that the country would be rich and militarily powerful, and the growing population of 46 million would be fed. The last uncharted African rainforest of the Congo basin, home to Stone Age hunter-gatherer people like the Aka Pygmy tribes, would be flattened. Animal species such as chimps, gorillas and bonobos would be decimated. And Kabila would be remembered as an African hero. The four want it known that Kabila's human rights record is as shocking as Mobutu's.

They wanted the famous Jane Goodall but she was very lucky. She was quietly woken from her sleep by one of her Tanzanian assistants who had been warned of the kidnappers' imminent arrival. Goodall slipped away into the night with her Tanzanian staff; she knows the trails in the hills like the lines on the palms of her hands. She left Tanzania that same night and did not return to Gombe for two years. Her Tanzanian staff would continue the behavioural life history studies of the chimps alone. Goodall had to press her case hard with the Tanzanian government before they would let her return, and for two years white scientists were considered a liability in this region.

Years later Jane Goodall still feels the effects of that night in 1975. Today she makes frequent but random, unpublicised trips to Gombe and rarely stays more than three weeks at a time. These precautions are a direct result of the kidnapping. Trained African researchers now follow the lives of the chimps full-time. The publicity generated by Jane Goodall's life with chimpanzees has encouraged many young people to wish to follow in her footsteps, which in general terms has been a good thing for primate conservation and evolutionary philosophy, but it may not have been good for Jane Goodall specifically.

Originally Goodall wanted to be alone to study wild chimps, and in the early years her days in the forest were spent with chimps almost exclusively, until her world expanded with the arrival of Hugo Van Lawick and her son, Grub. But after the films were released and a line of students turned up, her forest idyll evaporated. She became responsible for the academic guidance and general well-being of groups of ten to twenty naive students. In 1968 tragedy struck: Ruth Davis fell to her death while following chimpanzees. It took six days to find her body. She was buried at Gombe. The arrival of students in the jungle made the whereabouts of these white Americans and Europeans obvious to various African factions. Ironically, if Goodall had not become so famous her private world with the chimps would not have been destroyed.

It was said that women were attracted to Louis Leakey like moths to a flame, and this was prophetically true in Dian Fossey's case. In 1963, before Fossey became Leakey's gorilla girl, she was working in a health centre for crippled children in Louisville, Kentucky. At school she had wanted to work with animals, but she did not get high enough grades to train as a vet; disabled children were next on her list. Though devoted to the children she cared for, Fossey never lost her love of animals.

At fourteen her mother, Kitty, a diminutive blonde, took her tall, dark daughter to see the doctor as she was worried by her size. Fossey was already 6 feet 1 inch tall. The doctor assured Kitty that although Dian was an oddity she wasn't abnormal. Fossey never felt she belonged. She considered herself to be fundamentally flawed and never believed she could be attractive to people. Fossey's parents divorced when she was three years old and when Kitty fell in love again Fossey felt betrayed. She always hated her stepfather.

Fossey was an unhappy child; she felt unloved by her mother and instead diverted all her affection towards the horses she rode at the local riding school.

When Fossey first read about Goodall's work in the *National Geographic* magazine she was a single, thirty-one-year-old career girl, living in Louisville, Kentucky. There was no man in her life. She would not date anyone who wasn't taller than her and this self-imposed caveat narrowed down her opportunities to settle down and start a family, even though a large part of her longed to fall in love and marry. Fossey was finding her life in Louisville increasingly oppressive and restrictive, and grew bored by the people around her and irritated by her lack of autonomy.

As Fossey thumbed through the pages of the *National Geographic* and gazed at Van Lawick's photographs of Goodall with the chimps, she became fixated. In Fossey's eyes Goodall was an exemplary role model: a lone woman can live and work in the wilds of Africa! The inspired Fossey decided to borrow some money and treat herself to an African safari. Olduvai Gorge and the gorillas of Mount Muhavura were on her itinerary. Unfortunately when Fossey arrived at Olduvai she embarrassed herself. In her enthusiasm she ran towards one of Mary's digs but misplaced her footing amongst the rocks, tripped, broke her ankle and fell head first into the dig, before vomiting over the giant fossilised giraffe Mary was in the process of unearthing. Mary showed concern for their clumsy visitor and gave Fossey orange squash to drink as her ankle swelled up, but Louis was unimpressed and ignored her.

In a great deal of pain, the determined Fossey stuck to her plans. Hobbling along on crutches she made her way with a guide to the Parc National des Volcans, which straddles the volcanoes on the borders of Uganda, the Congo and Rwanda. She and her guide tracked up the side of Mount Muhavura for six hours, Fossey stopping regularly to rest her tightly bandaged ankle. Fossey's guide led her to Kabara Meadow, a clearing where gorillas could sometimes be seen and where, at this time, Joan and Alan Root were camped. Friends of the Leakeys, the Roots were conservationists and at this precise time they were trying to make a wildlife film about the eternally shy mountain gorilla. They were dismayed to see a disabled American tourist and her guide struggling towards them.

Dian Fossey became fascinated by the Roots' lifestyle. She had never met anyone like them; habituating wild animals in order to make a film about them was a dream to her. She told the Roots how much she loved animals, especially her dogs and horses back home, and said she was desperate to see a mountain gorilla. Relenting, they took pity on her and led her into the thick vegetation. Following a gorilla trail, the Roots introduced Dian to her first wild gorilla. She felt a tangible sense of belonging that had for so long eluded her. These animals were her kin. At this point Fossey must have decided somewhere deep in her heart that she would make this cold mountain forest her home. In the end she would never want to leave it, and it would not let her go. Eventually it became her final resting place.

Back in Louisville, with the safari just a memory, Fossey was once more a frustrated and trapped occupational therapist earning $5,000 a year. Her trip had opened her eyes to the opportunities of life, but she now had an $8,000 loan to pay back with interest and only photographs of Africa to cling to.

Dian Fossey was destined to be unlucky in love. She nursed her broken heart after the tall man of her dreams, Alexie Forrester, had declined to tie the knot, not once but twice. He was a Rhodesian whose father, an Austrian count, had escaped Nazi persecution during the 1930s and found peace in Rhodesia. There he started a very successful tobacco plantation. Alexie's mother, Peg, was Irish Catholic and had connections in Kentucky through the Catholic Church. Dian Fossey's priest had told her about the Forresters and had shown her a photograph of their three strapping sons. Dian liked the look of Alexie. He was 6 foot 6 inches tall and had strawberry blond hair. She was not a bit worried that he was seven years younger than her, and she decided to visit their home in Salisbury, Rhodesia after her safari and before flying home. Fossey had grown up in a small flat in San Francisco, so she was impressed by the Forresters' expansive plantation and old-fashioned colonial lifestyle. Alexie wanted to know if Dian could plough when he first heard of her, and on her arrival she wanted to prove to him that she could. After saying hello for the first time Fossey enthusiastically jumped up into Alexie's tractor and ploughed his field, hitting a stone and blunting the blade just before completing the job. Nonetheless, Alexie was very impressed. Alexie and Dian liked each other and when Dian left Africa for Kentucky they made plans to see each other again. In 1964 Alexie arrived in Kentucky to see Fossey and to attend university. By early 1965 they were madly in love and planning to marry. At the same time, however, Dian's estranged father, George Fossey, had walked back into her life. Dian was pleased to rediscover her father, yet she was angry with him for deserting her in the first place; it was an emotional reunion. Dian's stepfather had tried to control Alexie and Dian's marriage by telling them to get married straight away and not wait until 1968, when Alexie would graduate. But Alexie did not want to rush things. He found both Dian's biological father and her stepfather hard to understand and people had suggested that the age gap between him and Dian could cause marital problems later in life. With such additional complications presented by friends and relatives, Alexie decided he wanted to have a job before he married Dian, so the marriage, which had seemed imminent, was postponed indefinitely.

Three years after her safari, in 1966, Fossey noticed that Louis Leakey was booked to give a lecture on the life of early man in Africa at the State University. Feeling unwanted and desperately searching for something, Fossey queued for front-row tickets to the lecture, and afterwards jostled with students to speak with Leakey. Fossey was relieved that he remembered her, and had brought along a project she had put together detailing her safari. She had written to the Roots on her return and Joan had replied, sending Dian some photos of the gorillas. Fossey pretended to Leakey that she had taken the professional-looking snaps; he was impressed.

Leakey was interested in Fossey, as he was already looking for someone new to study the gorilla, though at first he was not totally convinced that she was right for the job. Some years later Fossey spoke of this incident. She said Louis's blue eyes twinkled with mischief when he asked her if she would like to be his gorilla girl; she answered in her Southern drawl, 'Of course, of course, my God.' Maybe he did not find her very sexy at their first meeting? Fossey had at this time an outbreak of boils on her face. He eventually decided to put Dian's commitment to a test. This story remains a potent part of her mythology and therefore it needs emphasising that it really did happen.

Leakey told Fossey she had to have an appendectomy before going into the field. He told her she could suddenly fall ill with appendicitis in the jungle and die, so she must have the operation before he would take her seriously. Although Mary had once become very ill with appendicitis, this was a ruse Leakey used to test Fossey's commitment, but how was she to know? She was initially stunned by the request, but determined to be rid of the said organ. She has since said that she felt that if an appendectomy presented the choice between working with gorillas or staying in Louisville, 'then anything, appendix, ovaries, spleen, get them out!' She was very poor at the time and couldn't afford the cost of the operation, so she had to embark on an insane and melodramatic assault on the accident and emergency rooms of various hospitals in Kentucky. Feigning appendicitis, Fossey would turn up doubled over in 'agony', grabbing her waist and crying out for help, though she kept forgetting which side was supposed to hurt. Fossey had to go through a series of examinations at a number of hospitals before her acting technique was perfected and she finally convinced one surgeon to perform the operation. Within two weeks of meeting Leakey in Louisville she was waking up in a hospital bed after the operation. At the moment of consciousness Fossey said she groggily thought to herself, 'My God, are the gorillas worth this?'

Hospital doctors are today aware of patients faking pain and begging for unnecessary operations, not because they are swamped with nutty women who wish to live with apes at any cost, but because it is often described in behavioural psychology as a feature of Munchausen's syndrome. But thirty years ago, after a number of humiliating hospital encounters Fossey got away with it. Her tenacity impressed Louis. He wrote and told her she hadn't really needed to undergo the operation, which at the time perplexed her. But later she said whenever she had a stomach pain while at Karisoke, her Rwandan study site, she felt some comfort knowing it wasn't appendicitis because, 'my appendix was in a garbage can somewhere in Louisville, Kentucky'.

Leakey left for Nairobi and Fossey waited with bated breath, resigning from her job and selling many of her belongings. As the weeks and months passed she became anxious, writing to Leakey regularly, demanding an explanation for the delay. He wrote back trying to calm her; he was on the case and she would have to be patient. Leakey went back to his old patron Leighton Wilkie. In November 1966 Leakey wrote the following letter to Wilkie:

> You will, I am sure, remember that many years ago you afforded me invaluable help in the way of a grant, to enable me to get the Chimpanzee Research, by Jane Goodall, started.
>
> As you are doubtless aware, she has made a magnificent job of her research, and at the moment, is writing

up her scientific results in the form of several books. She has also published a number of scientific papers, as well as two popular articles which have appeared in the *National Geographic* magazine.

I am sure that you are aware of some of the results, either through the *National Geographic* magazine articles, or through the television film that was shown last year in the States, and when her first book is published, we shall send a copy to you at once. You will, therefore, be aware that Jane has established that chimpanzees, under entirely wild and natural conditions, regularly make and use primitive tools. Indeed, as a result of this discovery by her, anthropologists have been forced to abandon the old definition of man as 'the primate that makes tools to a certain and regular pattern'.

Knowing your interest in primitive tools and tool-making, as well as in Early Man, I am writing to ask whether you would like to receive two or three examples of these exceedingly primitive tools that chimpanzees make, since we now have some available. I am also wondering whether you would be willing to help me launch a similar piece of research in respect to the Mountain Gorilla. As you are aware, it is always the initial launching of a research scheme of this sort that is so difficult. Once it has got underway, and has been shown to be a successful project, the considerably greater funds that are required to maintain the research can usually be obtained with less difficulty.

You presumably know of the study of the Mountain Gorilla by George Schaller. Personally I feel it was magnificent as far as it went, but it was very incomplete. For some time now, therefore, I have been looking for a suitable research worker to continue where Schaller left off, and in view of the greater patience of women, I have been looking for a suitable girl to do the job, and follow Jane's example.

I have now found what I believe to be an indubitably suitable candidate, and I am anxious to get her launched into the field to study Mountain Gorillas as soon as possible, and I am wondering whether you will be willing to help us with a grant in aid. If you are willing to consider doing so, I will send you more information, and you could also perhaps meet the girl in question, as she is American, and is still in the States.

Because Goodall had been so successful Wilkie decided to fund the gorilla project and he donated $3,000. Nine months after having her appendix removed, Fossey was once again flying to Africa. But she was not as well briefed as Goodall had been prior to her study. Fossey had no anatomy training or tips on gorilla observation, as Goodall had been given with regard to chimps. Instead Leakey would encourage Fossey to jump blindly into the deep end.

Fossey visited Leakey at his home on arrival in Nairobi. When Mary met Fossey she looked her up and down and asked her if she thought she really could outdo George Schaller, the current world expert on wild mountain gorillas following his one-year stint at Kabara. Fossey felt stunned by Mary Leakey's remark. Schaller's research would be a hard act to follow, but there were many gaps in his work. Although he had made a census of population numbers and range of territory, he had not habituated and observed the social interaction of gorilla family groups. Gorilla family bonds were yet to be observed and this area of study would become Fossey's preserve.

Leakey asked Alan Root to help Dian set up camp at the Kabara

meadow site. In return he could make a film about her work for
National Geographic. Alan agreed. A tent was bought, along with
binoculars, an old Land-Rover, tins of meat and fish, salt, pepper
and herbs to season locally grown vegetables, a live hen and a cockerel. Fossey was now ready for the drive to the Congo. It is
questionable how exploitative Leakey was of the women he sent
into the bush. She celebrated her thirty-fifth birthday on her first
day at Kabara, thrilled that Leakey and Root were organising her life
for her in this way, but Leakey should have taken more notice of the
political climate in the region. The Tutsi and the Hutu were murdering each other and white people were fleeing the area. It was
1967, and Fossey was walking into danger.

Root stayed at the camp for a few days, giving Fossey a crash
course in gorilla lore. Even though a gorilla is a massive beast, it
becomes invisible in the dense rainforest. If you are not a skilled
tracker they will hear you coming and flee and you are left none the
wiser. The terrain Fossey had to work in was very different from
Gombe. Except for the rainy season, every day Goodall was bathed
in warm, bright sunlight, whereas Fossey would be shrouded in
mist and soaked with rain. Gombe was a forest on a hillside, and
though it certainly wasn't easy for Goodall to explore the chimp's
world, at least she did not have to pull herself through impreg-
nable vegetation up a mountainside where the air was thin and the
mud thick, as Fossey had to. At first, Fossey would march purposefully along a gorilla trail in the direction from where the animal had
come, rather than the direction it was going.

A shy and patient man, Alan Root helped Fossey to feel at home
as much as he could, but after a matter of days it was time for him
to leave. She remained in camp with her three African assistants,
who were living in Schaller's old cabin. Her Swahili was hopeless
and their English non-existent; Fossey felt more relaxed talking to
her chickens. When Alan left she had to hold on to her tent pole to
stop herself from running after him. She was frightened of the

unknown and scared of the loneliness, but she was determined to outdo Schaller and stick it out for two years and then go on to study the orang-utan for Leakey.

If they didn't vanish first, the gorillas were sometimes hostile to Dian, and she was charged many times. Leakey had implored her not to turn and run, no matter how terrifying the situation, but standing her ground against two massive charging adult gorillas that were screaming and exposing their large, ivory fangs, was too much. Fossey would stand still as they rushed at her and at the point of impact she would throw herself into the bushes, the terror of the moment making her soil herself. Her instincts told her to mimic their sounds of contentment – a low monotone *aaar*, *aaar*, *aaar* – and to nibble at leaves the way they did. It seemed to work, and over time she habituated three separate family groups as the animals became accustomed to her. The Kabara field site covered the 25 miles of forest between Mikeno and the Karisimbi Mountains. Louis was pleased with Dian's progress and organised National Geographic to draw up a contract for Fossey in just the same way it had for Goodall.

Fossey wrote to Leakey regularly, telling him of her successes but never of her problems, such as her close shaves with charging gorillas, the persistent coldness and wetness of the mountain, and the cultural and language barriers between herself and her African staff, which she felt were insurmountable. Fossey feared that if she shared her insecurities her research grant would cease and she would have to return to Louisville, which in her mind had become a fate worse than death. She would suffer in silence.

Fossey had a crippling fear of heights because of a car accident as a young woman. Clambering about the mountainside after a family of gorillas, 10,000–14,000 feet up, caused her phobia to take over, and at times she would become frozen on a precipitous climb as she tried to follow the gorillas around the volcano cones. Fossey would try to keep this handicap a secret, forcing herself onwards and upwards, refusing to give up. Robert Hinde, her future supervisor at Cambridge University, remembers being shocked by Dian's psychological handicap; he was forced to grab her with both hands and pull her up a cliff face at Karisoke after a panic attack had frozen her rigid. He had great admiration for her when he realised the terror she felt on the climbs.

Fossey had been at Kabara for a few months when things changed dramatically. Politically the area was in chaos and embassies were battling to move their nationals out of the region. President Mobutu had closed the borders and ordered a radio broadcast to the Congolese, stating that all white people were enemies of the state and should be killed. But on the mountain it was like being on another planet. The world's problems were passing Fossey by, so on 9 July 1967 she was surprised by the arrival of armed guards who made up a story that for her own safety she must pack up camp and return down the mountain. Fossey protested but eventually complied, not realising she was actually being kidnapped and taken prisoner.

Her father, George Fossey, ended up raising the alarm in the United States. From news reports George reasoned the fighting was too close to her camp for safety. He contacted the National Geographic, leading to frantic phone calls regarding her safety, while Leakey tried to organise a helicopter-lift. They were unaware she was no longer in Kabara.

Fossey was kept for sixteen days at a farmhouse in Rumangabo, an enemy of the people. She was lucky not to have been killed. In the years following she hardly spoke of her imprisonment; when asked she clammed up the way she had done as a girl. But she did tell Biruté Galdikas and Louis Leakey that she had been held in a cage and repeatedly raped, spat on and urinated on, and that later she was put in a cage with some white men all of whom were murdered. Fifty kilometres from where Fossey was held eighteen white people were eaten alive.

Fossey knew she had to escape; she told her captors that she needed to pay for her jeep and that she would return if she were allowed to do this. Amazingly the drunken soldiers agreed to let her go temporarily to settle her debts, as long as one guard travelled with her. Fossey made her way, with her beloved hen and cockerel, over the border into Uganda to 'get the money'. Fossey stopped at a hotel, the Traveller's Rest, and ran inside and hid in a back room. The guard followed his prisoner into the hotel. At this point Baumgartel, the hotelier, a well-known character and known to Fossey, told the guard to leave, as Dian was staying put. The soldier retorted that he would be shot if he returned without Fossey, but Baumgartel stood firm. Ugandan soldiers escorted Dian's Congolese kidnapper back across the border. Her abductor swore if he ever saw her again he would shoot her on sight. Fossey was the last white person to get out of Eastern Congo alive.

Baumgartel urged Fossey to return directly to Nairobi, but only after Baumgartel agreed to look after her pet chickens would Dian board the plane Leakey had sent for her. Leakey admired Fossey's spirit, and wrote to his friends praising her character. Leighton Wilkie was so appalled by the 'blood-chilling' account that he immediately sent another $3,000. Fairfield Osborn of the New York Zoological Society, also shocked, wrote back to Leakey saying, 'These young women go out alone to far places, obviously relishing the risks involved. Do you think they are trying to prove they are better than men?'

Leakey nursed Fossey back to health in Nairobi. He made an appointment for her to see a doctor, and took her out to the best restaurants for dinner and on drives through game parks. It was probably during these five days that Louis fell in love with her, although it would be another two years before they finally made love. Leakey felt terribly sorry for what had happened, while Fossey tried to brush the experience aside, afraid of being sent back to Louisville, protesting that all she wanted was to return to the

Virungas. She had a hunch that gorillas were living on the Rwandan side of the mountains, in between Mount Karisimbi and Mount Visoke. She would name her new camp Karisoke – surely she would be safe there?

It wasn't long before Leakey's death that his third trimate, Biruté Galdikas, stepped up on to the pedestal. Galdikas says of herself that she was always, even as a child, single-mindedly ambitious; she felt she was destined to achieve something great with her life. Over the last thirty years Galdikas has remained focused entirely on the research and welfare of wild orang-utans.

She was born in West Germany in 1946 to Lithuanian refugee parents who were escaping from the Eastern bloc after the end of the Second World War. Galdikas's family settled in Toronto when she was a toddler; as a little girl she had always been interested in human ancestry and stories of Stone Age man. When she was six-teen her parents and younger brothers and sister moved to California. Aged nineteen, Galdikas attended the University of California in Los Angeles, where she chose to study anthropology.

In one of the first lectures Galdikas attended, the professor men-tioned – in passing – the young British woman living with chimpanzees. The lecture moved on to other topics, but Galdikas became stuck in the moment. The professor's aside had a reso-nance for her alone out of all the students in the hall; she says in that moment she saw her own future. She knew, in time, that uni-versity lecturers would talk about her in the same way that Jane Goodall had just been described.

While at university Galdikas firmed up her plans to study wild orang-utans. She has said that she chose orang-utans rather than the other apes for a comparative human study, because their facial expressions are so like ours. By the time she met Louis Leakey, Biruté Galdikas had already, but unsuccessfully, tried to raise funds to support herself in a study of wild orang-utans. In 1969 Leakey came to give a lecture at UCLA, and one of the students asked him of the relevance of primate studies in relation to human behaviour. Leakey declared these studies were imperative. He patted his jacket pocket and, playing to the gallery, declared, 'I've just received this morning a telegram from Dian Fossey – wild gorillas are sitting at her feet, untying her shoe laces!'

By 1969 Goodall had been with the chimps at Gombe for ten years and Fossey had been with the gorillas for three. *National Geographic* and other publications printed articles about the women's research. Intimate pictures of Fossey being groomed by gorillas and images of Goodall and her young son observing chimps had a fairytale element to their appeal. The photos disguised the hardship and the local African helpers imperceptibly. Both Goodall and Fossey were now stars. Galdikas knew all about them and was very much in awe.

Galdikas approached Leakey after his talk, in the way that Fossey had, and she told him of her wishes, like Fossey. He seemed cautious, but arranged for Galdikas to meet him again the next day. She visited Louis where he was staying with his old friends the Travises and the eccentric scientist tested her abilities, just as he had with Fossey. Leakey asked Galdikas if she would have her appendix removed, and she said she would – and her tonsils if he thought it was necessary. But she did not, for good measure, throw her ovaries into the organ cocktail, as Fossey had offered.

Using ordinary playing cards, Leakey performed a couple of tricks on Galdikas in order to test the quickness of her mind. Leakey laid a pack of cards out face down on a table and asked her to tell him which were black and which were red. Galdikas replied, 'I don't know which are which, but half of these cards are slightly bent and half are not' Leakey was pleased with her observation. He told her that, like Goodall and Fossey, Biruté had passed the card trick, though apparently all the men Leakey had tested had failed. A card trick like this tests whether people see what is actually there or 'see' what they expect to see and are therefore misguided by preconceptions. After telling Galdikas about his faith in women's special abilities, such as their eye for detail, it was agreed that she would, if the money was forthcoming, become Leakey's third ape lady.

Through the 1950s and '60s Barbara Harrisson was considered by most to be the world expert on the orang-utan. Then married to author Tom Harrisson, she was based with him in the former British colony Sarawak, Borneo. Infant orang-utans, orphaned when their mothers were killed and eaten by local tribes of Dyaks, were given to Barbara Harrisson. She turned her home into an orang-utan orphanage and had five infants and juveniles living with her. She was the first person to try and rehabilitate orang-utans, but she was not successful and all of her youngsters were eventually housed in zoos. Leakey thought it important for Galdikas to meet Harrisson. He arranged for the two women to meet in New York to talk orang-utans. Harrisson also briefed Galdikas on the importance of social hospitality in Borneo and she enlightened Galdikas on the question of the multifaceted nature of a simple sarong. Today wild orang-utans have been almost entirely exterminated from Sarawak's diminished forests where Harrisson once worked.

It took two years for Biruté Galdikas's grants to arrive from National Geographic, the Wilkie Brothers and Louis's Leakey Foundation. In the mean time Galdikas fell in love with and married her brother's friend, Rod Brindamour, high-school drop-out two years her junior. Conveniently Rod also nursed a desire to travel to exotic climes. Leakey was thrilled when Biruté announced she'd married, and he knew it would help her to stay in the field and complete her Ph.D. if she had the support of a man. Leakey proposed that Brindamour become Galdikas's site manager and photographer. Because of the temporary delay in getting Indonesian government permits Leakey suggested that Galdikas and Brindamour study the little known bonobo in Zaire instead, but Galdikas declined the offer. She was fixated on orang-utans alone. Just before leaving for Borneo Leakey suggested to Galdikas that she have a clitoridectomy as Kikuyu women do. Leakey assured her that if she had an operation to remove her clitoris she would abstain from sex with Brindamour and therefore avoid getting pregnant. Leakey did not want motherhood to prevent Galdikas from studying the orang-utans. Galdikas says she tried to put his mind at rest by telling him she would use a more conventional birth-control method.

By September 1971 Galdikas and Brindamour were at last en route to Kalimantan. Louis arranged for them to stop off first at the National Geographic offices in Washington for Brindamour to take the society's crash course in photography and collect essential equipment. As Fossey and Goodall had proved, a photographic record was essential. The newly-weds then flew to Kenya to spend time with Louis and Mary at Olduvai Gorge, and then moved on to Gombe to receive field observation training from Goodall.

Galdikas found Mary Leakey a difficult character; she couldn't see why Mary had to be so mean. On the other hand she thought Louis was a lovely old 'paternal' white-haired genius. She had no idea he was such a philanderer. If some women show an overriding love for apes, Louis showed an overriding love for women. When Galdikas first met Louis he had less than three years left to live. His body was disintegrating fast, but he was still notching up sexual conquests.

When Galdikas and Brindamour met up with Mary at Olduvai Gorge, Mary dismissed the primate studies outright, telling Galdikas that Fossey was mad, Goodall was manipulative and Louis was a womanising, profligate time-waster. Galdikas was bewildered and disturbed to hear Mary talk this way, but she was a newcomer to the Leakey's extended family. It would take her time to feel at home amongst these larger-than-life characters. In some ways the young Galdikas was naive, just as Fossey was defensive and Goodall stoical.

One evening at Olduvai camp, over a bottle of whisky, Mary warmed up and said Galdikas wasn't like the other primatologists. When she and Brindamour left camp Mary stood with her Dalmatians and waved to them until they were out of sight. Galdikas's mother, Filomena Galdikas, runs the Californian office of her daughter's Orangutan Foundation. In the late 1970s, after Louis's death Mary Leakey and Filomena found themselves at the same anthropological conference in Los Angeles. Filomena introduced herself to Mary and thanked her for all her help in guiding Galdikas through her primate studies. Mary abruptly told Filomena not to thank her as she had nothing to do with it. At that point the conversation came to an abrupt end, with Mary walking away from a silenced Filomena. The two women never spoke again. Gratitude and acknowledgement for the primate studies was the last thing Mary wanted.

In late September 1971, Galdikas and Brindamour moved on to Nairobi. Louis continued to test Galdikas, trying to boss her around by continually asking her to undertake little chores for other people. When she refused to run any more errands, Louis giggled with delight; she'd passed another test. After she emphatically told him 'No!' Louis seemed to respect her all the more. When Louis waved goodbye to them at Nairobi airport, Galdikas took the image of his beaming face with her as the last memory of her 'spiritual father'. She never saw him again.

By 6 November 1971 Galdikas and Brindamour finally arrived in Kalimantan's rainforest. They had been in Kalimantan, Borneo just ten months, and in constant contact with their first orang-utans, a female named Beth and her infant Bert, for eight months, when they learned of Louis Leakey's death. One evening, as they sat near their tiny transistor radio, their vital link to Western civilisation, Galdikas and Brindamour listened to the Voice of America news station. Isolated in a rain forest, they learned of the death of a 'prominent anthropologist' and grieved.

In 1977, five years after Louis's death, Mary Leakey unearthed her greatest fossil find, the Laetoli footprints. The Laetoli plateau is located 30 miles north of Olduvai, in the shadow of the once mighty Sadiman volcano, where the land is a montage of earthquake gullies, acacia trees and lava boulders. Some 3.6 million years ago, some of the volcanoes that created the Great Rift Valley were still bubbling, and during the dry season the Sadiman volcano started to erupt. Over a period of weeks, Sadiman continued to spit out ash as the dry season turned to rain, the falling water mixing with the warm volcanic ash to turn it into cement. One day Sadiman gave a final, angry eruption and spat out a thick, wide covering of ash that sealed the ground and any footprints that had been left, in the cement. In the 1970s Mary Leakey decided to dig there, feeling sure the area would yield fossil treasure.

Tick typhus infected her team and puff adders, buffaloes and elephants threatened the safety of the palaeontologists. Despite these distractions, buried in Sadiman's ancient cement Mary Leakey found the tracks of twenty different species of animals, from elephants to sabre-toothed cats. The indentations from the splattering of raindrops were fossilised all around. Within this area of the plateau, the fossilised bones of these same creatures were found in deposits a little younger than their equivalent footprints. Mary was delighted, but what she really wanted was ancient human footprints. Eventually she found what she was looking for.

The Laetoli footprints must be the most extraordinary hominid fossils ever found. Preserved in ash for 3.6 million years, after Mary Leakey excavated all 89 feet of them they looked just as they did on the rainy day when they were made. There are three bipedal tracks of footprints, from three people: one large, the male; one smaller, the female; and a child's track. Were they out foraging for food? Was this the first attempt to form a nuclear family? It has been suggested so. Or perhaps the female and child, while foraging, had strayed too close to another community's range, and as with chimpanzees, the larger male was kidnapping the female with her child in tow?

Whatever was happening in the lives of these early hominids is uncertain, but what we can tell from the tracks is that these three individuals were purposefully going somewhere when Sadiman erupted. They were obviously frightened by this explosion and trying to distance themselves from Sadiman's destructive force. The whole of their bodies would have been covered in the ash and, as the rain fell, they would have been caked in a thin layer of soggy cement that would have matted their hair and hindered their vision.

At first Mary Leakey was confused by the relationship between the tracks. The three tracks were so close together that they could only have been made by hominids holding on to each other and huddling close for comfort from the common threat as they walked along. When Alan Root came to film Mary with the footprints, he gave the right explanation. Having watched chimpanzees and gorillas with Dian Fossey, Root was sure he could easily explain the tracks. The larger individual was in front, the child walked slightly to the left of him holding his hand and the female held on to the male's hips and walked behind him, stepping into his tracks. Root had seen chimps behave like this many times. Chimps play follow-the-leader and step into the footprints made by the individual in front and, when scared, hang on to each other, just as we do. The comparative ape and early human study that Louis Leakey initiated helped to enlighten people to the nature of these fossilised footprints.

From the tracks we can see that whenever the male faltered, the female and juvenile did likewise. They followed his every move. The female's head was down to keep the rain and ash out of her eyes; stepping into the male's footprints would have helped prevent her from being left behind. As the threesome battled against the elements a primitive three-toed horse, a hipparion mare, cantered close by the hominids with her foal, also trying to escape the volcano's range.

By his early twenties Richard Leakey was already becoming a professional rival to his father, though his style was different from Louis's. Talented but cautious like his mother, Richard Leakey was about to become a hugely successful palaeontologist. He would go on to discover more *Homo* and *Australopithecus* skulls than his parents. He had chosen to excavate on the shores of Lake Turkana in northern Kenya, and his team would locate a new hominid, his famous 1470 skull, named after its field number. Found at Koobi Fora on the shore of the lake, 1470's brain capacity is equal to that of *Homo erectus*, yet the skull is dated a million years earlier. Richard's 1470 fossil gave yet more credibility to Louis's theory of the bushy *Homo* lineage, with several hominid species living alongside each other just as today several species of ape and monkey live in parallel.

One of Richard Leakey's greatest triumphs was the 1984 discovery by his team of an almost complete skeleton of a *Homo ergaster*, popularly known as Turkana Boy, who lived 1.4 million years ago before he collapsed and died, face down, at the edge of the Nariokotome river that leads into Lake Turkana. His body sank into the mud and his flesh rotted away, some of his bones were damaged when his body was trampled by prehistoric hippos and giraffe, their footprints remaining preserved in the mud. Over the last few years Richard's talented Hominid Gang have sieved through tons of mud searching for all his bones, and, tarsal by metatarsal, Turkana Boy has been pieced back together. It is a remarkable find. *Homo ergaster* was tall, with a large brain and a slender figure, similar in build to the tall, slim people found living in that region of Africa today, and he had adapted to living on the open plains. Although Louis Leakey was delighted by his son's growing success they had become competitors and would remain so, only managing a partial reconciliation shortly before Louis's death.

After a well-publicised stint as head of Kenya's Wildlife Service between 1989 and 1994, when he put a stop to the poaching of elephants for their ivory tusks, Richard Leakey is still discovering hominid fossils. But the arduous nature of palaeontology is even tougher on him now. In June 1993 he lost both legs below the knee in a flying accident that some suspect was an attempted assassination, and he now survives on one kidney, donated by his brother Philip. In 1998 Richard was asked to return to his role as head of Kenya's Wildlife Service. He gladly accepted. His wife Maeve and daughter Louise continue to discover fossils

Possibly the most momentous early hominid fossil discovered in Africa does not belong to a Leakey, but to Ron Clarke, who in 1994 was searching through boxes of fossilised bones at Witwatersrand University in South Africa when he realised they were not monkey bones, as labelled, but the bones of early man. The fossils had been found in South Africa's Sterkfontein caves near Johannesburg. Apparently the hominid tumbled into an inaccessible shaft and died, which explains why its remains were not eaten. Clarke returned to the cave and has since uncovered an almost complete skeleton, including skull, hands and feet. Known as StW3 or Little Foot, the remains are dated at 3.5 million years old, the oldest hominid bones ever found. It would have been Little Foot's immediate ancestors who left the footprints at Laetoli.

The *Australopithecus* hominid, which has been compared to contemporary bonobos, lived alongside archaic chimpanzees in ancient tropical forest that once covered South Africa. Clark isn't yet sure if Little Foot is male or female, as much of the skeleton is still imbedded in rock, but this ancestor walked on two legs and also swung through the trees. Over time, as Clarke carefully recovers more of

Little Foot from its stony grave, we shall understand more clearly how and when we separated from apes, though it will be Goodall's long-term chimpanzee research at Gombe that will help us to visualise how Little Foot behaved.

A recent hominid fossil find would have intrigued both Louis and Mary, but it came too late for both of them. Up until very recently it was thought that our ancestors did not mix or interbreed with the Neanderthal. It has been predicted that because the skeletal structure of the Neanderthal is much heavier and the pelvis wider than that of our ancestors it would have been impossible to interbreed. It has further been suggested that the Neanderthal's length of pregnancy was probably longer than ours, perhaps ten or twelve months. But a newly uncovered skeleton of a four-year-old boy has changed all this. In the Lapedo Valley, 80 miles north of Lisbon and 19 miles inland, a Portuguese archaeologist, Joao Zilhao, has found the 24,500-year-old fossilised remains of a child, but in a form never seen before by science: he was a hybrid. The boy had both *Homo sapiens* and Neanderthal morphology. The Cro-Magnons, who migrated out of Africa 100,000 years ago, and went on to become modern humans, lived alongside the Neanderthal whose remains are found in Europe, until 30,000 years ago when the Neanderthal died out and Cro-Magnon took over. From carbon dating we know this little boy was born 4,000 years after Cro-Magnon had migrated into what was Neanderthal territory. This find seems to prove that modern people are descended from both Neanderthal and Cro-Magnon ancestry and therefore we carry Neanderthal genes. The boy had been buried in red ochre and his body decorated with shells.

Human DNA holds all the information needed to make a person as well as a great deal of junk data that is no longer required. This junk DNA, deep inside our genes, contains sequences belonging to our ancestors – the DNA of *Homo antecessor, Homo ergaster, Homo erectus, Homo habilis,* and *Australopithecus afarensis, Proconsul* all the way back to *Altiatlasius* and beyond. In evolutionary terms it is more economical to keep the information filed away than to jettison it. Over time, some of this ancient DNA mutates and degrades, but nonetheless it's all there within our genes, layer upon layer, just as the fossilised remains of our forebears are found through layers of ancient soil.

Leakey had met his very first ape lady, Rosalie Osborn, in London's Natural History Museum in 1954. They fell in love and theirs was a love that endured. Leakey arranged for her to travel to Nairobi and work as his secretary, a cover for them to be together. Mary knew Louis had taken lovers before, but she quickly realised that Rosalie was much more of a threat to their marriage than the other women. Mary took to the bottle and their sons became distressed at their parents' arguments, especially Richard, who was ten. He fell from his horse and was badly concussed, and seemed to lose the will to live because he thought his father would leave them. Louis realised he had to end the romance with Osborn.

Leakey didn't send Osborn back to London, but to Rusinga Island, Lake Victoria, where Mary had found *Proconsul*, so Osborn could look for ape fossils. Leakey then sent Osborn to Mount Muhavura in Uganda to study the gorillas first studied by George Schaller. Osborn was very committed, making an excellent start, but after four months the funding ran out. Louis tried to raise money to send her to study the Gombe chimps, but at that time he was unsuccessful. Instead he arranged for Osborn to go back to England and attend Cambridge University to study zoology. She eventually returned to Nairobi some years later, where she found work as a biology teacher. She never married and always remained deeply in love with Louis Leakey.

With Vanne Goodall in London on 1 October 1972, eighteen years after Rosalie first met him, Louis Leakey died, aged sixty-nine. Richard Leakey would not allow Osborn to attend his father's traditional Kikuyu funeral, but she later visited the grave in secret, and continued to do so. After a year had passed the Leakeys had still not placed a tombstone on the grave. They had decided they wanted to use quartzite from Olduvai Gorge, the stone *Homo habilis* had used to make the hundreds of axe and arrow heads Louis had collected over the decades. But the Leakeys were all busy with their careers. By the time they organised matters and drove to the grave with their stone, Rosalie Osborn had taken it upon herself to do something and a headstone was already in place.

The Leakey family was stunned. To this day Osborn's headstone remains in place, with the acronym ILYEA carved into the bottom of the stone – The letters stand for I'll Love You Ever Always. Osborn always signed her love letters to Leakey thus. Most people think the loving sentiment on Leakey's tomb stone is from Mary, but for Mary this was a final, indelible reminder that she shared her husband with many other women. In response, Mary made her sons promise they would not bury her next to her husband, but cremate her and scatter her ashes at Olduvai, which they have done.

If Leakey had taken his teacher's advice and become a bank manager, if Goodall had remained a secretary – our world would look very different today. Olduvai might not have been excavated, the Gombe chimps might never have been studied and the closeness between modern ape and our ape-like ancestors might not have been explored as well as it has been. Also, without Jane and the subsequent Jane Goodall Phenomenon, we would have had neither Fossey nor Galdikas, and neither would we have had the many hundreds of other younger women who have followed in their heroines' footsteps and embarked upon long-term field studies of wild primates. Perhaps this should be the epitaph written upon Leakey's tombstone.

3

The Secret Life of the Female Baboon

Primates were studied before the 1960s, but only intermittently and mostly by male scientists. The studies that preceded Jane Goodall's research taught us that the natural order in the primate world was one where males were aggressively dominant over passive females. But with the influx of women primatologists, many of them inspired by Goodall's example, these beliefs were to be challenged in the 1960s and then overthrown in the 1970s. Significantly, women would unveil the female primate.

In the early 1960s primatology was formally established as a unique area of research. Certain universities in Britain and America started primatology courses within their anthropology and psychology departments. From the mid 1960s to the early 1970s, after the broadcast of the National Geographic Films depicting Goodall and Fossey, many young women were now choosing to study the sciences rather than humanities at university. Although Louis Leakey had presented Jane Goodall and Dian Fossey with opportunities to become field primatologists, it was Robert Hinde, an eminent animal behaviourist at Cambridge University, who was responsible for turning them into qualified scientists.

Robert Hinde was the man Leakey turned to when it was time for Jane Goodall to compose her doctoral thesis. A few years later Leakey would make a similar arrangement for Dian Fossey. Hinde is a tall, intellectual man, who doesn't suffer fools gladly. Like most traditional scientists, he will not let ideas and unsubstantiated theories take the place of hard, empirical data.

It was essential that Goodall gained academic credibility if she wanted people to take note of her discoveries and insights. In 1961 she was reluctant to leave Gombe, but she wanted to please Leakey and have a Ph.D. to her name. Goodall had already met the scientific establishment head on when she submitted a paper on the use of tools and the hunting she had observed in the Gombe chimps to Nature magazine for publication. It was returned to her with 'he' or 'she' crossed out and 'it' written over her type in pencil. At this time it was not considered scientific to acknowledge animals' gender in writing, and it certainly was not scientific to give them names instead of numbers. But, as Goodall says, 'primatology should not be treated as a hard science'. She crossed out the editor's corrections and reinstated 'he' and 'she' and the paper was published the way she intended. This was her first victory. Today all primates studied are given names and the members of the different lineages are given names with the same initial letter. For example, chimpanzee mother Flo gave birth to Faben, Figan, Fifi, Flint and Flame.

Under Robert Hinde's guidance, Goodall worked for her doctorate, spending her time in Cambridge writing up her results for a dissertation in the field of ethology (the natural social behaviour of wild animals). It was Hinde's job to help Goodall re-evaluate her subjective observations and start to question why the animals were seen doing these things.

Goodall's field notes were purely descriptive and she had

anthropomorphised the feelings of the animals, imputing intent. At one point in her dissertation Goodall had wanted to say that Fifi, daughter of Flo, had felt jealous. Hinde told her she could not say this because she could not prove it, but she could say that if Fifi had been a human child we would say she was jealous. Hinde remembers Goodall saying she felt so frustrated by the experience that she threw all of her notes into the corner of her room, tore at her hair and cried.

But Goodall's relationship with Hinde consisted of a two-way exchange of information. Hinde admits that he had not previously appreciated the importance of the study of individual animals until Jane Goodall, Thelma Rowell (a Cambridge student of his), Dian Fossey and other women primatologists whom he later supervised opened his eyes to the relevance of each animal's idiosyncrasies. Adaptations and selected changes to a species of animal evolve very slowly over time. Some changes may be anatomical and others behavioural, but certain individual animals who become more reproductively successful than others will pass their adaptation on to their infants.

Hinde was interested in how Goodall had documented distinct differences in character between her chimps. These personality traits made each chimp behave differently in any given situation. As evolution takes place on the ground, through the action of an individual as well as through accidents of nature, the differences between the animals' personalities are highly significant, with one animal's reaction influencing the behaviour of its friends and cohorts and its infants. A species does not evolve en masse; one individual can make a difference, a subtle change that will, over time, permeate laterally and lineally. It was observed that some female primates were more successful as mothers than other females. Some females gave birth to a great number of infants, most of which would survive to mature and successfully breed themselves, whereas other mothers did not become grandmothers so confidently. The behaviour of individual female primates started to become more significant to the scientists.

Jane Goodall received her Ph.D. but she was still at heart the keen field biologist Louis Leakey had wanted. The paradigms of the times, such as the popular notion of male dominant hierarchies and passive females, Goodall neither sought out, nor did she happen upon them; the lives of non-human primates were far more complicated than anyone at this time could have predicted. In 1965 with a Ph.D. under her belt, Goodall returned to Gombe and continued to observe the chimps in the unbiased way Leakey wanted her to do.

Although she would later become famous for her field research, in the early 1960s Thelma Rowell had been Robert Hinde's research assistant. She started her scientific career studying captive monkeys, undertaking with Hinde a number of mother and infant studies of Cambridge University's colony of macaques. In the late 1950s and 1960s the importance of mother–infant bonds was just being unravelled. John Bowlby, the respected psychologist and author of *Attachment and Loss*, wanted some research to be done on the psychological effects on infant monkeys of their physical removal from their mothers. He had already observed young children parted from their mothers, and had noticed that they experience a series of behavioural developments after separation. At first, the child panics, cries and protests; as the separation continues the child's protest turns to despair; and finally detachment takes over and the infant no longer seems to care. Bowlby thought children needed to form a lasting attachment to their mother, or their primary carer, in order to achieve sanity and confidence in adulthood, and believed even short-term maternal deprivation could cause long-term problems in personality development. Bowlby wanted to change hospital visiting hours, so that mothers admitted to hospital could still see their children and, conversely, children in hospital for treatment could have their mothers stay with them.

The infant monkey remained with its captive colony while its mother was removed for short periods of time, just as if a child was staying at home with its family while its mother went into hospital. The baby monkeys were observed passing through the same series of emotional states as children did, eventually becoming detached and anti-social, and monkeys separated from their mothers were seen to be less successful within the troop at finding food and making friends. It would be another twenty years before Sarah Hrdy's theory of infanticide would add a socio-biological reason for an orphaned infant's insecurities.

Signing herself T.E. Rowell, she submitted a paper on her findings to the *Zoological Society of London Journal*. The society was impressed and invited T.E. Rowell to come down from Cambridge and give a talk to the fellows, but when it was discovered that T.E. Rowell was in fact a woman there was some embarrassment. She could present her lecture but she was told she wouldn't be able to sit with the fellows for dinner because of her sex; instead she could sit behind a curtain, out of sight, and eat her meal. Thelma declined the invitation; after her lecture she shared a pork chop at a girl-friend's flat instead.

At the U.S. University of Wisconsin at this time the comparative psychologist Harry Harlow was also experimenting with infant macaques. These experiments were sadistic to the individual animal but they have, in retrospect, enlightened us to the fragility of the primate psyche – and to the twisted nature of Harlow's mind.

Harlow bred monkeys for laboratory experimentation, and was particularly interested in deconstructing the mother–infant bond. Forcibly parted from their real mothers and placed in solitary confinement, the tiny animals were given a choice of two 'mothers' to hold. The wire mother was a mass of wire with a bottle of milk fixed into its construction; the cloth mother consisted of a piece of wood covered in sponge and cloth, with a light bulb under the fabric giving off warmth, but had no bottle of milk.

Harlow was trying to prove that touch was a more important sense than hunger. He noted time and again how the insecure monkeys would cling to the cloth mother twenty-three hours a day and only jump over to the wire mother when hunger forced them to do so. Their empty stomachs filled, the monkeys would quickly return to the warm cloth. But as the infants grew into adulthood they were bereft of any social skills, which he had not anticipated. Temporarily released from isolation and reintroduced to groups of monkeys of the same age, the monkey would huddle into a ball and cover its face with its hands. Infant monkeys and apes inherit the social status of their mother, and if they have been deprived of a mother they can only assume the lowest rank.[1] They cannot mate successfully as they are unsure of the correct social behaviour at such times. Furthermore, these 'orphans' are more likely to contract disease, more prone to accidents and more likely to die before adulthood. It became evident that the physical attachment relished by an infant primate when carried and groomed by its mother is necessary for life. It is during the infant's early years that all the ground rules of social behaviour are learned. Without them a primate remains lost and confused.

Harlow wanted to continue to experiment with his subjects. When the females reached sexual maturity they were artificially inseminated in a trap he called his 'rape rack'. In evolutionary terms, unskilled mothers are failures. When a disturbed female gave birth she had no idea how to care for her infant; immediately removed, a new generation of infants to be nurtured by cloth mothers had been born.

Harlow added some cruel additions to his cloth mothers. Hidden spikes and electric shocks tested how much abuse a youngster could take. At this time the Cold War affected much scientific work. America's space programme needed research into long-term isolation for primates. Space is a cold and lonely place for both humans and monkeys and there were plans to send both into orbit for long periods of time.

Like Robert Hinde, Harry Harlow trained great numbers of students, though 90 per cent of them were men, whereas 95 per cent of Hinde's students have been women. Although Harlow was respected in his time and was awarded the American National Science Medal, his reductionist approach to the animals was anathema to most women's sensibilities.

Throughout generations of systematic abuse of monkeys, Harlow remained convinced that his monkeys were much happier and healthier than wild ones because they were free from disease and predation. Crisp and tidy experiments are possible in a laboratory setting, where data can be amassed and papers written. A field biologist has a much more holistic approach and cannot expect to test their hypotheses straight away. Sitting, waiting, and watching a story unfold is not a particularly efficient way to conduct science, but it does not involve cruelty and the results show the natural behaviour of the animal in question.

A metaphorical comparison might be suggesting to an anthropologist that she study the behaviour of patients in a mental institution to learn about human behaviour. Some basic human behaviour will be evident in mental patients but other behaviour will not be typical of the human species en masse; it will be behaviour peculiar to institutionalised humans. It is in this way that captive primates are, in certain aspects, behaviourally different from wild ones.

Over the years Goodall noticed that orphaned Gombe chimps developed psychological states similar to the detachment and loss felt by a motherless child. It wasn't long after Goodall's arrival at Gombe that she observed how chimp mothers were attentive to their infants; it is almost a lifelong love affair between the chimp mother and her child. Becoming an orphan is as socially devastating to a young chimp as it is to a human child.

Chimps migrate through their territory, moving from one flowering tree to the next. At Gombe the animals passed through Goodall's camp some frequently, some more infrequently, coming out of habit, to socialise and to supplement their diet with provisioned bananas. The camp became a meeting-place for the chimps.

Merlin was three years old when he returned to camp without his mother, Marina. Merlin was emaciated and motherless, and it seemed obvious Marina had died, for chimps will stay with their mother permanently until they are at least six years old. At camp the chimps gathered around the infant. They knew him and by their behaviour it seemed that they hadn't seen him for a long time. After the fuss had died down, his six-year-old sister, Miff, sat and groomed him for some time. At this point Miff effectively adopted her little brother. When it was time to go back into the forest she looked over her shoulder at Merlin and waited for him to catch up.

Goodall and some other students followed Merlin to study his progress. As the weeks passed Miff would groom him as much as Marina would have, similarly letting him ride piggyback and sharing her food with him. In times of danger, though, Miff was not big or strong enough to protect her little brother; she had only just found her own independence from Marina.

One day some adult males were displaying, competing with each other over their social status within the group. At a time like this there is much bluff aggression. Branches are swung around in displays of strength, and sometimes the situation escalates and blood is shed. At such times younger and smaller individuals should get out of the way, just as they would to avoid a predator.

Chimpanzees are as political within their communities as humans. Having power and powerful friends is the bottom line. Miff and the other chimps climbed into the trees, but little Merlin misinterpreted the situation and instead approached the displaying male, Humphry, pant-grunting in submission as though he was at fault and craving reconciliation. Merlin was in fact pestering

Humphry at a time when one wrong move could have led to Humphry's social downfall. He ran at the infant, grabbing hold of Merlin's arm and dragging him some distance along the ground. Merlin was hurt and did recover, but this behaviour was evidence of his psychological detachment from his society. When Marina was alive Merlin would have recognised the situation and, frightened, would have taken his mother's lead, jumping on to Marina's back so she could carry him into the trees.

Merlin failed to learn from his mistakes. He would continually try to ingratiate himself with adults at inappropriate times and suffer rejection, often being bullied in the process. He didn't know how to play with chimps of his own age and would become aggressive and defensive, frightening prospective playmates away. Before Marina died, Merlin was fast learning the art of termiting. The tools he made were often too short to reach the termites, but he had the correct idea. Two years after her death his use of tools was no better. He sat with Miff, an avid termite-catcher, but his skill was stunted and he never managed to improve on what Marina had taught him. Like young children, young chimps have short attention spans, but in this respect Merlin was more like an adult. He could sit alone and concentrate for an hour on his termiting; even though his technique never improved he seemed quite happy in his isolation.

Unable to find enough to eat, Merlin did not grow at the rate he should have. His bones stood out, his eyes were wide and staring, and his coat was patchy where he had pulled out clumps of hair. In the rainy season, when the temperature fell, his face and hands turned blue with cold. Goodall was surprised that he had lived for two years without his mother. When he was five a polio epidemic swept through the Gombe chimps, and Merlin was one of the first fatalities.

The 1966 polio epidemic ruthlessly claimed its chimp victims. Goodall had watched, mystified, as the first of the newborn chimp babies succumbed to the fatal disease. The infants would scream in pain and would lose the use of their limbs while their mothers tried to give comfort. When Goodall found out that there was a polio outbreak in the nearest town, everything started to make sense. Chimps are as susceptible to human diseases as humans, and vice versa. Jane panicked; she was, at this time, pregnant with Grub and neither she, Hugo nor their field assistants – let alone the chimps – were protected against the disease.

Goodall sent a desperate message to Leakey, who immediately flew in enough polio vaccine for both humans and chimps. Jane and Hugo administered the serum to the chimps in their bananas, but for fifteen of them the vaccine arrived too late. Eight died and seven were left crippled. Faban, Flo's eldest, lost the use of one arm, and Mr McGregor lost the use of both legs.

Bald Mr McGregor was one of the older male chimps. He did not hold high rank and probably never had. He was a slow purposeful type, not quick and ambitious like Humphry. When Flo, who was for many years the most sexually alluring of all the Gombe females, was in oestrous, McGregor was the last in line to mate her and during intercourse Fifi was able to push him off her mother and send him tumbling backwards down a hill. On one occasion, when McGregor spied colobus monkeys and purposefully led the chimps towards the prey, Goodall saw an irate male colobus chasing McGregor away! Colobus monkeys are much smaller than chimps and are the favourite species of monkey that chimps like to prey upon. It is exceptional for a colobus to confront an adult chimp in such a way.

During McGregor's last days, as he suffered the effects of polio, he was a grotesque sight. With both of his legs rendered useless he pulled himself along with his strong arms, but his backside was rubbed raw in the effort. Paralysed from the waist down, he didn't feel the pain of his sore backside but neither could he feel the use of his sphincter muscles and he became incontinent, with swarms of flies eventually feasting on his flesh and excrement. Goodall would spray fly killer around his posterior to try to give the dying chimp a moment of peace from the irritating hum of a thousand flies.

For some time Goodall had suspected that Humphry was McGregor's younger brother, and the chimp's behaviour during McGregor's illness confirmed her hunch. The other male chimps became very hostile to the ailing McGregor. They didn't seem to like the way he shuffled around, nor the flies, nor his smell. He was weak and they had an instinct to batter him, and groups of males led by Goliath would attack the helpless chimp as he cringed from their blows. Jane and Hugo would put themselves between McGregor and the other males if he was on the ground, but if he had managed, using his strong arms, to haul himself into the trees, Humphry would come to his defence. Humphry would display at the other males as they approached McGregor, and he even attacked the alpha male Goliath. As McGregor couldn't travel with a group to forage, despite pathetic attempts to follow them, Humphry would feel torn in two, wanting to follow the group but also feeling the blood tie to his brother, often staying back to keep him company.

On his last day, after all the other chimps, including Humphry, had left camp, Jane and Hugo gave McGregor a meal of eggs, his favourite food. While he was occupied they shot him. They did not allow any of the other chimps to see his dead body. When Humphry returned he looked everywhere for McGregor, and he continued to search for him for another six months. At the time Jane was severely criticised for putting McGregor out of his misery as well as for administering polio vaccines to the chimps. Other scientists thought she was interfering with the natural course of an animal's life and it was suggested that Goodall was following her feminine subjective emotions rather than adhering to objective scientific observations. But for Goodall her decision regarding McGregor and the dispensing of polio vaccines to the chimps was simply the correct moral action. In addition, if her study group had been wiped out by polio her observational research would have come to an end. Jane Goodall befriended wild chimps in a completely new way. As she was not a product of the scientific system, she was completely receptive to the chimpanzee behaviour she observed. She cared about the animals as though they were her friends, and never considered them to be unthinking, unfeeling research subjects. Only now, as we move into the twenty-first century and the fate of wild chimps becomes so precarious, is the ape's emotional and intellectual intelligence finally being appreciated.

Today, the latest psychological research into human and chimp personalities has shown that humans and chimpanzees have five basic character types. Human factors identified are: conscientiousness, openness, extroversion, neuroticism and agreeableness. Psychologist Lindsay Murray of the University of Chester has identified three factors to chimpanzees' characters: excitability, sociability, and confidence. These three factors result in five chimpanzee personality types: the excitably confident type, which are always adults; the sociably confident, which can be adults or juveniles; the excitably timid, almost always adult females; the sociably timid, usually juvenile chimps, and lastly, the sociably placid chimp, which is always a juvenile. It is now understood that an individual chimpanzee can have a personality almost as richly layered and complex as a human personality can be.

Back in the 1960s Thelma Rowell had wanted to remain within a university environment. She was not initially romanced by the idea of field work, as Goodall had been. Even with Robert Hinde's support, though, Rowell was unable to break into the old-boy network. In the 1960s women did not enjoy equality within university hierarchies, and forty years on career opportunities in academia are still somewhat biased in men's favour. Rowell has noted, 'In academia in the 1950s and '60s women academics were thought of as jokes or as eccentrics. The women themselves were realists and knew the network was closed to them because of their sex and you couldn't progress within the system unless you networked. At a monastic institution like Cambridge, as a woman, you couldn't even dine with your supervisor but your male colleagues could and this meant they could share thoughts and interests and network their way through the hierarchy. It is better today, but there is still bias towards men and men are still promoted by men over women because they are men. You find this more in Britain than in the U.S. In the 1980s at Cambridge most of the examiners were men and most of the women got second-class degrees, whereas men would be given mostly first-class or third-class degrees. In 1985 I decided to make a comparison with Berkeley University in the U.S. I found there that gender made no difference to grades. This was still true for Cambridge in the early 1990s. I expect it still is.' After Rowell gave her talk to the Zoological Society and her research on captive macaques came to an end she concluded that the life of a university academic would always elude her. Many women field primatologists had discovered that promotion within an academic institution would be denied to them, so field work beckoned. In a jungle a woman could be boss.

Rowell, now married to her zoologist husband, followed him to Uganda in 1962 where he had a post as a university lecturer. She decided to continue with her own primate field studies and embarked upon a six-year study of the olive baboon in a forested area of the Queen Elizabeth Park. As Rowell was a traditional university-taught primatologist rather than a scientifically unbiased self-taught researcher, like Goodall, she took certain beliefs with her to Africa. At university she had been taught the perceived wisdom of both Sherwood Washburn and his student Irven DeVore, baboon experts whose theories dominated primate studies. Washburn viewed modern man as a more civilised version of 'man the hunter'. It was the origins of this 'hunter' that lived in Africa's prehistoric savannah that intrigued him.

Louis and Mary Leakey's discovery of the 25-million-year-old *Proconsul* species of prehistoric primate on Rusinga Island in Lake Victoria, Kenya, in 1948 changed evolutionary thought and encouraged scientists to use monkeys as models for modern man. Washburn needed a student to study the 'sub-human' behaviour of baboons, as he wanted to see the origins of the hunter within us and male dominance hierarchies discovered in wild baboons. Irven DeVore was dispatched to Nairobi.

DeVore's study was very short and he had problems in habituating the baboons he studied in Nairobi National Park. He felt frustrated because they tolerated his wife but not him, so he provisioned them with peanuts to try to make them stay close by. It is tempting for researchers to feed their primates in order to achieve access but at the same time this gesture affects the creatures' behaviour; you are no longer studying truly wild animals.

DeVore found it much easier to observe the adult male baboons rather than the female baboons. The male baboon is considerably bigger than the female. He is an impressive creature with a large fluffy mane of hair and sharp canines that he shows off with conspicuous yawning. Occasionally the males bluff aggression, or actually fight. At first glance it could be said the males seem more active and more interesting than the dowdy females. DeVore had arrived in Africa under the influence of Washburn's teachings. He expected that the males would be aggressive and dominant over females, and when he met the baboons that is exactly what he saw. This is the intellectual trap that Leakey didn't want his trimates to fall into; it is easy to see what you expect to see – even if it's not there.

During DeVore's provisioning, rivalry between the animals intensified as they jostled for peanuts. The males would compete with each other for food, pushing the smaller females and juveniles aside. Washburn and DeVore presumed the males always pushed the females around this way, and did not make a connection between this aggressive male behaviour and the provisioning of peanuts. DeVore did not observe the natural situation where senior females, with their amassed knowledge of their environment, lead their group, including the large males, to food and water, even in a dry season. Most days baboons calmly follow an older female as she guides the group to a patch of root vegetables, where they will sit down and proceed to dig up the plants and eat.

To understand why Washburn and DeVore were not able to see the truth we have to appreciate the influence of scientist Baron Solly Zuckerman, a British anatomist who in the 1960s became the chief scientific advisor to the government. In the 1930s, Zuckerman spent only nine days observing wild baboons in Africa, shooting a few females for anatomical studies back home. He then turned his attention to a captive colony of baboons at London Zoo. During the 1930s people interested in primates only travelled to the tropics to go hunting or to capture primates for circuses or private collections. All primates studied at this time were either stuffed in a museum, belonged to a private owner or were represented by their bones, and most of the accepted 'facts' on primate behaviour were taken from the studies on captive populations. Captive animals behave differently from wild ones, especially if the ratio of females to males is unrepresentative of the gender split found naturally in the wild.

In the 1930s and 1940s the colony of baboons at London Zoo consisted of an unnaturally large number of adult males, with just a few females, whereas in the wild a community of baboons consists of a stable core of females and their infants, with the males in a minority. On reaching sexual maturity the males leave their mother and their natal group and find a new community to join, the inexperienced males existing on the periphery of the community as they try to ingratiate themselves with a senior female. Only with her approval will they be welcomed. In this way they transfer from group to group trying to find security and females to mate with.

The exception to the rule is the hamadryas baboon. The male hamadryas has a harem of females that he herds and dominates using the threat of aggression. He is usually only aggressive towards his females if he fears they will try to mate with another male, physically punishing them if they stray too close to a new male. He will also fight with the new male if the stranger invades his territory and tries to take his females from him. The baboons at the zoo were hamadryases, and the percentage of females to males was so small that it rendered the colony socially unstable.

The baboon colony at the zoo was in continual upheaval as the captive males, with nowhere to escape to, had to stay and fight with the other males over status, food, space and access to the bullied and underrepresented females. Anyone who watched the baboons could see that the animals were often very aggressive with each other and that the few females had a particularly difficult time. Deaths were frequent as weaker animals were killed in fights.

At the time neither Zuckerman nor anyone else knew any better, and assumed this exaggerated male aggression was quite reasonable. Zuckerman thought the baboons' behaviour was similar to that of man, and believed he had discovered that chauvinism had an evolutionary force at its root and that male dominance was the natural order. This male bias was adopted in the academic world and subsequently taught to generations of students, including DeVore.

But Zuckerman's theories on primate behaviour and male dominance were not the only views. Also in the 1930s, the American psychologist Clarence Carpenter from Columbia University undertook the first field study of primates in their natural habitat, studying the wild howler and spider monkeys in Panama over a number of months. He concluded that howlers lived in largely cooperative and not particularly hierarchical societies. His views were much closer to the truth, but his research was overlooked in favour of Zuckerman's research; it did not have the same appeal. Male scientists of the 1930s, 1940s and 1950s preferred to think of primates as having male-dominated social structures, which seemed to give credence to male hierarchies in human society. In the 1960s, as more women chose to become field primatologists, conflicting reports on primate behaviour developed. Women did not observe the male dominance of passive females, as had been reported by senior male scientists.

Although DeVore had difficulty in getting close to the animals, baboons are generally easier to habituate and study than most other primates. They are noisy, showy and adaptable and for the most part live in open grassland rather than forest. For protection from predators they retreat to high rocky outcrops or the branches of tall trees at night. They are active in the daytime and not usually camouflaged in thick foliage, but conspicuously squabbling in large groups in open savannah making them easy to observe. It is no surprise they were seized upon as the model for man's hunter-gatherer origins as well as other innate human behaviour. Today, when extinction threatens so many primate species, the savannah baboon's adaptability is serving it well.

In spring 1963 Thelma Rowell habituated the Ugandan forest baboons to her presence and began her study, spending two weeks with her baboons and two weeks back in Kampala with her husband. She worked entirely alone and slept in a small tent. Rowell observed her baboons and looked out for the male dominance hierarchies. She also waited for the aggressive male coalitions to show themselves, but she saw none. Rowell's male baboons were mostly peace-loving and were often subordinate to females. Males were not permanent members of the community, but as young adults would arrive and hope for acceptance. Rowell observed that older males would try to see them off, but if a female, especially a senior female baboon, liked the visiting male he would become her special friend and with her sponsorship would be allowed to join the group. The female baboons were hierarchical and political animals in their own right.

Rowell was stumped. What was wrong with her baboons? She felt it was inconceivable that her observations were right and that other senior behaviourists such as Zuckerman, Washburn and DeVore were all wrong. This type of professional conflict is exactly what Leakey wanted to avoid. Had Jane Goodall studied baboons she would not have questioned her data, as Rowell did; her academic naivety would have led her to her own conclusions. After Clarence Carpenter's study in the 1930s, Rowell was only the second primatologist to challenge the assertion that the 'natural order' sees male primates dominating the lives of female primates.

Rowell was also confused by the absence of DeVore's 'army model', a type of behaviour seen in some other mammals such as the cape buffalo and the musk ox but not observed in baboons. DeVore claimed the adolescent male baboons would move on the periphery of the group, deflecting attacks against weak and vulnerable baboons sheltering at the core. High-ranking males and females would form the next layer in this protective circle, with nursing females and infants encased in the middle. He described coalitions of males as 'troops' and individual males as 'fighting machines'. We now know this macho behaviour was imagined. In reality, when preyed upon baboons have the philosophy of 'every baboon for itself'. The long-legged adult males take flight, while females carrying infants lag behind and become easy prey for a leopard.

During the Cold War it must have been comforting to think that one could see the origins of male military defence in our relatives. Without sufficient data, DeVore's theory was welcomed and Washburn went on to publish widely on the origins and biology of conflict in non-human primates and our direct ancestors. Washburn's man the hunter, he believed, had evolved into man the warmonger in a world where woman the gatherer remained insignificant.

At the same time, just fifteen years after Pearl Harbor, Japanese primatologists were observing the fuller picture when they saw transitory male Japanese macaques playing second fiddle to a dominant related sisterhood of females. The females formed the social core of the monkey society, and their roles extended beyond that of being a mother. But Japanese researchers would have to wait until the 1970s before the West was ready to listen. The nefarious theory of male-dominant hierarchies stuck. Society at large could easily relate to the paradigm of male aggression and dominance over females in our primate cousins. Today DeVore says it was 'bad luck' that he didn't notice the centrality of the female's social position with baboons.

But back in 1966, Thelma Rowell was unsure of her next move. She did not feel brave enough to take on Zuckerman, Washburn and DeVore, but she did want to write up her findings. Rejection by the male-dominated academic world made it even more difficult for Rowell to speak out over their major scientific error, and even undermined her belief in her own observation. Rowell decided to blame the passivity of her baboons on the forest environment they lived in. Rowell claimed that life in the trees as opposed to life on the open savannah somehow encouraged peace and harmony and that was why the army model and male dominance were missing from her baboons. The subsequent myth of the arcane forest baboon that grew out of her research made Rowell regret her professional diplomacy. As the years passed and more women primatologists studied baboons, the false impression of macho male dominance was challenged by others, especially Barbara Smuts working at Gilgil in Kenya and Jeanne Altmann working at Amboseli, Kenya. By the mid-1970s Rowell felt confident enough to dissent, asking the polemical question, 'Is our own species more than usually bound by hierarchical relationships, at least among the males, who have written most about this subject?'

Another mistake that Rowell has tried to challenge is the belief that male dominance is rewarded by reproductive success. It is a widely held belief that powerful males have more access to fertile females and therefore father most of the infants. Although this has never been definitively proven, it is a popular theory, especially among male scientists.

In the study of animal behaviour before the advent of DNA testing, paternity was determined by who a researcher saw mating with whom. At this time Rowell watched ambitious young males challenge the status of older males who in turn would fight to retain their rank. She thought probably 50 per cent of infants were fathered by senior males and the other 50 per cent fathered by younger, lower-ranking males. More often than not the female would decide whom she wanted to mate with and would go and seek him out. A newcomer always held attraction for the females. Often the males would be so busy pushing each other around that they would fail to notice 'their' female creep off for a moment of passion with a new healthy young male. Even after dark, when the males are sleeping, females have been observed to mate with strangers. Although they both risk a battering, the urge to be together is a powerful one.

The arrival of DNA testing still hasn't proven whether the alpha male has more infants than the beta male, and whether the beta has more infants than the third in line, and so on, but it does seem to show that the five highest-ranking males are likely to have more progeny represented in an arbitrary sample of infants tested. Overall the data is still suggesting a fifty–fifty split, but male primatologists refuse to concur.

Thelma Rowell and some other women primatologists believe that men have a personal stake in the issue and desperately want a higher rank to be equated with a higher number of matings. Male scientists seem to think that if the hard work of achieving social status fails to bring sexual benefits then there's no meaning to life. On this point Thelma Rowell told me: 'There is a sex difference here. Male scientists fight furiously over their belief that the highest ranking male has the greatest reproductive success. Whether this is to do with their subjective hopes or fears, I don't know. But only half the DNA studies done on this can supply data to support this belief. The other half show that it was a subordinate male, who you never saw mating, who sired most of the infants. Men just want it to be true and will stretch their data to make it fit that picture.' This type of male primate behaviour can be seen in the mating strategies of male mangaby monkeys. The mangaby is a forest-dwelling semi-terrestrial monkey. There are two types of male mangabies: dominant, large males that mate confidently and openly with the females; the other male type is also adult but remains the size of a juvenile. These smaller, slimmer males hang around the females and seize opportunities to mate surreptitiously behind the dominant males' backs. Until a full DNA survey is undertaken with a wild group of mangabies, we will not know which male reproductive strategy works best.

Another woman primatologist who has studied baboons and focused her attention on the female baboon is Barbara Smuts. She was born into an intellectual household; her parents were historians who many years later returned to academia during their retirement. Smuts's father was seduced by evolutionary biology – recently Barbara and her father published a paper together – and Barbara's mother, at seventy-four, is publishing a scientific book on child development. Smuts says her mother has been a significant role model. In addition, as a child growing up in Michigan in the 1960s Smuts was a fan of Jane Goodall, watching the National Geographic specials on television and reading the magazines. As an animal lover it was amazing for her to realise that one could have a job doing something one liked; work didn't have to be drudgery. Barbara would pass a butcher's shop on her way home from school and she would ask for eyes, hearts and other body parts to dissect. She would sit at the kitchen table poking the organs while her tolerant mother prepared the family dinner.

Single-minded about her ambition and encouraged by her family, she first went to Harvard and as an undergraduate was advised to study anthropology because it was a 'soft' option for girls. Smuts went on to Stanford University to read psychology, and in 1975 headed for Gombe with the aim of working alongside her heroine. She had decided she was going to study the behaviour of adult female chimps for her Ph.D. After being released by her kidnappers Barbara Smuts was back in America in the latter half of 1975, recovering from her ordeal and unsure of her next move. It was impossible to continue to study the Gombe female chimps, even if she felt brave enough to return, and to this day the lives of the female Gombe chimps remain less well documented than those of the males. Smuts instead turned her attention to Kenyan olive baboons that were living outside a small town called Gilgil, near the larger town Nakuru. Back in Africa but now in Kenya, a politically stable country of a different terrain, with the glorious view of the double peaks of Mount Kenya in the distance, Smuts embarked on a new study of sex and friendship in baboons. She didn't just want to observe the behaviour of the female baboons, but wanted to ask questions. In the early days of primate studies, being the first person to observe an animal performing some action or showing some social behaviour was all that was required. But Edward O. Wilson's 1975 theories of sociobiology radically influenced biology and zoology and changed scientific approaches, and a more sophisticated study was now needed.

Sociobiology is the study of the evolution of social behaviour, examining how behaviour between individuals has a selective force behind it. It was believed that everything happens in an animal's life for a reason and the communication and relationships between individual creatures denote the sociality of the group. Field work was no longer purely observational but developing a more theoretical agenda.

British field primatologists were mostly trained in zoology and therefore had a Darwinian grounding in evolution. But many American field primatologists came from an anthropology background and were not trained in maths, chemistry and biology but in history, sociology and geography. Edward O. Wilson's neo-Darwinist theories caused more of a jolt to American primatologists than to British ones. It was much harder for American anthropologists to apply evolutionary ideas to behaviour than it was for British zoologists. The American scientists had to take on a more strictly scientific approach to keep up with the evolution of the science. In the 1960s, people were surprised to see co-operation amongst wild animals and many studies centred on co-operative behaviour. Wilson was very interested in the effects competition (especially male competitive behaviour) has on the individual and ultimately on the group. His influence meant that from the mid-1970s 'cooperation' was focused upon. In the 1990s field primatologists went back to re-examine co-operative behaviours.

Wilson theorised that all behaviour co-operative or competitive was a strategy or a game. Costs and benefits had to be assessed by the individual at a primal level before it acted on its instincts. Wilson believed there was probably a gene for most of these behaviours or strategies, and twenty years later the research substantiates his beliefs; for example, it seems primates have one gene for selfishness and another gene for altruism. From the mid-1970s onwards the world of primatology did not just want to hear that female baboons have been observed inviting unrelated, unknown adult males to join them as protective friends for their infants, but wanted to know why.

Barbara Smuts was one of the first scientists to question why females did what they did, rather than just noting them doing it. She wanted to know why some females chose certain newly arrived and unrelated fully grown males over other males and not only befriended them but, more importantly, trusted them with the lives of their infants. Help in minding the baby was a good thing for the female, though not so beneficial if the male baboon became proprietorial over her social life. Could the males actually be more trouble to the females than they were worth? If sex came into the equation, Smuts questioned the costs and benefits to the female baboon.

Jeanne Altmann and her husband Stuart Altmann were also studying baboons in Africa during the 1970s. Jeanne Altmann was asking the same questions of the female yellow baboons in Amboseli as Barbara Smuts. Jeanne Altmann had started her professional life teaching mathematics, then married a professor of primatology, stopped work and had two children. After a few years she went back to work as Stuart's research associate. After helping him collect data for his doctoral thesis in Puerto Rico, from 1963 she worked with her husband in Kenya's Amboseli National Park. Jeanne Altmann was soon to become a highly influential self-taught primatologist.

From her experience in the field, Jeanne Altmann became aware that different field biologists had varying techniques for collecting data. Most scientists, like Zuckerman, Washburn and DeVore, recorded what caught their attention, such as the large squabbling males, but Altmann quickly realised that, for a study to be truly scientific, each individual animal – male, female, young, old – had to be observed for an equal amount of time. Something that didn't immediately catch your eye could be nonetheless highly significant, such as the more subtle social behaviour of the female baboon. The existing subjective sampling meant that researchers' data could not be fairly compared to another study sample as the idiosyncratic and personalised data created discrepancies. More and more scientists were undertaking field research and a universal standard was needed to establish the same methods for all biologists, allowing scientists from different sites to make comparative studies and to collaborate with each other over time and between species.

Over a few years Altmann's protocol for data collection was implemented, so that today field biologists take her original standards for granted. She advised, for instance, that if the study animal did nothing during the set period of sampling then that lack of activity should also be recorded because it was also of interest. Because of Altmann's ideas, field work developed a common standard and a researcher could no longer only take note of two fighting males. Looking back to 1974, when her recommended methods were first being absorbed, it seems amazing that a maths teacher could influence primatology so radically. Altmann's paper 'Observational Study of Behaviour: Sampling Methods' was widely read. Altmann also gave lectures on the subject. Men senior to her changed their ways; Sherwood Washburn's students and Robert Hinde and his students listened and re-evaluated their own findings. Hinde encouraged his students to adopt Altmann's methods. With the acceptance of Altmann's sampling protocol another woman was forcing powerful men in the world of primatology to think again.

Altmann's second outstanding contribution to primate study has been her analysis of the life of the female baboon. Once her children were of school age Jeanne Altmann had returned to work at the University of Chicago, and while juggling her various incarnations of wife, mother and researcher Altmann realised the nursing female with infants was in fact the most interesting animal in a primate society. It was here, with the mother–infant bond, where ecological pressures were at their greatest. This relationship was the palpable interface of evolutionary impact.

The amount of time in any given day that a mother can commit to her infants, to herself and to her social position affects her immediate well-being and the evolution of the next generation. There is never enough time, and the female must choose certain activities in preference over others. The budgeting of time fascinated Jeanne Altmann, who coined the term 'maternal time budget' as she noted how much time female baboons spent on the myriad of daily activities. The female baboons instinctively weighed up the costs and benefits of each task and therefore how much time to dedicate to it: how much time to groom and suckle the baby, to forage for food, to groom powerful friends, to be groomed, to search for and defend a good sleeping site, for example. A high-status female could afford to spend longer being groomed than grooming others and less time defending her sleeping site. The result of Smuts's and Altmann's research is that we now recognise that modern woman's concerns have grown with her in evolutionary terms. All contemporary primate mothers are presented with a clash of interests.

Jeanne Altmann was one of the first Western scientists to appreciate that baboons were matrilineal monkeys whose social rank was handed down directly from mother to daughter through countless generations. In baboons it is the male who leaves the group and finds his own way in life, living by his wits with no one to help him. In a matrilineal society it benefits high-status females to have daughters. In patrilineal societies it pays high-status females to have sons. Most monkeys are matrilineal, but the South American spider monkey is patrilineal. In this species, at the onset of sexual maturity, the female leaves her natal group and her future is uncertain. Meg Symington, who studied these monkeys, noticed that high-status female spider monkeys were more likely to have sons, low-status females were more likely to have daughters, and middle-ranking females had an even sex bias in their offspring. Apes are patrilineal and some scientists, such as John Manning, who studies sex ratios at Liverpool University, would say humans are also a patrilineal species. From sex ratio studies on humans there does seem to be a trend towards high-ranking, dominant women being slightly more likely to have sons. Altmann noticed her high-status yellow baboon females were twice as likely to have daughters and low-ranking females were more likely to have sons.

Altmann has also found from a more recent DNA study of paternity that in multi-male breeding systems, such as baboons practise, male social status does make a difference to reproductive fitness in males. In many cases, higher status does translate into more frequent mating and at Amboseli during this particular study, Altmann's alpha male fathered most of the offspring. However, Barbara Smuts's olive baboons at Gilgil seem to be different from all other savannah baboon populations studied so far. At Gilgil the highest status males do not mate with fertile females more than low-status males. Smuts and other scientists have yet to establish why this is, but both male (Timothy Ransom and Robert Seyfarth) and female scientists have consistently reported this result. Any attempt to unify the sex lives of primates remains elusive, even if the importance of social status is understood.

Inherited biological status is only the starting-point for an infant primate. Maturing individuals can better their status through their choice of mate and their choice of friends. Consequently, for the sake of her infants and for subsequent progeny, a primate mother will not be a good mother if she devotes insufficient time to social networking. A primate mother who devotes all of her time to the immediate care of her infants will only provide her infants with half of the benefits they could receive from her. In the case of human mothers those women who have a job and time for leisure and for friends present their children with a positive image and, in practical terms, with an income that provides their children with social standing as well as psychological and financial benefits. Women like this are good mothers. As long as they spend time with their own children every day, spending time with other individuals is no bad thing. What is significant is the amount of time a mother spends with her developing infant. Its demands are awesome and, as Bowlby pointed out, this is where the conflict occurs.

Having children and a career is often referred to as 'having it all', as though this is the desirable lifestyle of the fashionable woman. But this term misses the point. The 'all' is an immense burden to an individual woman and globally most women have no choice.

Women have always worked and cared for dependants, their designated role a complicated one that often attracts criticisms. In the West the press regularly carries stories of this *bête noire*, the working woman who leaves her children to 'run wild'. These children, we are told, are disaffected and problem cases, often left 'home alone' while their disgraceful mothers socialise or run off on holiday. There will always be conflict when an individual juggles her dual roles of working woman and mother.

Jeanne Altmann's term for this among her baboons is 'dual career mothering' and the 'work' the maternal female baboons engage in is primarily the search for food and tending to powerful friends. By drawing our attention to this female predicament in human and non-human primates Jeanne Altmann did women a great service. Latch-key kids, child-minders, nannies and 'home alones' are not just phenomena of the industrial age, but have a place at the dawn of time. Dual career mothering was a struggle our pre-hominid ancestors negotiated and is one our primate cousins also deal with today.

I hear many of my women friends talking wistfully of the 'quality time' they plan to spend with their children, though I know of none who have achieved it. But one non-human primate mother does have quality time with her baby. The polyandrous saddleback tamarin monkey from South America is the exception to the rule. It was presumed to be monogamous, but when Ann Wilson Goldizen studied the creature she found that the female openly mates with two male partners. When the saddleback tamarin is in oestrus the two males calmly take turns to mate with her, and there is no jealous fighting between them. After the infant is born, the mother often only comes into contact with the baby when it requires suckling. At all other times the two males share the other roles of parenting between them, even though only one of them can be the real father. It is easy to see the benefits to the female saddleback tamarin – but why do the males have this sexual strategy?

Each male has a 50 per cent chance of seeing his genes reproduced, so the co-operation between the males benefits them both. In the long term the chances are that both animals will end up fathering infants, and with so many carers present all the infants should safely reach sexual maturity and breed themselves. The two father figures are often siblings, and playing the role of an uncle to an infant that shares 25 per cent of your genes is almost as good as it gets. Rather than the male's behaving altruistically or opportunistically, he is actually playing a genetically selfish game.

Human psychological studies have shown that women are flexible in their approach to time; routines can be modified and new tasks absorbed. They can juggle their duties and fit more activities into their day than men. This is because for their and their children's sake they have to. With each new baby born, a woman has to meet the ever-changing demands of her existing family and her community. Until very recently men have not expected to share in the traditional female role of raising children. Evolution has not selected men to be able to juggle time in the way that women can. Because of this difference in behaviour between the two sexes, many women living in developed countries feel it is easier to deal with all of the tasks themselves rather than delegating duties to their male partner.

Elder siblings realise a new baby creates direct competition for resources, such as their mother's precious time. A very young child's survival instinct can manifest itself in the form of jealousy and sometimes bullying towards the younger sibling can occur. Some women, usually in developed countries, have access to contraception and can, if they want, choose to remain childless, but most women eventually become mothers. Those who adroitly budget their maternal time have this skill because it has been sexually selected for over millions of years.

Game theory – the study of dilemmas and power struggles between cells, individuals and groups – explains relationships such as the mother–infant bond as a game of competition. The infant demands an exhaustible amount of resources from its mother in the form of physical bonding, transportation, protection, sustenance, education and the learning of social skills. But if the mother gave every last minute of her time to her infant, she would neglect herself and suffer the consequences. Her downfall would involve the downfall of the infant too. Within reason she must put herself first within her maternal time budget.

The arms race between the mother and her baby begins in the womb, where the conception of an infant is a game between the father's and the mother's genes. At this stage, male and female strategies are genetically programmed. The male's reproductive imperative is the fathering of an infant every time he has the opportunity, as death or injury may prevent him from having another chance, while the mother's strategy is to allow only the very healthiest embryo to develop, as the physical costs to her will be enormous. Specific chemicals attached to our DNA control the development of the placenta and the development of the embryo. The placenta is a spongy organ which enables the developing embryo to feed off the mother's blood. An embryo is the name given to an unborn baby during the first few weeks of pregnancy. As the father wants reproduction irrespective of later costs to the mother, the developing placenta is controlled by his genes, but the growth of the embryo is controlled by the mother's genes.

The mother's body automatically tries to rid itself of the alien tissue that has attached itself to her insides and is using her blood to grow. She feels sick, she feels exhausted, she grows pale. Chemicals from the father's genes in the placenta send out hormones to make it grow bigger and increase the blood supply, but in reprisal hormones from the mother's body try to cut off the blood flow. Over 50 per cent of the time her body wins the fight and very early in the gestation spontaneous abortion occurs. Often the developing foetus is reabsorbed into her body and the nutrients are recycled. The termination of the pregnancy is not a conscious decision by the woman; mostly she will be unaware of her internal fight for the perfect baby.

The majority of babies born are completely healthy, but this is only the beginning of the fight for the mother's body's resources. The infant is like a parasite. A mother uses on average 125 calories a day gestating her infant, but when she begins to lactate the calorific cost to her will increase twenty times. Jeanne Altmann noticed that her lactating baboon mothers were far less social than at other times. As they needed to conserve energy, they would limit socialising to their immediate family. Altmann also noticed that nursing females stood a higher chance of dying than non-lactating females. Weakened, they were more vulnerable to parasitic and predatory attack.

All functions within the body, such as the menopause, can be interpreted as games. In evolutionary terms a successful mother could be one that keeps producing babies until the day she dies, such as the female turtle, who doesn't become sexually active until she is sixty years old, returning to the sandy beach of her birth – and her mother's birth – and at night laying hundreds of leathery eggs that she buries deep in the sand. The more babies the better, surely? Turtle hatchlings probably never see their mother; from day one they are alone and very few will make it to adulthood. The female leatherback turtle uses her resources to produce thousands of eggs in her lifetime and if just two reach adulthood and breed she has been an evolutionary success. A primate mother uses her resources to care much more deeply for a much smaller number of progeny. All primate infants, especially human babies, have long childhoods and are dependent on their mothers for all their needs for years.

A forty-five-year-old woman with six children, all at different stages of development, needs to see these youngsters into adulthood. She has invested much of her own time and energy in them, especially the eldest, who now may reward her by helping to care for the younger siblings. If she continues to produce babies she runs a higher and higher risk of eventually dying in childbirth and her remaining children may find it difficult to survive without her. Her menopause prevents this from happening. No other creature lives on for years after the reproductive potential has ceased; the female human primate's menopause is unique. Evolution has cast women to play multiple roles. The nursing of infants is just a small part in the epic.

Women have the ability to see their commitments through to the bitter end, a trait understood and used by Louis Leakey. They have an instinctive fear of loss of status, with or without children, because it's easier for a woman to become a victim than it is for a man. If women seem to particularly hate losing status it is because they have much to lose, unlike the serene female turtle who is never held responsible for her infants.

Baboon mothers negotiate their time similarly to women. They have to socialise and win powerful friends otherwise their social insecurity will be a hindrance. Food sources will be kept from them as senior animals feed first. Baboons with high status are always fatter and a high-status infant has the social confidence to push a low-status adult away from the food resource. Bullies, keen to impress others with their strength, will pick on low-status females and their infants, raising their own status as they push others down. A baboon mother needs help with her childcare, just like a human mother, or her other duties will suffer. This is where male friends enter the equation. A bargain is struck – childcare in exchange for group acceptance and sexual favours. It's a fair deal and on the surface everyone wins. Infants find themselves with a strong protector in addition to their mother, the mother now has the time to develop her social climbing and the new male is presented with a role that allows him to join the baboon troop. If he is very lucky he may also be allowed to father the female's next infant.

Both Barbara Smuts and Jeanne Altmann noticed that a purely

platonic friendship could develop between a male and a female baboon. It was a new observation. It seemed at first that in this type of relationship the costs were low to the female and the benefits high. Smuts hypothesised that the male's subtle social sophistication came with age and experience and that juvenile males had to learn that social finesse rather than bullying worked better with females. These relationships between adult males and females were symbiotic. Only later, at the female's invitation, might the relationship become sexual. New men or soft lads appeared to have developed first among male baboons who seemed to be in touch with their feminine side, serving as caring and reliable minders for infants that were not their own.

Some years later Barbara Smuts would be disillusioned to discover that these soft lads were impostors. They remained soft lads for only as long as the strategy worked, and if it failed to earn them breeding opportunities a more sinister approach would come into play. It's not only females that weigh up the costs and benefits of their actions. The 1989 movie *When Harry Met Sally*, written by Nora Ephron and directed by Rob Reiner, took a frivolous overview of this type of behaviour in humans. Can men and women be friends without getting into bed with each other? Women seem to yearn for this type of relationship, but among heterosexual men and women it is very hard to achieve as sex keeps appearing on the agenda. Some heterosexual women specifically seek out and build adoring friendships, without sexual coercion, with gay men. Could this desire for platonic, mixed-gender friendships explain why women were first to recognise elements of this behaviour in baboons?

Robin Dunbar from Liverpool University has also studied the baboon. He says the sexual life cycle of a male baboon echoes that of a man. A male baboon's prime lasts for only a short period in the animal's life – at most four years, between the ages of nine and twelve years. It is a struggle to get to the top of the male hierarchy and only a few make it there. Others have to make their best years work for them in second, third or even tenth place. The transient period at the top is marked by stress and instability, and the subsequent decline usually lasts longer.

While sexually maturing, a young male, age four to six, will venture forth to avoid inbreeding, leaving the security of his natal group behind. He will wander alone until he finds another group to accept him. Not yet sure of himself, he will try and win the confidence of a group of females. His imperative is acceptance by the group so that eventually he can mate with as many females as possible. It is anti-social for him to rape females, and many of the females have friends of both genders who will help fight off unwelcome advances. The best way to have sexual relations with an oestrous female is to become her friend, and the best way to become her friend is to become her babysitter. But this can take time; a new baboon has to learn a subtle and patient approach to the females, as they already have lots of male friends and do not like noisy displays. By age 8 a male baboon is fully grown and has full extended canines; he is entering his prime.

Once a dominant female and her infant have given a male the seal of approval, he can ingratiate himself with other males of a similar status. This way they can offer mutual help in fights, keeping the pressure on the male whose status is immediately higher. Together, in their self-interest, males form coalitions to oust higher-ranking males, and they slowly work their way up the ladder. Male chimps behave in a similar way, but they are also bonded by blood, which makes the knot even tighter (in chimpanzee society it is usually the female chimp who must venture forth alone to avoid incest). Baboon males transfer from group to group. It has been suggested that they move on if they think they would have a better chance of mating elsewhere.

Another strategy a new male baboon might take, if he is very strong, is to arrive and challenge the resident dominant male outright. There may be no contest and the resident male will step down; otherwise they will fight. If the resident male's cohorts come to his aid he may be able to defeat the newcomer. But not for long. On the second or third challenge the newcomer, with the aid of his new-found male friends, will take a position at the top of the pyramid, with his male cohorts sharing in his success. While at the top and in his prime, he will feed first and be groomed incessantly, and is likely to mate with more females than any other male during the same period. Much of his time is also spent stressfully trying to defend his position from other younger males. When his turn to make way for the next young gun finally arrives, he steps down to find that his friends – female and male – matter all the more. As he settles into old age he competes and networks less and spends most of his time and energy looking after his last progeny, making sure they succeed in his place. Other older baboon males aged about 15 years will spend time away from the group alone. They seem to prefer a quiet life once their body is stiffening and their canines are worn. They only return to the group for possible mating opportunities. Sometimes these older males form a dispersed group together.

Recently a more cynical hypothesis to the male baboon's strategy of befriending an infant has been voiced. Although the male baboon seems altruistically to offer cuddles, piggyback rides and protection to infants, usually his hidden agenda is to mate with the infant's mother. The male is gratuitously advertising his parental skills, saying, I'm a nice guy, I'm kind to your baby, why not mate with me? But if the female turns her attention to another male or seems unreceptive, male baboons have been seen to lose interest and discard her bewildered infant. He may also bite and rape her. After mating with a consenting female he may also neglect his new infant friend in favour of another infant, his ambition being to mate with as many females as possible.

Isaac Burns, a male sociobiologist acquaintance of mine, uses his research to benefit his private life. While he packed for a trip to a chimp reserve, I noticed packs of sweets and photographs of his sister's children among his bottles of malaria tablets and insect repellent. When I questioned their inclusion he told me he wanted to have regular sex while in the jungle. Convinced the basics to human sexual politics are the same as the politics of beasts he studies, Isaac admitted these simple items were his tools for sexual conquest. Women, apparently, always fall for it. The sweets would be a tempting luxury for sugar-bereft women primatologists after six months in the field and by lovingly showing them pictures of children he could show he was a nurturer and a soft lad, making himself doubly irresistible. He emailed me to let me know it had worked a treat.

Had Isaac Burns been aware of the Chicago-based research of Alan Hirsch from the Smell & Taste Treatment and Research Foundation, he might have liberally covered himself in baby powder too. Hirsch has researched the effects of smells on the female sex drive by measuring blood flow to the vagina while women sniffed at different scents, and observed that the smell of baby powder, candy-coated liquorice and cucumber sexually excited women the most. Hirsch also discovered the cosmetics industry may have misjudged women's tastes, because men's colognes have proved themselves to be a distinct turn-off when compared to the allure of a cucumber.

Jeanne Altmann received due recognition once she published her sampling paper in 1974. Altmann's 1979 dissertation was titled 'Ecology of Motherhood and Early Infancy' and by the time she received her Ph.D. she was already editor of the leading behavioural studies journal, *Animal Behaviour*, the first woman editor of any major journal dedicated to behaviour or primate research. Altmann methodically continued to observe her baboons and she is still at it.

With help from a long line of African, American and European students, the Altmanns have collected an extensive database on the baboons. They chart each animal's family tree and identify individuals using a collection of mug shots that help new students to recognise the animals by face. The animals' territorial range, sleeping sites and water sources have all been mapped. As in most field sites, the primate's predators are probably the same as ours and dangerous animals are always a concern. Tragically a past student was killed (and partially eaten) by a lion.

At dawn Amboseli in Kenya offers researchers a fabulous view of Mount Kilimanjaro, just across the border in Tanzania. Quietly researchers watch and wait as the African sun climbs up in the sky, waking the yellow baboons with its warmth. The baboon that rises first and decides the day's route across the savannah is constantly changing, and a researcher needs to be there at sunrise to see who is stepping forwards as the new leader and where the baboons will go. To stroll along among a habituated but wild troop of baboons is poetry in motion. When it's time for the next person on the baboon-watching rota to take over, the researchers can contact base on their walkie-talkies and communicate their whereabouts.

Over the years Jeanne Altmann has seen infant baboons mature and become political animals, intent on improving their own status as the new generations of infants are born. The work has never become dull and repetitive. The eternal scrapping and the animals' individual personalities are highly entertaining to the observer. One becomes emotionally involved and ends up caring about their lives. Altmann still believes the mother–infant bond is the crucial real-life interface of evolution at work and the maternal time budget remains a popular area of study, especially with women researchers.

Jeanne Altmann is expanding her research. She has recently started to anaesthetise the baboons, using a blowpipe and dart, to get close to the baboons without them realising (if traps are used, the animal remains conscious and is traumatised) so she can measure the size of the sleeping animal and take a blood sample without causing too much stress.

Some of Altmann's latest research has been on the social status of the oldest females in the baboon group. When Altmann was a young woman with small children, it was the life of the baboon mother that fascinated her. Thirty-seven years later it is the quieter life of the older female that interests her. She has studied the strategic lives of female baboons from the simian cradle until the grave, and has said of her career, 'A characteristic of my pleasure in the work and, I believe, our success, is collaboration. We respect and give intensive training to our Kenyan and American staff and students. They are essential to the data collection, the analyses, and the development of new ideas; some stay in the field and in collaborations, others move on but do with a model of science and work that I think is important. I am not and don't want to be Indiana Jones, though I am pleased with my successes and pleased when they are recognised and modelled.' In 1999 Altmann turned her attention to a completely different species. She has been collaborating with her graduate student, Karen Ryan, on kin selection, and together they have observed that monogamous male mice bias behaviour towards females according to very small differences in their kinship to these potential mates.

Today, thirty-five years after she left Britain for Uganda, Thelma Rowell has returned to Britain and now resides in an isolated village in the Yorkshire Dales. She too has turned her attention to another species of mammal – soay sheep. When I asked her why, a smile lit up her face. 'Sheep are everyone's paradigm of silliness.' Thelma is questioning the pedestal on which we have placed primates, she says; because our nearest relatives look like us we have arrogantly presumed them to be more intelligent than all other animals. But we know through the long-term observations of elephants by women biologists such as Cynthia Moss and Joyce Poole and from studies of dolphins and hyenas that primates are not the only intelligent animals. Bottlenose dolphins have been studied by an

American team working in Western Australia, which includes
Rachel Smolker, Janet Mann, Richard Samuels, and Barbara Smuts.
Spotted hyenas have been observed by Lawrence Frank, Kay
Holecamp and Laura Smale. The primate species we know most
about – baboons, macaques, chimpanzees and humans – are all
gregarious, sociable and long-lived. Thelma thinks an important
factor leading to intelligence is sociability, and sheep are social.

The woolly grazing forms of sheep dotted on a hillside make a
comforting image of country life, but it is an unnatural life for the
sheep. They are most probably all females, all of the same age, and
the flock may consist of a number of smaller groups that have only
just been herded together into that field. These same-age females
are not together long enough to form relationships, but are purely
breeding machines. Likewise most scientific sheep studies are looking at ways to increase the growth rate and the rate of reproduction,
such as the recent genetic emergence of 'Dolly' the clone. Rowell is
being intentionally subversive. She wants to ask the same type of
social organisation questions of her sheep that she asked of her
baboons. Rowell's naturally small soay family consists of three ewes
and three ram lambs, bought at market and now living in a field
beside her cottage. Every day she wanders out into the fresh
Yorkshire air to observe them.

Rowell's sheep seem to have the ability to form a culture in the
same way that primates do. They recognise other individuals and
can form and sustain relationships with each other as their mutual
roles within the group change over their lifetime. The sheep have
assumed an order of rank, with the eldest female dominant. Rowell
has seen mutual grooming, intervention in a fight from a third
party and reconciliation following a fight. Such forms of behaviour were only recognised in primates fifteen years ago. But finding
an editor willing to publish her findings is proving as hard today as
it was for her to publish her original work on baboons.
Challenging the accepted paradigm of animal behaviour is a role

Thelma Rowell has grown used to, but it has not become any easier.

Barbara Smuts, like Rowell and Altmann, is now adding the behaviour of another species to the equation. Since 1997 Smuts has been struck by the parallels between chimps and Western Australian bottlenose dolphins. Although these marine mammals are 60 million years distant from primates in evolutionary history, they have developed a similar social sophistication to chimps, displaying intelligence and individual personalities akin to primate behaviour. Barbara Smuts plans to make a comparative study between chimps and dolphins to try and establish the ecological pressures common to both species that could have affected the evolution of their intelligence. Smuts is intrigued to discover that, like male chimps, bottlenose dolphins sexually harass fertile females, while groups of male dolphins form coalitions similar to those of chimps to compete with other groups of males for access to females. Smuts believes that social intelligence paves the way for certain sexual strategies that chimps and dolphins appear to have in common. Smuts says that within hours of beginning her dolphin observation she was able to see similarities between them and chimpanzees. Smuts, Rowell and Altmann are all using the behaviour of another animal species to help them to better understand aspects of primate behaviour.

Today Barbara Smuts, who as a girl was a devoted fan of Jane Goodall, has evolved into a popular role model for the latest generation of female primatologists. Her insights into female primate sexual oppression at the hands of male primates has made a lasting impression, especially on young women researchers. She has inspired many of them to explore this phenomenon themselves. The life of the female primate is no longer a secret of nature. We know now that she is a multi-faceted animal – competitive, politically strategic, sometimes a victim and sometimes dominant, but never just dull and passive.

4
Wise Monkeys

Primatology is a well-established area of research in Japan. It arose independently of Western influence and was always much more holistic in its approach to the animals than Western science used to allow. Japanese scientists wanted to know how macaque monkey societies worked as an integrated whole rather than researching individual behaviour. Ironically, methods used to study animal behaviour in the West have become progressively more holistic since the development of Jane Goodall's influence, while in Japan today the recently arrived Darwinist doctrine is encouraging a more rigorous scientific analysis of evolution. East and West are meeting halfway.

The need to translate Japanese research into English was for many years a hindrance to the exchange of ideas and knowledge. For the most part, the West ignored the Japanese research, and vice versa. When some of the first Japanese research was translated in the late 1950s it seemed very naive and far too anthropormorphic for the West to take seriously. Western scientists, with their inheritance of Darwinian interpretation, tended to look down their noses at the Japanese scientists, whose cultural inheritance influenced their own subjective style of science.

With neuroscience, psychology and palaeontology constantly supplying tantalising further pieces of factual information for the jigsaw of human evolution, it has become evident that the similarity between ourselves and non-human primates is no illusion. The fossil record shows that we started to take a separate evolutionary path from our primate cousins relatively recently, approximately 5 million years ago. Anthropomorphism is no longer scientifically sinful, but is seen by some as pushing the boundaries of science in the right direction. Ignoring the obvious similarities between ourselves and non-human primates, the West first had to discover and 'prove' how much like the other primates we are, whereas the Japanese instinctively knew it to be true from the start and never doubted their assumptions.

A handful of Japanese women primatologists are now making their name in this area of research. There has never been a Japanese version of Louis Leakey to encourage women to undertake field research, so Japanese primatology meetings are not dominated by women, as might be the case in London. In Japan, for the moment, male primatologists greatly outnumber women. Umeyo Mori and Mariko Hiraiwa-Hasegawa are two Japanese women, who, despite chauvinism, have succeeded in becoming well-respected primatologists. Their lives and research show how Eastern insights and Western theories have come together.

Neither North America nor Europe has an indigenous non-human primate, other than the small wild population of macaques found on Gibraltar, which were probably first established on the Rock when Roman legionaries, returning home from North Africa, abandoned their pet monkeys there. Asia is home to numerous Old World primates, including most of the sixty species and sub-species of macaque, as well as langurs, siamangs, the snow monkey, the loris, the tarsier, the gibbon, the orang-utan and the proboscis monkey. There are no Japanese species of ape. The macaque is the most geographically widespread Asian monkey, various sub-species of macaque inhabit Japan, and the macaques that live in the forested, snowy Japanese mountains known as the Shiga Heights are the most northerly living non-human primates. They have red faces, hands and bottoms and pale grey fur, which becomes especially fluffy in the winter.

Japanese macaques are culturally seen as wise and devious, just as in the West we identify the fox as sly. The absorption of the fox into Western fairy stories and fables mimics the similar role of the macaque in Japan. Cultural stereotyping is powerful; most people in the West would be surprised to hear that foxes are not sly, that they do not as a rule raid farms and kill chickens and lambs, that their main food is earthworms. The Japanese stereotyping of the crafty macaque is much nearer to the truth. The macaque is well adapted for a terrestrial lifestyle, nimble on its hands and feet and able to manipulate objects with dexterity. The Japanese have always seen their monkey as a close human relative that mirrors man, and their culture is full of references to monkeys displaying human-like behaviour. When Japanese scientists started to look at monkey behaviour they brought their own cultural apparatus with them, interpreting the natural world as following sociological rules that have formed the origins of their own society. Culturally it would seem to the Japanese that monkey and man have comparable roles in the great scheme of things.

It did not seem unscientific for the Japanese to describe some of the behaviour of the individual monkeys as wise, selfishly devious or, at other times, as kind towards others. Up until very recently Western scientists would have thought this description of monkey behaviour to be a gross over-interpretation. Before sociobiology emerged in the West, it didn't seem possible that animals could think and feel similar things to us. Anyone who claimed this might be considered an over-sentimental fool. In Japan the macaque has been credited with a mischievous spirit and devious nature for centuries. The Buddhist religion has a monkey king at the heart of it.

In 635 a Chinese monk, Xuan Zang, travelled to India in search of the Sutra, the Buddhist holy book, and on his return to China he translated it, thus influencing the shape of Buddhism as we know it today. It is told that Xuan Zang travelled with a monkey king, who protected and guided him on the journey, and many legends surround their lives. While Shakespeare wrote his poetry in England, in China this allegorical fable was written up into the classic novel, *Monkey King* (known in Asia as *Journey to the West* by Wu Cheng'en) has remained popular for well over 400 years. The magical character of the monkey is a mixture of wisdom and rebellion, representing a basic human struggle against oppression, which in this case would originally have been the feudal emperors. Armed only with his quick wit and a supernatural iron bar, the monkey is invincible, and even the might of Buddha cannot suppress the monkey's spirit.

The Christian West sees itself as special and humans as superior to all other life forms. We prize the natural world because it exists independently of man, the Garden of Eden, a place for us to cleanse ourselves of our sins. In the Japanese tradition, as in much of the East, there is one world and humans are part of that natural continuum. Beautiful natural areas are treated as organic works of art, which may occasionally need human organisational skills to rearrange the landscape.

With such an anthropomorphised monkey in their midst, up until very recently there was little room in Japan for Darwin's theory of evolution. Darwin's ethos – that all evident behaviour and appearance of an animal has been selected for because of the drives of survival and reproduction – was not embraced in Japan. Traditional centuries-old fables and legends about the wise but devious macaques could never be questioned or quashed by Darwinism. The monkey tales were imperceptibly carried from Japanese culture into Japanese science.

Professor Kinji Imanishi is the respected grandfather of Japanese primatology. During the early 1940s zoologist Imanishi had decided to ignore the Second World War and instead, unaccompanied, took to hiking through Inner Mongolia to study the behaviour of wild animals. He observed wild horses and gazelles, as well as the Mongolian people, and during this period developed his theory of an animal sociology. Worried that he might die during the war and his theory would be lost to the world, he wrote *Sebutsu no Sekai* (*World of Living Things* 1941). After the war finished he worked for the Research Institute for Humanistic Studies at Kyoto University, and in 1951 published *Prehuman Societies*. These two books are his legacy to primate studies in Japan, and have had an immense, and some would say detrimental, influence.

Post-war Japanese primatology studied the social anthropology of the monkeys. From the early 1950s the life history of individual monkeys was written down, and it was observed that males transferred from group to group so that the cohesion of a troop resided with a sisterhood of monkeys. It quickly became evident to the Japanese scientists that the monkeys were matrilineally hierarchical; the females were dominant over males and the high-status female monkeys gave birth to automatically high-status daughters. Sons also depended on their mothers for social rank, even more than on their own ability to dominate others for resources. It was also observed by Japanese male scientists that male rank had no bearing on numbers of offspring.

In the West nearly all the field reports described a hereditary father-to-son dominance, where high rank was equated with high numbers of progeny. In Japan scientists had seen that low-ranking males had just as much chance to father infants as high-ranking monkeys, and these findings corroborated the Japanese mythical notion of the matriarch, especially the spiritual nature of mother–son bonds. Because of cultural and linguistic barriers it was not until 1965 that Jeanne and Stuart Altmann oversaw translation of the Japanese research. Had it been assimilated into Western thought earlier we might well thank them and not Western female scientists, such as Jeanne Altmann and Barbara Smuts, for shining the spotlight on to the social significance of the female primate.

After 1965, as more and more translated Japanese research filtered through into the West, a gradual assimilation took place. Western research projects started to say the same things as the Japanese and the Japanese scientists started to be seen as a group of researchers who could make the West look scientifically old-fashioned. The threat from the East was similar to the effect Jane Goodall was having on Western science. It was time for change.

For a while white Anglo-Saxon male scientists possessively defended their outdated corner by critically inferring that women and Orientals did science in a certain way that was not theoretically grounded. But there is no replacement for factual observation, such as that of Jane Goodall and the Japanese. Over the last thirty years, as Western primate studies have entered a theoretically postmodern era of science, where an individual animal's subjective feelings and thoughts have become an issue, the Japanese research has become a crucial addition. In the West we have now officially endowed non-human primates with human-like sensibilities. Just as with people, we can now speak about non-human primates' individual personalities, Machiavellian or emotional intelligence and their interpersonal relationships. This is the approach Jane Goodall instinctively chose when observing her chimpanzees, and this style of acknowledging emotion has been used to study the whole of monkey society by the Japanese for centuries.

Despite the Japanese love for the natural world, very few conservation charities are working in Japan. The environmental lobby has yet to get a grip of the Japanese people's imagination. The Japanese view the animals that swim in their waters and walk on their land as theirs to do with what they choose. As a result of agricultural encroachment in the mountains, the macaques that live there are forced to compete for space with the invading farmers. Every year 2,000 monkeys are trapped and shot; the macaque birth rate cannot sustain a cull of that size. Mariko Hiraiwa-Hasegawa has told me that cultural moods and attitudes in Japanese society lag approximately ten years behind those of the West. Feminism has only just arrived in Japan, while the inherent sexism in Japan means that only a tiny percentage of university places are taken up by women. Most scientists, including primatologists, are still men. When Hiraiwa-Hasegawa first entered university in the late 1960s only 3 per cent of science and medicine students were women. Today about 12 per cent of Japanese university students are women.

For many years, Chie Nakane, Professor of Cultural Anthropology at Tokyo University, was the only female role model for women primatologists such as Mariko Hiraiwa-Hasegawa or Umeyo Mori. Chie Nakane was a tough-talking, cigar-smoking, no-nonsense lady who did nothing to help younger women climb the professional ladder of academia. Similar to former British prime minister Margaret Thatcher, Chie Nakane had fought hard to reach the top in a man's world. She was a pioneering scientist, the first Japanese woman to travel alone to India and Tibet to study the social systems of the local people. In the 1960s she visited London University to lecture students. Today Chie Nakane is in her early 70s and holds the honorary position of Professor Emeritus at Tokyo University. But Chie Nakane was a one-off; she would have inspired younger women although she didn't actively encourage women to follow her. Against the odds Umeyo Mori and Mariko Hiraiwa-Hasegawa are two very different women scientists who stand out in Japan.

Mrs Umeyo Mori is the *grande dame* of primate studies in the

East and Japan's first woman field primatologist. She never knew her father, who was killed two months before she was born in 1941. Umeyo's mother, Satsue Mito, a schoolteacher, became the bread-winner, and in 1947 settled in Ichiki, a small village where Tohichi Kanchi, Umeyo's maternal grandfather lived. After the war, Japan suffered a desperate recession, and Satsue Mito was lucky to find both a job teaching at the village school and a new family home just 200 metres from the shore. Umeyo Mori told me that all her mem-ories start from this time; she cannot recall anything that happened to her before she moved to Ichiki.

The young Umeyo Mori would look through her bedroom window, out across the water to the uninhabited island of Koshima, where a troop of wild macaques lived. Most Japanese children had to work, the majority of people were very poor and many were starving. As a little girl Umeyo would be sent out to walk along the shoreline to collect shellfish for the family's supper. She had a gen-uine interest in wildlife and would sit for hours watching birds nesting and raising their young.

Umeyo Mori was six years old when her lifelong fascination with the Japanese macaque began. At low tide you could walk across to the island, but when the strong sea current flooded in a boat was needed to carry one across to the monkeys. But it was hard to observe the animals as they foraged in the undergrowth. Mori, accompanied by her mother and grandfather, took advantage of the long, open sandy beaches and left food to encourage the animals out of hiding. Umeyo's grandfather had visited the island and provisioned the macaques with sweet potatoes since the 1920s. He had written to the local authorities describing the island of Koshima as a natural treasure and he suggested it be turned into a nature reserve. Tohichi Kanchi also suggested the monkeys could be observed by scientists, as they would eventually accept anyone who fed them.

In 1949 Professor Kinji Imanishi was studying wild horses with his male students Itani and Kawamura in the village of Toi just over the hill from Mori's village. On his way back to Tokyo he decided to visit Ichiki to see the Koshima macaques. This was eight-year-old Umeyo Mori's first and unforgettable meeting with the eminent Imanishi. Imanishi's writing was very popular; Japanese philosophers and scholars of humanities taught his theories, partly because Imanishi appealed to Japanese people's nationalistic senti-ment. In 1951 Imanishi sent Itani back to Ichiki to provision the Koshima macaques and to watch them at close quarters. Ten-year-old Umeyo Mori watched the scientists as daily they placed sweet potatoes at designated sites and waited for the monkeys to accept their presence. Because Japanese macaques live in close proximity to humans and often try to scavenge from people and from Buddhist temples, ethically the Japanese scientists had no problem with the provisioning of food to allow them to get closer to the monkeys.

In the West a long-term field study is seen much more as a woman's preserve, but in Japan it is commonplace for groups of male primatologists to embark upon ten years or more of monkey studies. Toshisada Nishida founded the chimpanzee field site in the Mahale Mountains in Tanzania and Takayoshi Kano has stud-ied bonobos at his Wamba site in Ziare since 1974. The Koshima macaques have been systematically studied for forty-eight years; it is the longest primate field site in the world and the majority of researchers there have been men. Tragically in 1968 Kenji Yoshiba was drowned in a freak storm when his tiny boat capsized just metres from the shore. Yoshiba had done much to establish a per-manent primate research centre at Koshima and he is remembered with affection.

For the first two years the only things the scientists could study in any detail were the monkeys' faeces, but gradually the researchers were in a position to recognise each individual monkey by face and to name them. Umeyo Mori's grandfather ran the village's only inn, and the scientists stayed there during their research. The natural history of the Koshima macaques became a family affair. Umeyo's grandfather and mother were already well-informed amateurs when the sophisticated scientists started staying at Tohichi's inn. Itani asked Mori's mother if she would help them with their research by continuing to collect data on the monkeys after the scientists had returned to Kyoto University. Satsue Mito was delighted to be asked.

Koshima is a tiny island with a circumference of only 4 kilometres. A small number of racoons live on the island with various bird species, but they were not systematically observed in the same way as the monkeys. Making notes on the monkeys' behaviour, Satsue would walk around the island with Umeyo. Satsue was pleased to help out and ended up spending forty years of her life watching the macaques. She was responsible for the basic data about the monkeys, such as their birth and death rates, shifts in status and the transfer from troop to troop of different monkeys.

Satsue Mito never married again, but she was not without male company; she found the scientists stimulating and the macaque research took up most of her spare time. Life in Ichiki was relentlessly parochial and dull, and the scientists brought an air of urban superiority and sophistication that was attractive to both Umeyo and her mother. When Satsue Mito was fifty-five years old she retired from teaching and started working full-time as a research assistant for the newly established Koshima Field Laboratory Primate Research Institute.

A sweet potato is a treat for a macaque, but the frustration of the animals trying to rub off and spit out the sand meant the meal became a bittersweet experience. One day a two-year-old female, named Imo, had a vision and went down into the shallows to rinse her potato. Some macaque species are good swimmers, but it was not common behaviour for the Koshima macaques to go in the sea, yet this smart female quickly figured out that the water would rid her potato of the gritty sand and transform it into a desirable feast.

It was 1953 and Satsue Mito was the first person to observe the now legendary female macaque, Imo (which means 'sweet potato' in Japanese) resolve the problem by washing the potato clean. Some years later National Geographic was alerted to the monkey's behaviour and it filmed *Monkeys, Apes and Man*, which stars Imo, by then an old lady, in all her glory, wading into the sea, meticulously washing the sand off her potato and then eating it.

The other monkeys in Imo's troop watched her actions with interest. They were quick learners. First her siblings developed the skill, then her playmates picked it up and soon afterwards Imo's mother became the first adult macaque to display this behaviour. Today it is believed by behaviourists that human children are influenced socially by their friends and their siblings more than by their parents. The skill spread from one individual monkey to another, although some monkeys never learned how to do it. The monkeys that technically lagged behind were the oldest females and the adult males. The dominant males, not so closely tied to the family network, didn't bother to learn the process but simply mugged younger, weaker monkeys for their clean potatoes and ate them, forcing the smaller, bullied macaques to go back and wash other ones.

When Imo became a mother her infants learned their mother's trick. First the youngsters would want the one Imo had washed and, like a typical primate mother, she would give it to them and then start over, until her infants finally acquired the skill for themselves. By 1962 the technique had spread through the whole troop and it had become the norm.

When grain was left on the beach for the monkeys the same problem occurred – sand would be caught in the monkeys' teeth. The animals would try to brush off as much sand as possible before putting handfuls of grain in their mouths. Again it was Imo who surmounted the problem. Grain floats in water, Imo discovered. If she picked up handfuls of grain and let them float in the sea, the sand would be washed off and the clean grains could be scooped up in her hands and eaten. The macaques watched wise Imo once more, and again picked up her ingenious discovery.

Watching the information spread through the group meant the researchers could observe a primate community change and culturally develop. Solving problems is the milestone in species evolution. Our hominid ancestors continually found answers to problems; bright individuals have enlightened their communities and the change has pushed us forwards. Our brains and our intelligence have expanded as our progression has speeded up. Surmounting obstacles shows intelligence and primates are the one group of animals who persist in problem-solving.

Umeyo Mori and her mother watched this learned cultural skill spread from Imo to all the other monkeys, thinking creatures equipped to adapt to beneficial new behaviours. As Umeyo's childhood was centred around the Koshima macaques it is no surprise that she picked up the skill of primate-watching from her mother. Satsue Mito's observational data are still being used by scientists today to help them to trace back the bloodlines of family groups using DNA samples taken in more recent times. Umeyo Mori's mother's observations continue to yield information on the cyclical population changes of the troops, but Satsue Mito has never received any formal recognition for all her hard work. Her name appears in no academic paper nor is she acknowledged in any Kyoto University books written on the lives of the Koshima macaques. It seems, in the end, that the male scientists did not want to share their scientific superiority with a simple schoolteacher from Ichiki.

The long line of scientists intrigued Umeyo Mori, and she spent more time in her grandfather's inn listening to the scientists' conversations than playing with local village children. Over their evening meal the researchers would talk animatedly of what the Koshima monkeys had been doing and of other primates they had already studied or planned to study. Enraptured, Mori found it exciting to stay up past bedtime, sitting at the dinner table listening to the researchers talk. They would mention India's hanuman langur, Tanzania's chimpanzees, the Cameroon's gorillas and mandrills and south-east Asia's gibbons. These were stories of adventure and discovery, describing exotic and inaccessible parts of the world. Umeyo Mori was hooked, though she didn't yet know it.

When it was time for Mori to go to college she chose to study statistics rather than primatology or any other related course. Primatology still seemed a man's profession and Mori already knew so much about the subject that she wanted something new that would be her own. Mori majored in mathematics, and after graduating her first job was at the National Statistical Institute. She worked there for several years, slowly progressing upwards in the hierarchy, as much as any Japanese woman could.

Mori should have been content, but on trips home she would meet scientists staying in her grandfather's inn and be brought up-to-date on the macaque research. She decided to make a career change. The Primate Institute of Kyoto University had been established and Mori took up a post as a research assistant in 1968.

As a child, Mori watched the behaviour of the juvenile monkeys as they played and fought one another. Their island had been her playground too, and she observed them as they learned by example and matured into adults. Now, as an adult, she would study the social development of young monkeys and make comparisons between human and monkey behaviour. Mori packed her belongings and left Kyoto to return to her roots and the Koshima macaques. She was coming home but in a different guise. She was no longer an amateur, like her mother, and her monkey observations would now be recognised.

Umeyo Mori started field research on the social play of juveniles and infants of the Koshima troop in 1968. That same year Akio Mori, a young student from Kyoto University, arrived to study the communicative behaviour of the Koshima troop. Umeyo and Mori would glance over at each other as they diligently followed their particular study animals around the island. Eventually they fell in love and married in 1972. They have two teenage sons.

Because of the provisioning the population of monkeys had grown considerably since the study had begun. The monkeys had become accustomed to receiving food; they didn't go hungry and as a consequence were breeding more successfully. There were now many infants and juveniles for Umeyo Mori to watch. She studied whom the young monkeys chose to groom and whom they chose to play with. She noticed that 90 per cent of grooming was under-taken by the youngster's mother and the other 10 per cent of grooming was reciprocated between siblings, but playtime was spent with peers of the same age. As the monkeys grew older Umeyo noticed that the mother's dominating role in the child's life eased off and 45 per cent of grooming was undertaken by other members of the group.

In 1974 Mori's paper, 'The inter-individual relationships observed in social play of the young Japanese monkeys of the nat-ural troop in Koshima islet', was published. It was well received and Umeyo Mori took her rightful place amongst the scientists she had once held in awe. A trip to Africa was arranged and from 1973 to 1976 Mori visited first the high mountains of the Ethiopian plateau to study the obscure gelada baboon and then in 1979–1980 the Cameroon to observe the drill monkey. She became the first Japanese woman primatologist to undertake field research abroad.

The gelada lives in groups of small harems. Usually there are five females and their offspring for every one male. At puberty the males leave the group to join gangs of young males who move around together, but as individuals they are always trying to dis-place a harem male and take his place. The females watch males fighting for their attention with confident interest. The bonded females are all related by blood; mothers, daughters, sisters, aunts are committed to each other and protect one another, all for one and one for all. They spend much of their time sitting in rows, one grooming another. This mutual hairdressing serves not only to free them of parasites, but also to reaffirm their friendships on a daily basis.

The harem male spends much of his time watching his females, trying to make sure that they do not mate with any strangers. If one of his females lingers too close to another male, he will charge and scream at her in punishment. At this point her sisters come to her rescue and often chase their harem male away. This is all very embarrassing for the male, as he needs to be perceived as a tough guy. He risks having his bluff called, his breeding females becoming monopolised by another male and his chance to father infants being lost for good.

Mori found it exciting to study a new species of monkey. She believed the origin of what we think of as a family set-up – one dominant father figure, a baby-producing mother figure at the centre and older grandparents on the perimeter of the family caring for grandchildren – could be seen in the gelada. She did not study the all-male groups of monkeys, who were waiting for their chance to find some wives, but the social relations of the one-male harems, observing new, strong males take control of a unit. After the fight was over and a new male had won, he benevolently allowed his now sexually repressed predecessor to remain within the group to care for his own progeny. The new male would quickly become sexually active, and eventually the females would accept the change and all new babies would be his.

Mori wanted to understand more about the relationship between baboon mothers and their young. In the mid-1970s Umeyo Mori and her Western counterparts such as Jeanne Altmann and Barbara Smuts became more aware of each other's work. Mori noticed that Western women primatologists also showed a primary concern for the female baboon with infants and reasoned that this was the case because women had sympathy for the female baboons and easily related to the animals' trials and tribulations in the same way that male researchers easily related to male monkey behaviour. At the same time Mori felt cautious of this type of instinctive empathy; it can affect methods of data collection and subtly influence results. If you think you understand a female baboon's actions because you can get inside the animal's mind and predict her next move, you may only see what you want to see, and not what is actually happening. Female baboons are not women; it's been 25 million years since we shared a common ancestor. Mori believed both a male and a female scientific perspective were important, but because men had dominated the field for the first twenty years a female bias was necessary to iron out the male-centric imbalance.

Mori says: 'I envy the female solidarity of Western female primatologists. Female solidarity is not strong in Japanese primatology or in other biological or ecological fields in Japan. Some women are working actively regardless of the sex bias against them, but others are struggling with unfair treatment of women. Japanese women need to co-operate with each other both in research and in achieving appropriate positions of work.'

Mori had two sons in her early forties. When they were young she was fascinated by their behaviour and studied them to compare her relationship with them to that of a baby monkey and its mother. Mori said that it wasn't until her boys started to learn to talk that she fully appreciated the difference. Up until this watershed, development of the two species runs almost in parallel. 'When my sons started to speak I realised they were going to become humans day by day and leave their monkey cousins behind.'

In the 1970s, after twenty-five years of provisioning, the Koshima troop had their provisions of sweet potatoes and grain reduced. In a quarter of a century five generations of monkeys had been born and adapted to life at the feeding stations. They no longer spent their lives migrating around the island following the flowering of different plants, and over the generations had lost their knowledge about indigenous seasonal foods, so their ability to forage for themselves was vastly compromised. They were spending more and more of their lives on the beach at feeding sites. Mori had watched this change in their behaviour, she had personal child-hood memories of how the macaques used to behave, she knew the scientists must slowly reduce the amounts of food provisioned.

Living unassisted in the wild is hard; it is much easier to lie about, waiting to be fed. The social life of the Koshima macaques had changed in a number of ways. There was now more time for grooming, socialising, playing, taking naps and nursing infants. The troop was expanding in numbers. Starvation was no longer a threat to the monkeys. Infant mortality had dropped and the mon-keys were living longer, healthier lives.

Provisioning caused all sorts of biological changes and inequal-ities between the status of the females. For some low-status females the onset of oestrus was occurring later and birth intervals were expanding, while for other, high-status females, who could domi-nate and take most of the provisioned food, there were even smaller birth intervals, so that high-status female macaques were starting young and producing many babies in their lifetimes. Female macaques are usually ready to reproduce at five to seven years of age, but after twenty-five years of having their natural diet supple-mented by nutrient-rich foods such as grain, soya beans and sweet potatoes some low-status females were not having their first baby until they were ten years old. All females could expect to live longer than their great-grandmothers.

For some female monkeys the delay in reaching sexual maturity hindered their integration into the adult females' hierarchy. As the juvenile period had been extended, fully grown female monkeys were still behaving like children and these older non-productive females' rank order was rather unstable. Umeyo Mori soon realised that females were integrated when they became mothers. Grooming interactions between these overgrown female juveniles and other 'normal' monkeys decreased. They became socially isolated and stayed together on the peripheral part of the troop, some of the females moving independently to other troops.

Back in 1952, when the Koshima monkeys had first been artificially fed on a regular basis by researchers from Kyoto University, there were 20 individual monkeys. By 1970 there were 120 monkeys and the population was still rising, even though the island could not naturally sustain a population higher than 100. Over a period of a few years, as the provisioning was radically reduced and finally stopped, the monkeys kept waiting on the beach for food to arrive. During this process of rehabilitating the macaques to their original lifestyle, many of the weaker members of the troop died. Birth rate decreased and infant mortality increased, with the oldest and the youngest and those holding the lowest status quickly perishing. The effect on females meant that high-status females had to wait longer for the onset of oestrus and birth intervals expanded and life expectancy shortened for all. The monkeys also had to relearn the locations of different food sources. Today the monkeys are still observed, but they are no longer fed; they are rarely seen on the beach in fact. Scientists now follow a troop through the centre of the island on foot, and it is the macaque's true nature that is observed.

In 1986 Umeyo Mori wrote up her research on the social development and relationships of Japanese macaque females. She never did lecture at Kyoto University; the college promoted her husband but promotion eluded her, so she left. Today she lectures on the comparisons between human relationships and non-human primate relationships at the private Nagoya-bunri College, Inazawa, Aichi. For example, the origins of human mother–infant bonds and sibling rivalry and male–male competition can be seen in non-human primates, such as the Koshima macaques. Two-thirds of Mori's students are female, a particularly high female to male ratio in Japan. She has certainly become a role model for young Japanese women. Today Mori is so busy lecturing that she has little free time. It has been years since she made the journey south to see the Koshima macaques. 'Now I don't have time to do field work on primates. I miss the monkeys so very much.'

Mariko Hiraiwa-Hasegawa represents the second generation of Japanese women primatologists. Her formative years are sharply contrasted with Umeyo Mori's idyllic childhood playing on the beach. Growing up in Tokyo, Hiraiwa-Hasegawa is an intellectual urbanite, an only child born to a sophisticated liberal family. Her father is an international banker and her mother a housewife. As a child she was encouraged by both of her parents, but especially by her mother, to achieve personal success.

Japanese economic affluence has not changed social attitudes to women's roles. Without a Louis Leakey to turn stereotypes around by encouraging and championing women's skills in primate field research, Japanese primatology has remained male oriented. Mariko Hiraiwa-Hasegawa graduated in anthropology from Tokyo University in 1970. She met her husband, Toshikazu Hasegawa, while they were undergraduates. Toshikazu was studying psychology. They married in 1977, but they have never had children. She says: 'I wish I had done that when I was twenty-two. I've now turned forty – it's too late to become a mother now. I just can't be bothered with all that!'

I first met Mariko Hiraiwa-Hasegawa in London, when she was on sabbatical from Senshu and Tokyo Universities where she now lectures in behavioural ecology. She explained that she chose to leave pure primatology behind her 'because of my utter disgust with the system in Japan' and has instead moved into the modern

Western neo-Darwinist disciplines of sociobiology and behavioural ecology, analysing the behaviour of human beings. For many years she had suffered from sexual harassment from a fellow academic, a man whose work she had respected but towards whom her attitude would painfully evolve into incomprehension, disillusionment and finally disgust. Just like Umeyo Mori has said, Mariko Hiraiwa-Hasegawa told me she has always felt 'jealous of the support Western women working in primatology can give each other'.

Hiraiwa-Hasegawa's first study was of the macaque that lives in the Shiga Heights in northern Japan. In the winter temperatures fall well below freezing point and thick snow covers the ground. During these chills the macaques have taken to bathing in natural hot springs, wading in like health freaks and sitting in the steaming pools, almost disappearing in the mist. They need to emerge from the hot water before the sun sets in order to give their thick fur time to dry before nightfall, otherwise they could freeze to death. These macaques have been filmed as they sit from the neck down in thermal water while heavy snow falls all around, some of it settling in peaks on the top of their heads and noses.

Hiraiwa-Hasegawa wrote up her research into the maternal care in macaques in 1983 and then took the opportunity to study chimpanzees for her Ph.D. She worked for three years at the longest running Japanese chimpanzee field site in the Mahali Mountains on the shores of Lake Tanganyika in Tanzania. When she went there in 1979 she also worked to achieve a National Park status for the mountains, helping to protect the creatures that live there.

Some of her time there would be spent alone with her African field assistants from the local Tongwe tribe; at other times her supervisor would arrive for a month or so.

Hiraiwa-Hasegawa wanted to approach her work through a Darwinian perspective. Today her specialist area is female sexual selection (identifying the traits a female seeks in the males she mates with and how this affects the future of her infants). Certain questions would need to be asked in certain ways. But because she couldn't study individual behaviour owing to Imanishi's legacy and her supervisor's anti-Darwinian stance, her three years in the Mahali Mountains turned into a difficult period which resulted in her hating both her supervisor and chimpanzees.

Hiraiwa-Hasegawa's admiration for her supervisor waned. By the end of her time in the Mahali Mountains she was ready to reject much of her academic experience and start over instead. Hiraiwa-Hasegawa didn't mind the basic living arrangements at camp. There was no power, so they cooked on a log fire, drank rainwater caught in buckets and washed in the lake. The nights were pitch black, and candlelight could not penetrate them. They slept in traditional mud huts with corrugated iron roofs which would echo like an endless express train in a tunnel during the rainy season.

Tropical twilights are brief and often spectacular. Sunset is sudden; night falls and it remains dark until 6.30 a.m. Dawn lasts only thirty minutes and by 7 a.m. the sun has risen quickly in the sky. The fast enveloping morning light would wake Hiraiwa-Hasegawa. After a breakfast of tea, rice and dried fish she would go out into the forests to try to locate one of the study groups of chimps. May to October is the dry season, when it is hot during the day and cold at night; the rains come from November to April, when the nights are hot and sticky.

Two years had passed and Hiraiwa-Hasegawa had seen some 'nasty' chimp behaviour and suffered much personal distress. Most of the time the male chimps would ignore the male field researchers – it was too much trouble to fight with them – but Hiraiwa-Hasegawa is a petite woman. To an opportunistic chimp she would have seemed a pushover. A chimp, especially a male chimp, is forever trying to show off his physical prowess to the rest of the troop, or when his back is turned he may become the next victim. The chimps were habituated to humans and a few males had challenged Hiraiwa-Hasegawa a number of times, running at her and pushing into her. If the animal thinks itself superior, things can become dangerous.

Mariko Hiraiwa-Hasegawa had to stand her ground but she was scared. Wild adult chimps seem like a different species from the adorable baby chimps the general public are used to seeing on television. Hiraiwa-Hasegawa began to hate them. The species name for the chimpanzee is *Pan troglodyte*; Pan was the Greek god, half man and half goat, who amused himself by causing misery to others, and the Greeks used the term troglodyte to describe ancient African tribespeople who lived in caves. According to the Greeks neither Pan nor the troglodytes could be trusted.

In November 1981 the rains had come, her supervisor was back in Japan and Hiraiwa-Hasegawa was running the site with Mohamedi, her African field assistant. They were in camp one day when they suddenly heard screaming. They jumped up and ran through the trees towards the noise.

At this time there were two communities of chimps in the vicinity that Mariko Hiraiwa-Hasegawa was observing – M group and K group. M was dominant in size and aggressive to K, encroaching into K's territory. Members of K group were defecting to M rather than staying to fight a losing battle. On this occasion Ntologi, M's alpha male was attacking a lone female from K with four of his male allies. Wantendele and her three-year-old son Masudi had been foraging for leaves together well within their boundary, but they had been ambushed.

By the time Hiraiwa-Hasegawa arrived at the scene Wantendele and Masudi were covered in blood. The males had bitten off one of Wantendele's fingers and had kicked, scratched and battered them both. The female chimp tried to protect Masudi as best she could, stretching her hand out to the males in the classic chimpanzee 'begging' gesture, arm outstretched, palm facing the sky. Mariko Hiraiwa-Hasegawa said the males were focused solely on the female and her son with a killer look in their eyes. It seemed like nothing could distract them from their murderous intentions.

Hiraiwa-Hasegawa and Mohamedi instinctively felt they had to stop the carnage, picking up long pieces of sugar cane and hitting the males. But the males ignored them. Chimps are considerably more muscular than humans and probably didn't feel the sugar cane across their backs.

The blood bath had drenched their minds with lust for a double killing. Hiraiwa-Hasegawa felt desperate; she couldn't allow nature to take its course and needed to bolster the offensive. With her approval Mohamedi picked up a large, heavy stone and threw it at Ntologi. It was not a direct hit and brushed past the chimp, but it was an impressive enough gesture to break the spell. Accuracy in throwing is a trait of *Homo sapiens*; our particularly mobile wrist joints combined with primate binocular vision allow for a precision that chimps do not have. With the flick of a wrist a stone in a man's hand evolves into an effective weapon. Chimps can lift heavy weights and throw things in fights but not with the same accuracy.

Ntologi began to scream in submission, and the other males lost their confidence. The males looked for leadership to Ntologi, who decided to retreat into the trees and was followed by the others. If they hadn't lost confidence but had decided to attack Hiraiwa-Hasegawa and Mohamedi using their vastly superior brute strength, the chimps could easily have killed and feasted on them both and no one would have known what had happened.

Hiraiwa-Hasegawa stayed with Wantendele and Masudi, sitting on the floor and watching them. The female cried and cuddled her child, and they held tightly to each other for another thirty minutes. Eventually Wantendele got up, unsteady on her feet, and took her traumatised son into the trees towards what was left of K. This account is reminiscent of the climax of William Boyd's novel *Brazzaville Beach*, where a Jane Goodall figure feels morally compelled into shooting dead a group of murderous chimps she has been studying. Boyd endows his chimps with self-knowledge, and his narrative depicts apes that are enough like humans to pay the ultimate price for the ultimate sin. But Hiraiwa-Hasegawa's real-world experience did not have the convenient catharsis of a novel.

It was hard to ascertain the chimps' true motivation for attacking the female and her juvenile son. At three years of age Masudi was a little too old to be the victim of infanticide, even though a greater proportion of infanticide victims are male. When mature males kill infant males they are effectively killing a possible competitor. Some males have even been seen to engage in cannibalistic rites when the infant could well have been their own baby. Mariko Hiraiwa-Hasegawa had already seen male chimps forcibly take biologically unrelated young babies and crush their heads in their jaws and eat them while the bereft mothers fought to save their infants.

If a lactating female loses her baby, she is quickly brought back into oestrus and becomes an attractive possession worth fighting over, but Wantendele was not in oestrus at the time of the attack and killing Masudi would not induce oestrus either; she was not receptive and therefore at that time was of no use to a male. On that occasion Wantendele and Masudi appeared to represent the enemy; they came from K and had to be exterminated. Chimps, like humans, are xenophobic and commit genocide.

A year later, after Hiraiwa-Hasegawa had left the mountains, she heard that Wantendele had given birth to a new baby and soon after had been attacked again by Ntologi and his cohorts. Without Mariko or any other human or chimp to help defend her, Ntologi and the other males ate the infant alive. Subsequently Wantendele and Masudi defected to M, and Wantendele mated with her baby's murderers and started her reproductive life over again with her oppressors. A female chimp's biological function is to stay alive, reproduce and give long-term care to her infants. Staying with K was not helping Wantendele as there were no strong males to protect her or her babies; she had no other choice.

Mariko Hiraiwa-Hasegawa related to Wantendele's suffering. Both male chimps and male humans often use bullying strategies to coerce females into having sex with them. Women often remain in relationships with violent men, which can be partially explained by the feelings of vulnerability experienced by lone women without a male presence. He may undermine your confidence or even beat you, but staying with him will at least prevent the possibility of a group of other males from harassing or attacking you. The meagre scattering of battered wives' refuges cannot offer adequate protection from male behaviour selected for during our prehominid ancestry.

Barbara Smuts has become a good friend of Mariko Hiraiwa-Hasegawa. She has also studied aggressive sexual coercion of females by males in baboons, chimps and humans. Even if a female does not want to reproduce with a certain male who has been systematically aggressive to her, she will concede rather than face another battery. If she mates with him and no one else he can be sure that her baby is his and that he has reproduced his genes. If the female reproduces a son, he may be aggressive like his father; battery and rape may ensure she has grandchildren. However brutal, reproduction is the bottom line.

Wantandele accepted this fact. If she had a baby with Ntologi he would not murder the infant. He might, of course, hit her from time to time to prove his superior status but it would be unlikely that he would attempt to murder her and very unlikely that he would murder his own progeny. Ntologi would realise his offspring needed a mother to care for them; the baby's survival was paramount to both parents.

Masudi followed his mother into M and took up his lowly position within the male hierarchy. Wantendele's status with M was low for some years until she was old enough to win some social status for herself. Resident females resent the arrival of another female. They already have to compete with each other for the best sperm and a share in the food. When a female transfers from her birth group to another community, it takes her years of complicated politicising for her and her progeny to achieve worthwhile social status. Powerful friends make all the difference to our lives.

Ntologi had the longest tenure as an alpha male chimp of any ever observed. He remained in control of M from 1979 to 1995. Prime ministers and presidents do not stay in power that long. Only tyrants in non-democratic regimes exert such control because, like Ntologi, they rule by force. Eventually he was beaten in a fight by his second-in-command, Kalunde, and was forced out of M to live in exile. Kalunde couldn't afford to have Ntologi around; after sixteen years at the top he still had a great many supporters. The chimps were used to being led by Ntologi, and many younger ones had never known anything else.

Not all new alpha males force their predecessor from the group. Quite often the ex-leader steps down and ages gracefully, caring for his youngest children. But if Ntologi had remained within the group he could possibly have seized an opportunity to win back his power by using his many allies to help him in a fight. Kalunde couldn't afford to take the chance, but even in exile Ntologi was still a threat. One day Kalunde, backed up by his allies, sought out Ntologi and murdered him. It was the only way Kalunde could feel secure at the top. The brutal accounts of the lives of these chimps are Shakespearean in their tragedy – Ntologi and Julius Caesar had much in common.

When Biruté Galdikas heard how Mariko Hiraiwa-Hasegawa had put a stop to the murder of two wild chimps, she made a point of writing to Hiraiwa-Hasegawa's supervisor, saying that she thought she had done the right thing in saving the lives of Wantendele and Masudi. Hiraiwa-Hasegawa appreciated Galdikas's gesture of female solidarity, but other Western scientists, such as zoologist Tim Clutton-Brock from Cambridge University, were critical, arguing that she should have remained detached and calmly observed the incident. But to sit through something like that demands a cold, detached disposition. Some scientists have that quality in abundance; others do not.

Mariko Hiraiwa-Hasegawa told me: 'While I watched the chimps in the Mahale Mountains, I started thinking more and more about human behaviour. Their intelligence, their nice and their nasty behaviour were so similar to human behaviour, but at that time I had no evolutionary training to understand the implications for human nature. I was very idealistic about people, especially about men. Between the ages of twenty-five and thirty-five I would travel in Europe during the summer holidays with two of my girlfriends. When in Italy we would be persistently courted by Italian men. I didn't know how to handle the situation, I had to learn how to say, "I'm *not* interested in you." I was a late developer with regard to men's behaviour.'

In 1986, when Harvard University Press published Jane Goodall's scientific tome *Chimpanzees of Gombe: Patterns of Behaviour*, it succeeded in dispelling most of the rumours that Goodall's research was not theoretical enough. It was well received and Goodall at last had the respect of the scientific peers who had so often dismissed the Leakey women. Mariko Hiraiwa-Hasegawa was asked to review Goodall's book for the PSGB's magazine *Primate Eye*. Now free of her supervisor and more and more aware of how contemporaries were using a Western neo-Darwinist theoretical framework for their research, Hiraiwa-Hasegawa had grown critical of the traditional Japanese style of observational science. Pure observation was out and the hypothesising of explanations for animals' actions was increasingly fashionable in Japan. Anything less than that had become objectionable to Hiraiwa-Hasegawa. She gave Goodall's book a bad review.

Having found a theoretical framework to cling to, Hiraiwa-Hasegawa accused Goodall of lacking one on which to hang her observations. She now says that if she were to write a review of that book today she would not be as critical, but in the mid-1980s she was fighting hard to reinvent primatology in Japan and wanted to establish behavioural ecology as the guiding force for research. And personally she was reinventing herself. Hiraiwa-Hasegawa wanted to be rid of Imanishi's legacy and her university supervisor's influence.

When the theories of sociobiology and behavioural ecology grasp you, it is similar to finding religion. Critics of sociobiology would say too much interpretation can result from sociobiological theories, but as a late developer Hiraiwa-Hasegawa was glad of the guidance provided by evolutionary psychology. Men's behaviour was beginning to make sense. If others around you are not using a neo-Darwinist doctrine it becomes logical for the converted to criticise its absence. Hiraiwa-Hasegawa has just finished the first translation into Japanese of Darwin's *The Descent of Man*, and she intends to continue to translate all of Darwin's works and have them ready for the great man's bicentennial in 2008. Hiraiwa-Hasegawa doesn't want young Japanese women intending to study primates to be without this knowledge.

Hiraiwa-Hasegawa's main criticism of Goodall's book was the absence of any analysis of 'fission-fusion', a term used to describe the grouping of separate units within a larger community. Chimps, bonobos and dolphins behave this way, as do hunter-gatherers such as the Kalahari's !Kung San people. In basic situations where limited resources of food or water are widely scattered, it would be a disaster if the whole community arrived at one fruiting tree or at one meagre watering-hole. How would they share the resource fairly? Fights would break out; the powerful and strong would take control and everyone else would stay hungry and thirsty. This would be fatal for the weaker individuals, who would not have the strength to walk to the next fruiting tree or watering-hole.

Fission-fusion takes care of this type of disaster. It allows smaller groups to disperse amicably, with individuals mutually confident that this separation is only a temporary measure. They accept they will regroup again, always coming to each other's aid in times of need, as in territory disputes. Male-dominated fission-fusion is found only in chimps, dolphins and humans.

Male chimps have adapted their behaviour to the pattern of fission-fusion. Individual thuggish males can dominate small groups of between two and ten individuals that contain no immediate male competitors, bullying weaker males into being their allies and controlling the reproductive lives of the isolated females. Males relish being big fish in small ponds. Female chimpanzees must transfer alone from the community of their childhood experiences to another group of unrelated males in order to avoid incest. Older sisters have been observed welcoming younger pubescent sisters arriving at a group. They remember each other well from their shared childhood. This sisterly bonding is very helpful, as they will support each other in fights.

But it is quite possible that a young female may enter a group where she has no immediate relatives. If during the fission-fusion process she finds herself in a small group with one other female and three other males, she will probably be beaten into submission by the stronger male who wants to mate with her. He demands respect and the other chimps will not defend her as she is not their kin and there has not been time for her to form relationships with them. The others have more to gain in retaining the status quo rather than sticking their necks out for a low-status female. If the young female is strong and offers help to the other chimps in times of stress, she will ingratiate herself and will probably find the gesture reciprocated.

I asked Hiraiwa-Hasegawa what she thought of the Great Ape Project GAP, a philosophical movement in the West campaigning to give apes civil rights. It is supported by Jane Goodall and many other well-informed professionals but it is not without its critics. Hiraiwa-Hasegawa told me, 'Chimps are very intelligent. It is difficult to find the boundary between humans and apes. But if you decide to treat chimps as special animals, then why not macaques? I spent five years studying macaques. You feel you can read their thoughts. They are loving to orphans, they play nicely with babies. They are very protective of babies, they are very human-like, very loveable creatures, more so than chimps, so the Great Ape Project should include macaques.' Both Umeyo Mori and Mariko Hiraiwa-Hasegawa have an overriding love for their very own Japanese macaque.

Hiraiwa-Hasegawa's defection in 1986 from non-human primatology to the study of human behaviour is a common one. 'I found it hard to study chimps because they are so clever. You have to approach primate studies differently from all other animals. The individual animal's intelligence, social tradition and individuality do not reflect reproductive success like it would in fallow deer, for instance. Female mate choice and sexual selection in deer and sheep is very straightforward by comparison.' Primatologists tend to move on to human evolutionary psychology or stick with their animals and become conservationists, like Goodall, Fossey and Galdikas did.

Today Mariko Hiraiwa-Hasegawa and her husband both teach undergraduates evolutionary psychology and behavioural ecology at Tokyo University. Since 1996 Hiraiwa-Hasegawa has been researching the cognitive abilities of men and women and the lives of Japanese murderers. In America and Europe much research has been done on the differences between men and women, and Hiraiwa-Hasegawa wanted to simulate these experiments in Japan. Were the Japanese like everyone else; is the battle of the sexes universal? Yes. Like men elsewhere, Japanese men were shown to be much better at three-dimensional perception than Japanese women, but the women were much better at location and object memory.

It is thought that these gender differences have been selected for because men hunt game and need to follow the prey throughout the undergrowth. Often a successful hunter has to use three-dimensional perception to 'see' through the trees to relocate the prey. Women, it seems, are more likely to get lost, whereas men have a better sense of direction. On the other hand, women's superior cognitive ability to remember objects and locations is probably connected with the job of looking for edible vegetables. While foraging you may have to hide your infant, and it is important to be able to find her again. Certain tubers will only grow at certain times of year in certain places. It is vital to remember these details.

Hiraiwa-Hasegawa also substantiated the universality of Leda Cosmides's cheater detection work. An evolutionary psychologist, Leda Cosmides has discovered that the human brain has a distinct process for detecting cheating and bluffing behaviour in those around us. This cognitive specialisation is an adaptation that has been selected for. Game theory tells us that co-operative behaviour is beneficial to most of us most of the time, yet we still have urges to push in front of everyone else by stepping on people who considered us as friends. This ambition or defection is at the root of evolution. Some of us are better at sensing such betrayal than others, but we all have an instinct for recognising covert cheating behaviour that is detrimental to us and we all, consequently, have the instinct to take revenge. As individuals, we must remain vigilant. Hierarchical behaviour such as this is endemic in the primate world.

Like many other primatologists who have moved into sociobiology and evolutionary psychology, Hiraiwa-Hasegawa is finding the study of human behaviour fascinating. She has always loved detective stories. Elmore Leonard, Dick Francis and Agatha Christie line her bookshelves next to the *Origin of Species*. Homicide and sexual jealousy are fast becoming Mariko's specialist fields.

Why do men murder? Human killers would share many of their motivations with alpha male chimp Ntologi. Retaining or gaining power may provoke murder. Richard III killed his competitors and as a result gained the ultimate power, the right to sit on the throne of England.

The general pattern for homicide shows that the numbers of women committing murder are negligible but the figures for men are quite different. Young men between the ages of twenty and twenty-five are more likely to commit murder than men of other ages. After the age of twenty-five the incidence of murders declines rapidly. A ninety-year-old man is very unlikely to commit murder. But Hiraiwa-Hasegawa found a difference in Japanese men. From 1990 to 1994 the peak for Japanese men committing murder was found to be among those aged in their forties.

Hiraiwa-Hasegawa was astounded. Why was Japan so different from other countries? She looked back through the records. Between 1955 and 1965 the peak occurred in the men's early twenties, but by the 1980s it was shifting to older men. She realised the same generation of men was still killing; they were just getting older. After the Second World War there was a massive baby boom in Japan. By 1946 thousands of babies had been born; this post-war generation had grown up together and, even with Japan's economic success, had to compete for resources. These men had to compete for wives, jobs, promotions, and school places for their children. They would even have to compete for a space to be buried in. Since 1955 the rate of homicide in Japan has been decreasing, the opposite of the situation in the United States. In Japan this is due to nationally improved standards in education and housing. Hiraiwa-Hasegawa hopes to continue this research by examining the post-war generations of all the other nations who fought in the Second World War.

When it comes to sexual jealousy Japanese men are the same as any other men. They are thrown into a rage if their wife or girl-friend sleeps with someone else. This type of male sexual jealousy over physical relationships is powerful.These feelings have been selected for to frighten women into remaining faithful, giving a man the confidence that any children born belong to him. Of course even with the threat of the old man's sexual jealousy women still take lovers – and this is known as female choice. One difference found by Hiraiwa-Hasegawa was that young Japanese men were more sexually jealous of their girlfriends' emotional relationships than of their physical relationships. She explained that the Japanese students interviewed were most probably virgins and only concerned themselves with emotions. Unlike their over-sexed American counterparts, Japanese students are physically late starters. For example 83 per cent of male and female Japanese university students are virgins; the figure is much lower in the US.

Together Hiraiwa-Hasegawa and I attended a Darwin Seminar at the London School of Economics, where philosophers Peter Singer and Jonathan Glover and game theorist Ken Binmore were speaking. Peter Singer, originator of the GAP is well known for his philosophy of animal rights, and in his lecture he spoke about the origins of a primate's moral sensibilities. He announced that, since we accept that humans have an ethically co-operative side to our nature, selected for because it serves us mutually, we must now rid our society of sexual double standards between men and women. Singer feels that for too long women have been publicly degraded by this male hypocrisy – female subjugation at the hands of men must finally come to an end. Singer was pleased to prophesise that DNA testing means that men will no longer need to bully women because they feel insecure about the paternity of their children.

But Hiraiwa-Hasegawa and I felt Singer was missing the point. Women and other female primates have always had a counter-attack to this treatment – female choice. Women are not just generic victims of patriarchy. On the surface sexual double standards seem to keep women in their place, but in reality men are kidding themselves. Female choice ensures male–male competition; men jostling for women's attention means that women get to pick the strongest seed.

Many women – and females in other primate species, such as langurs – trick the dominant male in their life into thinking their baby is his. Quite often a female will choose the sperm from one male and the paternal skills of another. I have a number of girl-friends who have done this, though thankfully their husbands, ex-lovers and children are all blissfully unaware. Only the woman knows the truth and she rarely speaks of it because she has worked hard to achieve the family of her choice and one comment could blow her world – and that of others – apart. DNA testing is perhaps the last thing a modern woman with ancient urges needs.

Could female choice explain why the maternal grandparents of a new baby often remark how much the baby looks like its father? They are keen to make the father feel confident in his paternity, and don't want him deserting their daughter. Could this genetic inse-curity also explain why maternal grandmothers always seem closer to their daughter's children than paternal grandmothers? The maternal grandmother knows her daughter is hers and that her daughter knows her baby is hers, but the maternal grandfather and the paternal grandparents can never be sure the grandchild is theirs. Could this be why women with sons often yearn for daugh-ters? If the matriline is broken, Granny cannot be sure her grandchildren are really hers. There is a powerfully deep, unspoken confidence between unbroken generations of women very similar to the tight bonds seen between matriarchal monkeys, who also know that they are blood relatives.

Mariko Hiraiwa-Hasegawa told me her research into human behaviour has 'helped me with my marriage in understanding my husband and in understanding men generally. Now I understand the basics of human nature I've found I can understand several of my own unconscious emotions. I used to have strong feelings of rage. I didn't understand them, but now I do. I was talking to a long-term girlfriend, a medical doctor, about my sexual jealousy research and how men are more sexually jealous. I explained this was a male reproductive strategy – men don't want to care for children that are not theirs and female choice is how women respond. My friend said that evolutionary psychology is going to take control of my life and all the pleasure will be lost. I explained that understanding does not mean you lose the feelings. We may search for enlightenment, but we are still animals. But it does help to take away some of the emotional confusion that causes pain.'

Much wrangling goes on between theoretical scientists, but it is well to remember that their neo-Darwinist theories come in and out of fashion. Goodall's original observational data is used by all desk-bound theoretical scientists studying chimps, because nothing can replace an observational fact.

5
Wild at Heart

The rehabilitation of apes can take years. The animals bond to their human carers and learn the rules of a captive life. To give the apes what should have been rightfully theirs from the very beginning – freedom – can be a double-edged sword, but some women have loved their apes to such an extent that they have put themselves and, on occasion, the animals through great personal hardship in their bid to set them free.

Today, Goodall fights to keep wild-born chimps safe and forever wild. But in recent years she has had a change of heart with regard to the rehabilitation of captive chimpanzees. From her years of observation of the Gombe chimps, especially mother–infant bonds, she now believes captive chimps must have space, but remain captive if they were cared for by humans from a young age. Goodall believes it is humans the orphaned youngsters have bonded to and human culture they have become a part of, and she thinks this is irreversible. The chimpanzee mother–infant bond is a powerful one and if broken the infant is lost. The mother–infant bond is well illustrated in the story of Flo and Flint.

*

At Gombe, Flo was the female chimp closest to Jane Goodall's heart. Their bond of familiarity enabled Goodall to sit quietly and observe mother and infant interaction. As Flo was a dominant female, the other chimps took her cue and eventually all allowed Goodall to enter their lives.

Chimpanzee families consist of a female chimp, her youngest baby and any older siblings. Even when grown-up, chimpanzees remain attentive to their mothers, brothers and sisters. There is a division of labour in chimp societies and the father does not play a direct role in the day-to-day care of his infants. Flo was a successful mother, probably approaching forty years of age when Jane Goodall first set foot in the Gombe hills, and already the mother of three surviving offspring, Faben, Figan and Fifi, when Jane first met her. As a highly successful and experienced mother, Flo was a very attractive mate for the male chimps, who do not like to waste their sperm on females without maternal experience.

As Flint grew up his personality emerged and he showed himself to be a demanding youngster. He wanted piggyback rides everywhere. When Flint was five years old and Flo was nearly fifty, she gave birth to Flame. Unfortunately Flo had been unable to wean Flint from the breast and was nursing Flint right up to the day her new daughter was born but Flint was jealous. Fifi, on the other hand, took great interest in her pretty baby sister and loved to cuddle and play with the infant.

Flo was utterly worn out by Flint's behaviour. He wouldn't grow up: when Flo walked along through the forest with her baby daughter clinging to her stomach Flint would pester his mother and have tantrums until she would allow him to do the same. Flo would carry the extra weight of Flint hanging on under her stomach, his little sister squashed between them. At five years of age Flint was old enough to explore with his older brothers, sister and other juveniles; he was old enough to make the first tentative steps towards independence but he wouldn't go. He could make his own bed to sleep in at night, but he would always push into Flo and Flame's nest at night. Weakened by the extra stress of such a difficult child, Flo contracted pneumonia and became very ill. She was so run down that she couldn't even climb a tree to make a night nest.

One day Jane found her lying on the ground, unable to move, and Flame was missing. What happened to poor little Flame is a mystery. She was probably taken and eaten, possibly by another chimp, possibly by bush pigs after Flo's grip around her weakened. But with his little sister out of the way and Flo making a slow recovery from her pneumonia, Flint was able to have his mother all to himself again. In the last years of their life they were inseparable. Fifi had turned out to be as sexy as Flo and had just become a mother for the first time to Freud. Fifi had seen her mother hold Flint upside-down by his ankle and tickle him under his arm, just as Flo had done with Fifi, making him roar with laughter. Fifi took a leaf out of her mother's book of maternal tips and played with Freud in just the same way. Goodall had never seen another chimp mother play with her infant in this particular way before. Tickling was common, but hanging the infant upside-down to intensify the pleasure was Flo's addition. Just as Fifi copied her mother's maternal behaviour, so perhaps had Flo. We shall never know what Flo's mother was like, but it is likely she taught a trick or two to Flo, who passed them on to Fifi. Fifi has gone on to give birth to seven infants, though the last baby caught a disease and died. But seven births and six surviving progeny is a record for an observed wild chimpanzee mother.

Flo continued to give Flint rides right up to the time she died. By then Flint was a hefty eight-year-old and poor Flo could hardly move under his weight. Flo's teeth were worn down to the gums and she walked at a slow pace, but if Flint ever got into trouble with a baboon Flo would muster up the strength from somewhere to bristle out her hair and, emitting threatening waa-barks, would run at the baboon, chasing it away.

The area over which the two of them foraged for food became progressively smaller as Flo aged and Flint became a heavy weight to carry, meaning there was less variety and quantity of edible food-stuffs available. Flo and Flint would often come into camp and Goodall would supplement their diet with bananas and eggs.

On a sunny autumn day in 1972 Flo's dead body was found lying face down in a stream, a grieving Flint close by. Goodall spent the night by Flo's body, no doubt sharing Flint's grief. She wanted to protect the corpse from hungry bush pigs, and reasoned that Flint's depression would be greater if he found his mother's remains being feasted upon. Goodall observed Flint as he watched the life-less shape of his mother then climbed up into a tree above his mother's body and looked at the remains of the final nest they had shared together. Flint stayed beside his mother's body for three days and nights. He had lost his reason for living. His elder brother, Figan, arrived and comforted Flint, encouraging him to travel with him. Under duress Flint left.

With Flint out of the way, Goodall and her students gathered up Flo's remains and her body was sent back to the United States for analysis, but a few days later Flint was back at the side of the stream looking for his mother. Fifi arrived and groomed and comforted Flint, but she could not reach his lost soul. As well as having the role of big sister Fifi was now a mother and eventually she had to move on to find food for herself and her infant; as Flint would not follow she had to leave him to his misery. Three weeks after Flo's death, Flint died. He lost his appetite and wasted away. He died curled up on the bank of the stream, just feet from where Flo's body had lain.

Flint's body was also sent to the United States for analysis. I visited anatomist and anthropologist Adrienne Zihlman at the University of California, Santa Cruz. Zihlman is the keeper of the bones of some of the most famous primates, and she opened a drawer to reveal, side by side, the skulls of Flo and Flint. Zihlman held Flo's skull aloft. Through Jane Goodall's detailed documentation of Flo's social life, both Zihlman and I felt we had known her well. I remembered watching film of Flo, and in my mind I saw her face, tatty ear and lolling bottom lip. I saw Flo grooming Flint, and sitting with her friend Olly. It was a sobering moment. Gently her skull was placed back next to Flint's skull and the drawer was again shut to the light.

From dating Flo's skull it seems that the chimp may have been as old as fifty-five when she died. Only a skilled chimp lives that long in the wild; Flo's brain was slightly bigger than the average female Gombe chimp brain. Zihlman went on to tell me that Goodall's hypothesis that Flo's indulgence of Flint inhibited his natural development is probably unfounded. Zihlman suspects Flint didn't grow up because he couldn't. From examining his skull, it seems that his brain was not growing correctly and he was mentally disabled. Flint would probably always have needed a mother to care for him. Flo had tried her very best, probably realising that Flint was not maturing probably and therefore fussing over him the way human mothers tend to their sick children. At eight and a half years of age, Flint should have been socially capable of living without his mother by his side. But at the end of her life Flo could no longer defend Flint from their inevitable tragic conclusion.

For a time Goodall relished being a part of her chimp's community. Jane Goodall has said she loved David Greybeard more than any other chimp she has ever known. Sharing his life allowed her to make discoveries about David's chimpanzee nature that transformed Goodall into a world-famous scientist. When she first arrived by the shores of Lake Tanganyika, there were plans to allow local farmers to cut back the forest and farm there, and had this happened the Gombe chimps would have become extinct. Because Goodall had seen David Greybeard exhibiting the supposedly human characteristics of meat-eating and tool-using, the area was turned into a national park that gave protection to the animals and allowed Goodall to continue her studies.

She had established an intimate bond with David Greybeard long before Hugo Van Lawick arrived in Gombe. David was the first chimp to invite himself into camp to feast upon the fruiting palm tree that gave shade to Goodall's tent. But she knew him by sight long before he did this; Goodall describes him as having 'a distinctive handsome face and well defined silver beard'. When David Greybeard first came into camp, so as not to frighten him away Goodall would hide from view inside her tent and peep out at the confident chimp.

One day Goodall was eating a banana when David turned up in camp. This time Goodall decided to sit outside her tent to watch him. He approached her, stopping a few feet away, then stood up and fluffed out all his hair, a magnificent and terrifying sight. The chimp charged at Goodall and she thought he would attack her, but instead he grabbed her banana and ran off with it, settling down on the other side of camp to eat it. This was the beginning of the infamous banana feeding. Goodall asked one of her African helpers to go to market and purchase large quantities of bananas to encourage the chimps to come to camp. Following six months of observing very little, the unexpected theft of a banana by a fully grown wild chimp was an epiphany that left Goodall wanting more.

The next day David wandered back into camp with Goliath, the alpha male, in tow, and eventually numbers of chimps visiting the camp grew. A few weeks later both Goodall and David felt confident enough to get a little closer to each other. Goodall held out a banana to David and he approached, making a quiet but threatening cough. He stood up, bipedal, hit a tree, reached out and gently took the banana out of Goodall's hand. Although David was a senior male, his status was not achieved through aggression. In fact the chimp was a peace-maker, calm and sympathetic to others and never bullying. If Goliath became angry David would make the peace by grooming Goliath and laying his hand on the others who were becoming over-excited.

When Hugo Van Lawick first arrived in 1962 neither he nor Goodall knew what was in store for them. Although Louis Leakey had told Jane's mother, Vanne Goodall, that he had found a husband for Jane, at this stage Jane and Hugo weren't even certain they could make a film together. Goodall was also unsure how Van Lawick and David Greybeard would get on. Van Lawick decided to hide in his tent at first, allowing David to become accustomed to the new equipment and smells in camp before showing the chimp his face. David casually strolled into camp and sat down as usual to eat his fill of bananas. When he'd finished he rose and purposefully strode over to Van Lawick's tent. The chimp pulled back the flap of the tent, stuck in his head and stared Van Lawick in the face. Satisfied, David then turned on his heels and knuckles and walked off into the hills.

By December 1962, when Van Lawick finished filming, he and Goodall had fallen in love. They decided to get married as soon as they could. Goodall had to return to Cambridge to finish her Ph.D. and it was decided that London would be the location for their wedding. After Van Lawick left camp, Goodall felt very lonely. She turned to David Greybeard for the comfort of contact. Louis Leakey tried to find companions for his protégées as he was worried they would begin to see their apes as a 'romantic sexual thread'.

One day, as David Greybeard ate his bananas, Goodall tentatively reached out to his shoulder and started to groom the chimp. At first the chimp pushed her hand away, but she tried again and this time he allowed Goodall to stroke him. Jane says this contact with a fully grown wild chimp, possessing a strength five times as great as her own, yet nonetheless allowing her to groom him as though she was another chimp, was the best Christmas present she has ever had.

When Goodall and Van Lawick married in 1964, Louis Leakey was unable to be there, so he sent a tape recording of his speech and his granddaughter was a bridesmaid. A clay model of David

Greybeard was placed on top of the wedding cake instead of the usual decoration. Large, blown-up photos of all of Goodall's favourite chimps adorned the walls at their wedding reception. If Goodall could have had the chimps as guests, no doubt she would have done so. While in London for the nuptials she heard that Flo had given birth, so they cut short their honeymoon to get back to Gombe and see newborn baby Flint for themselves.

The physical closeness Goodall had begun to enjoy with some of the chimps meant she often preferred their company to that of humans. David Greybeard's reputation spread quickly. Visitors would turn up at Gombe hoping to catch sight of the wild chimp that allowed humans to watch him while he ate bananas. One day Goodall was sitting on the hill above camp with one of her chimps named William when they both spotted people arriving down below, an incident described in *In the Shadow of Man*: 'I was sitting there with William, peering at them as though they had been alien creatures from an unknown world.' Goodall was entering the chimpanzee world and seeing humans as aliens. Her marriage to Hugo Van Lawick and the birth of her baby son, Grub, kept her from removing herself from humans altogether.

Although Goodall today believes that chimps must be uncontaminated by human contact otherwise they will be for ever spoiled, she did not always feel this way. A moment of physical contact between herself and a male chimp remains to this day the highlight of her career: 'There are certain people – and Dian Fossey was one – who want to always be in contact with their subjects. It's a danger and I fell into this at the beginning. I was so thrilled, and I wouldn't change this for a moment, when David Greybeard, who grew up fearing humans, actually allowed me to groom him. That was one of the highlights of my whole life, and I wouldn't want it different.' But after the kidnapping and her enforced departure Goodall realised her research could go on without her because others could do the work, her conclusion was thus: 'We must not interact with the chimps. We must try and see that their lives remain as uncontaminated as possible. We do not want to be part of their community; we want to be observers noting down what they do.'

Today Jane Goodall never attempts to rehabilitate the rescued orphans that now live at the Jane Goodall Institute's four African chimpanzee sanctuaries, in Uganda, Kenya, Tanzania and the Republic of the Congo (Brazzaville). The chimps there will be used as ambassadors in eco-tourism, and revenue raised can be used to help save the forests where the wild populations live. Goodall believes that once humans are imprinted on the mind of a chimp you cannot rehabilitate her and it is psychologically extremely cruel to try; the early years of a chimp's life are as crucial for that chimp's mental stability as a safe childhood is for a human. Most rehabilitant chimps have lost their mothers and had early trauma and Goodall maintains they are too disturbed to make it back into the wild. She also believes that captive apes do not know how they should behave. Much chimpanzee behaviour is learned by young wild chimps through their observation of their peers and elders, especially their mother. If they are taken from their mother and the wild at a young age they will have little remembered behaviour to guide them. Goodall thinks the only time a new chimp can be introduced to an established group is if the chimp is a female in oestrus; in the wild it is only females, in heat, who can successfully transfer to a new group.

The Jane Goodall Institute gave me an example of an ex-captive male called Poco who resides in their Kenyan sanctuary. Apparently Poco learned to walk bipedally while in captivity because his cage was so small. He is a senior and well-respected male among his community of chimps at the sanctuary and the young chimps look up to him in more ways than one. Poco is still walking on two legs and the young chimps are copying this behaviour. But it is unnatural for chimps to habitually walk on two legs. Jane Goodall believes captive or ex-captive apes, who have never lived like normal wild chimps, cannot create a genuine chimp society together as they do not know the social rules.

In November 1971, as soon as Biruté Galdikas arrived in her new Indonesian home in Kalimantan in Borneo, to research the lives of wild orang-utans, she also became involved in the exhausting and thankless task of trying to save the lives of captured baby orang-utans. As with other apes, infant orang-utans are 'orphaned' when their mothers are shot dead. As babies they are adorable; they do not bite as much as baby chimps and are easier to keep alive than baby gorillas. In the early days at Gombe Jane Goodall never tried to rehabilitate chimps and Dian Fossey only attempted it with infant gorillas twice. Jane Goodall (among others) has also been critical of Biruté Galdikas's rehabilitation work with orang-utans in Borneo. Goodall told me: 'Biruté is releasing more and more orang-utans into an area that they didn't originally come from. She is creating artificial crowding. I'm sympathetic – it is hard to see what else she can do – but at the same time I understand the criticism.'

Galdikas named her first infant rehabilitant orang-utan Sugito, in honour of Mr Soegito, the Indonesian forestry officer who guided Galdikas and husband Rod Brindamour through endless bureaucracy and protocol on their way to becoming the first Westerners to set foot in Tanjung Puting National Park. Although orang-utans and proboscis monkeys (found only on Borneo) were protected on paper before Galdikas arrived, orang-utan infants were openly traded. Galdikas and Brindamour insisted the forestry officials enforce the law and Mr Soegito gave his blessing to their Western need to do something. It was somewhat embarrassing for him to confront locals – many of them were members of his family – with pet orang-utans, but white Canadians found this task easier.

Female orang-utans have a baby approximately every eight years.

The young orang-utan hangs on for dear life to its mother for much of this time. In the absence of his mother, Sugito chose Biruté Galdikas to care for him and he refused to let her go. Galdikas's latent maternal instinct was triggered and they became the most important thing to one another. Their relationship would become far more intense than if he had remained with his biological mother.

Galdikas and Sugito's relationship was one of selfless compromise on the part of Galdikas and pathetic desperation from Sugito. Galdikas had to sleep with him glued to her, his strong fingers gouging her skin. The little orang-utan would intermittently urinate on her throughout the night. The warm rush on her skin would wake her, then the resulting coldness, stinging and stench would keep her awake. This continued for years. In the wild the baby would have lifted his rear away from his mother and his trickle of urine would have rained down through the branches of the tree they were nesting in without either animal getting wet.

If anyone tried to prise Sugito off Galdikas to give her a chance to bathe and wash off his urine and excreta, he had a panic attack. The little ball of red fuzz would metamorphos into a screaming, biting, randomly urinating and defecating beast. Anyone within firing range would be liberally covered with some sort of mess and more often it was easier for Galdikas to bathe with him attached to her. Sugito wanted no one near Galdikas when they went to bed. The baby orang-utan came between husband and wife more effectively than any extra-marital affair. Not surprisingly Sugito and Brindamour hated each other. If Sugito ever had the opportunity, he would ram his penis in Brindamour's ear. But in the little ape's eyes Galdikas could do no wrong – until she took on more ex-captive orang-utans.

Sugito was jealous of anyone, orang-utan or human, that Biruté paid attention to. As a juvenile he drowned a pet kitten and three younger ex-captive orang-utans that Galdikas was caring for. Sugito didn't want siblings and he didn't want his 'mother' attending to anyone but him. In the wild he would not have suffered such competition. The two of them, mother and child, would have been a single unit until the female became pregnant again when the child reached eight years of age. Upon pregnancy, the mother orang-utan rejects the juvenile's relationship, rebuffing it and sometimes even biting it to force it to venture forth on its own. A fruiting tree might support one adult female and one young orang-utan, but not one pregnant adult and one half-grown orang-utan. The immediate vicinity is just not big enough to support the two of them.

When Sugito started to drown other creatures Galdikas did not want to believe it was done with intent. She wanted these incidents to be tragic accidents, the results of experimentation on the part of Sugito. But when he was caught in the act for the third time and Galdikas looked into his eyes, she realised he knew exactly what he was doing.

In 1977 Galdikas invited Gary Shapiro, her first American primatology student, to come to Camp Leakey. Camp Leakey is Galdikas's base within the Tanjung Puting National Park. Shapiro was a psychology student at Oklahoma University, where he had been teaching chimps American Sign Language (ASL) and he became the first person to teach sign language to an ape in its natural environment rather than in a laboratory. Galdikas wanted him to ask Sugito why he had murdered the baby orang-utans. Shapiro did succeed in communicating with a number of free-ranging rehabilitant orang-utans at the camp, with Princess being especially receptive to him, learning to request food and cuddles with ASL, but Sugito refused to open up. He was not a good student and would just get up and walk away when Shapiro tried to teach him signs.

When Galdikas returned to California to submit her Ph.D. thesis Brindamour seized his opportunity to get rid of Sugito, expelling the nine-year-old ape who knew too much back into the Garden of

Eden. He took the animal miles away from Camp Leakey, and released him. Galdikas had expected Sugito eventually to leave camp voluntarily, as he was at the age when a juvenile leaves its natural mother, so she presumed that as he matured the call of the wild would encourage him to make his way into the jungle in his own good time. But when Galdikas returned she found that Sugito was gone. Galdikas never saw her Oedipal child again, and always resented that she was not consulted on his fate or allowed to say goodbye.

Biruté Galdikas has always tried to mix the rehabilitation of orang-utans with her observation of truly wild orang-utans, but over the years the decision about where to home all the 'orphaned' infants confiscated from the ape trade has become one of her biggest headaches. With the loss of rainforest habitat increasing to 5,000 tons of logs leaving Indonesian ports every single day, more and more orang-utans are being made homeless. Many rescued orang-utans are released into the Tanjung Puting National Park, which struggles to sustain its own resident apes, let alone all the newcomers.

Gundul was another male orang-utan that Galdikas struggled to rehabilitate. Once an ape has become used to being fed at a site it makes sense to them to return from time to time when foraging for food is hard. Gundul was a sub-adult male; he was in years a mature male but he had not yet developed the cheek pads, long call and extra weight of a fully fledged adult male orang-utan. The long call can only be made by a cheek-padded male, and it consists of loud, low grumbles that build into a bellowing cry and eventually end in a series of more grumbles. The noise can travel miles through the forest. Gundul made frequent visits back to Camp Leakey, and on one occasion in 1975 he raped Galdikas's cook.

When Galdikas and Brindamour first arrived in Borneo, a forestry official asked Galdikas if she was afraid. Afraid of what? she enquired. Afraid of being raped by an orang-utan. She was told that this had happened to women many times. Orang-utans will grab a woman and lift her up into the canopy where she cannot escape, only bringing her down when they have finished with her. Galdikas and Brindamour thought that the man was superstitious and naive. As young, ambitious field primatologists, they laughed at the absurdity of such a thing!

It is difficult to explain what motivated Gundul to rape. Male orang-utans are more sexually attracted to females who have already proven themselves to be successful mothers. The cook was a mother and Gundul had observed the woman with her children. Also, as Gundul had lived with people during the first year of his life, a form of cross-species misplaced imprinting might have occurred; this is known in captivity but much rarer in the wild. Perhaps Gundul was just sexually attracted to the woman. Certainly, when great apes are raised by humans they do not fear people and are confident in their superior brute strength.

Biruté Galdikas describes the Gundul story in her book, *Reflections of Eden*: 'One day, I went to the platform with a visitor from North America and one of the cooks. When Gundul arrived, he ate a little but seemed distracted. Suddenly Gundul grabbed the cook by the legs and wrestled her down to the platform, biting at her and pulling at her skirt. I had never seen Gundul threaten or assault a woman, although he frequently charged male assistants. The cook was screaming hysterically. I thought, 'He's trying to kill her.' I had a vision of Gundul tossing the cook off the platform into the shoulder-deep swamp water and drowning her.

'I attacked Gundul with all my strength, trying to jam my fist down his throat. I shouted to the visitor to take the dugout back to Camp Leakey for help. My repeated blows had no effect on Gundul; but neither did he fight back very aggressively. I began to realise that Gundul did not intend to harm the cook, but had something else in mind. The cook stopped struggling. 'It's all right,' she murmured. She lay back in my arms, with Gundul on top of her.

Gundul was very calm and deliberate. He raped the cook. As he moved rhythmically back and forth, his eyes rolled upward to the heavens.'

'Gundul was by far the most difficult ex-captive I ever received. His sexual assault on the cook was one of numerous incidents. Gundul was afraid of Rod and gentle with me, but he terrorised the local assistants, the Indonesian university students, and, to a lesser degree, local visitors. He stalked a young American female Ph.D. from MIT for weeks. Gundul had been raised from infancy by humans. His owner, a general, had treated him like a shaggy orange prince. It did not surprise me that Gundul was sexually attracted to human females and viewed most local men, men without uniforms, as servants.'

It's not just rehabilitated orang-utans who behave this way. A female primate keeper was almost raped at Chester Zoo in England by an orang-utan in the early 1990s. Nick Ellerton is curator of mammals at Chester Zoo, where he is responsible for a large collection of chimps and a small collection of orang-utans. Ellerton informed me that sub-adult orangs have a high sex drive with Kama Sutra techniques. Many captive males have to be vasectomised because there are too many unstable bachelor groups as there has been too much unsupervised inbreeding in the past, with certain individual orang-utans fathering so many infants that their genes now dominate the captive community.

One such bachelor was sitting quietly in his enclosure when a woman keeper entered the next-door enclosure to clear up. An orang-utan sitting motionless looks like a benign Buddha crossed with a ginger shag-pile rug; it certainly doesn't look like a monstrously strong, quick-thinking, fast-moving sex pest. The adjoining door between the two enclosures wasn't locked, and the animal wasted no time. Before the keeper realised what was happening, the orang had opened the adjoining door, grabbed hold of her and removed her trousers. Terrified, the woman froze. The animal was intent on raping her. Luckily Nick Ellerton was passing by and rushed in, grabbing a chair en route and smashing it over the orang-utan's head. The ape was momentarily stunned, giving Ellerton enough time to help the woman to safety and to lock the orang-utan in.

In 1996 a documentary was made at Camp Leakey about orang-utans and one of their many human admirers, the actress Julia Roberts. She wanted to meet wild orang-utans and left the safety of Hollywood for a tropical Indonesian wood and a close encounter with Hollywood's favourite great ape. But things very nearly went badly wrong when Kusasi, an adult male orang-utan and ex-rehabilitant, paid a visit to camp. Some ten years before, Galdikas had successfully sent him on his way and he had not been seen again until one day he reappeared as a cheek-padded male. Kusasi wanted to mate with all the females in the area of Camp Leakey. At present he is still there and when he achieves his goal he will disappear into the forest again in search of new females.

Kusasi took a fancy to Julia Roberts. The documentary crew was filming her walking along a path when the ferociously strong male orang-utan grabbed her. Roberts, terrified, struggled, but Kusasi wouldn't let go. The film crew had to stop filming and rush to her rescue, and it took five men to peel the ape's fingers off the actress and pull her free. If Julia Roberts had been alone when the orang-utan captured her she might well have been carried up into the treetops and raped.

Five years after arriving in Indonesia Galdikas and Brindamour had a son, Binti. At eighteen months of age Binti had four-year-old juvenile orang-utans as playmates. Galdikas noticed that, just as we like to know the sex of a new infant and ask the mother whether it 'is a boy or a girl', newly orphaned orang-utans at camp would be picked up and sexed by older rehabilitants. Any new orang-utan at camp would also want to sex Binti by peering into his nappy. There are some very endearing photographs of Binti and orphaned orang-utans having baths together.

Ex-captive orang-utans had the run of Camp Leakey and for Galdikas to continue her research a nanny was needed. Galdikas employed a seventeen-year-old Indonesian woman called Yuni to help care for Binti. Binti and Yuni loved each other and, while Galdikas was consumed with her thesis and the needs of rehabilitant orang-utans, Brindamour also turned to Yuni for love.

Brindamour and Galdikas achieved a great deal together in Borneo, but after seven years their relationship was over. Brindamour claims that Galdikas never told him she wasn't coming back. Galdikas says that in the end Brindamour was only willing to invest seven years to the lifelong quest to understand and save the orang-utan.

A number of issues rose their heads. Brindamour was ostensibly Galdikas's research assistant. He had helped her achieve her Ph.D., but his career was not progressing and he wanted to establish himself in Western terms. Another problem was that Binti was emulating juvenile orang-utan behaviour because these animals were his only friends. When Galdikas turned her head for a moment, Binti would quickly climb up a huge dipterocarp tree, just like an orang-utan, and she would fear for his safety. Like most Western parents, they both wanted Binti to go to nursery school and have human playmates, but this was not possible at Camp Leakey.

Another problem was that Galdikas had grown fat. After giving birth to Binti, she took a long time to lose weight again, and Brindamour blamed his philandering on Galdikas's extra pounds. After Galdikas had gained her doctorate, Brindamour told her he wanted a divorce. He wanted to return to North America, put Binti in nursery school, marry Yuni, go to school to study computers and start his life over as a computer systems analyst. Their idealistic hippy days avenging poachers and illegal loggers together were gone. Brindamour could take no more and wanted to return to the 'real world'. He told Galdikas, 'Every year it got harder and harder to come back to Kalimantan. I never want to return to Kalimantan as long as I live. Never!' Galdikas felt betrayed by Rod and Yuni's love affair. She still loved him and didn't want a divorce, but she agreed to his demands.

Of this time in her life Galdikas has written, 'The archetypal Western male, Rod, went to Borneo in search of adventure. He liked testing himself, pushing himself to the limit . . . and then a little more. He was the Marlboro Man with a mission, saving the forest and the orang-utans. But when you have the same adventure day after day, the exhilaration and the feeling of triumph fade. I went to Indonesia for so-called "female" reasons: I wanted to help. If I had to take risks, I did. But I wasn't interested in adventure for adventure's sake. My triumph came from feeling at one with the orang-utans and the forest; I exulted in the peace and the quiet. Because I wasn't looking for thrills, I never got bored. The more I knew about orang-utans, the more I would be able to learn. After seven and a half years I felt even more committed that when I arrived.

'Only after Rod announced he was leaving did I realise how much I had resented the fact that it always had seemed so easy for him. I whimpered, I suffered, I gritted my teeth. Now I discovered that Rod had hated it even more than I did. Never saying anything, never complaining, was part of his Western maleness. I had grown weary, but never enough to consider leaving. For me studying and rescuing orang-utans wasn't a project or a job, but a mission.

'Six months after Rod left, I took Binti to North America to enter nursery school. It was the most difficult decision I ever made. But I felt strongly that Binti should be educated in North America and that a boy should live with his father. I had been appointed to a permanent, part-time faculty position in the Department of Archaeology at Simon Fraser University, British Columbia, which meant that Binti and I would have at least several months a year in North America. During this trip Rod and I finalised our divorce.

Now Rod could marry Yuni, and Binti would have a stepmother who knew him and loved him. Whatever flashes of resentment I might have felt toward Yuni, I knew that I could trust her with Binti's welfare.'

Some years later the new Mr and Mrs Brindamour moved to Australia for career reasons and Binti moved in with the family of a school friend, who effectively raised him. Rod and Biruté decided it would be best for Binti not to disrupt his schooling by moving him to Australia. Binti is now in his early twenties and at a cross-roads, unsure which direction to take. He often works as a volunteer for his mum at Camp Leaky. Galdikas obviously felt that Binti had people around him who loved him, even if she was not there full-time, but she knew that without her the red apes had no one. Their need for her and her need for them were the most powerful, primal urges. Galdikas returned to Borneo alone.

From 1979 to 1980 Galdikas and Pak Bohap spent time together surveying Tanjung Puting Reserve and fell in love. Even though he has married the world-famous scientist he has never left Indonesia and does not speak English. He is a traditional Dayak farmer and apparently the only Indonesian man orang-utan Gundul seemed to revere. Galdikas and Pak Bohap have two children, Fred and Jane. The children visit North America regularly with their mother, and love to see their big brother Binti, but their early years have been typically Indonesian. Like Binti before them, Fred and Jane are at ease with orang-utans.

In *Reflections of Eden* Galdikas writes, 'I have looked into the eyes of a wild orang-utan and seen that orang-utan looking back at me. The experience is almost indescribable ... Communing with a wild animal of another species means glimpsing another reality. Perhaps the closest analogy would be visiting the parallel universe of the Dayaks.' Biruté Galdikas has immersed herself completely in orang-utan, Dayak and Indonesian life and is no longer the Western woman she used to be. Her work with rehabilitant orang-utans continues to this day. The orang-utans she rehabilitates are all wild caught in Indonesia; she has never brought a captive-born, tame orang-utan from America to Borneo for release.

Few American laboratory chimps have ever been rewarded with freedom after all the years they spent teaching us about ourselves. Lucy is one language chimp that did reclaim her birthright, retiring from experimentation and emigrating to Gambia, where she was taught how to live as a wild animal rather than as a surrogate American by Janis Carter. The relationship between Carter and Lucy is an epic love story between a woman and a chimpanzee. Carter gave up everything for Lucy. Together they transcended the mental and physical barriers to intercontinental rehabilitation. Together they discovered Africa, the cradle of life.

In the early 1970s psychologist Bill Lemmon ran Oklahoma University's Primate Centre. Like Harry Harlow, Lemmon wanted to unravel the female human maternal and sexual psyche by experimenting with female primates. Wanting to study a female chimp's innate maternal instinct, Lemmon's planned experiment was to test newborn Lucy's ability to correctly raise her own chimp baby, which would be conceived through artificial insemination when she was older. If Lucy was never to see another chimp and all her role models were humans, would she think of her baby as a human baby or as a chimp baby? Would she raise her baby like a human mother or like a chimp mother? And how would she communicate with her baby? Would she, for instance, teach her baby American Sign Language?

Maurice Temerlin, a clinical psychotherapist and one of Lemmon's favourite students, was chosen to take Lucy from her mother when she was just two days old. Maurice and his wife Jane Temerlin treated Lucy like a daughter and raised her with their son, Steven.

Lucy lived like a human. Maurice and Jane Temerlin were her

mother and father and Steven her brother. She had complete freedom of the house and consumed a health-food diet, Ski fruit
yoghurt being her favourite food. Behaviourists from Oklahoma
University such as Roger Fouts and Sue Savage-Rumbaugh studied
her and taught her ASL.

One day Roger Fouts was visiting the Temerlins' house to give
Lucy a lesson in ASL. While he sorted out his papers Lucy defecated
on the living-room carpet. The following conversation took place
between Fouts and Lucy in ASL:

Roger: What's that?
Lucy turned her head away, pretending not to hear him.
Roger: Do you know? What's that?
Lucy: Dirty dirty (Lucy's term for excrement).
Roger: Whose dirty dirty?
Lucy: Sue's.
Roger: It's not Sue's. Whose is it?
Lucy: Roger's.
Roger: No! it's not Roger's. Whose is it?
Lucy: Lucy's dirty dirty. Sorry Lucy.

When Lucy first met Sue Savage-Rumbaugh she bit her hand,
but they overcame this moment of hostility and eventually there
was enough trust between them for Savage-Rumbaugh to take Lucy
out for rides in her MG. Savage-Rumbaugh would drive Lucy all
over Norman, Oklahoma, the chimp pointing in the direction that
she wanted to go and Savage-Rumbaugh following her directions.
If she ever ignored Lucy's requests, the chimp would grab the steer-
ing-wheel and drive the car herself while Savage-Rumbaugh tried
to fight her off the controls.

Lucy became famous in the United States. She was featured in
Life, the *Los Angeles Times* and the *New York Times*. She loved to see
pictures of herself with her pet kitten in the newspapers. A large

spread on the life and times of Lucy also appeared in *Psychology Today*.

Temerlin was fascinated by Lucy's sexuality. The following extract from his *Psychology Today* article clearly depicts aspects of Lucy's American life.

In every way that we could we gave her the same enriched environment – books, attention, love, language – that we provided our son, Steve. Lucy has been as much in my heart and mind as my son or wife. Every day when I got home from the office Lucy would greet me with her soft, guttural 'I love you' sounds, and she would cover my mouth with hers in a chimpanzee greeting. Once while she was sitting in my lap we developed a body image love game. In rapid succession Lucy learned to touch her eyes, nose, ears and mouth whenever I asked her to.

When Lucy turned two and a half, she became disturbed by watching Jane and me make love, and ever after she would try her very best to stop it. Curiously, Lucy seems to interpret the sex act the same way that children do: Daddy is hurting Mother, or at least an aggressive and dangerous act is taking place. Lucy tried to stop it. She would grab my arms or legs (never Jane's) and pull with all her might. Once I lost my erection as I suddenly became aware that Lucy was slowly but firmly biting the calf of my legs. I did not need words to get the message: 'Stop doing that to my mother!'

When Lucy reached sexual maturity at the age of eight years, my relationship with her changed dramatically. The great intimacy we had shared, including close skin contact, mutual hugging and mouth to mouth kisses ended abruptly. She began to avoid me. If we started a game, she would stop abruptly and walk away. If I sat down next to her on the couch, she would get up and walk away without so much as a backward glance. Two weeks after she began to menstruate, Lucy's genitals enlarged to five or six times their normal size and turned a deep pink colour, signifying that she was capable of coitus and conception. When a wild female chimpanzee comes into oestrus, she will mate with any male who appeals to her and sometimes with all the males in the immediate area. During this time of oestrus Lucy totally rejected me. She would not hug me, she would not kiss me. She would not allow me to cuddle, or hug or kiss her. Yet during the same period she made the most blatant and obvious sexual invitations to other men. She would jump into their arms, cover their mouths with hers and thrust her genitals against their bodies. This behaviour disconcerted Fuller Brush men, Bible salesmen and census takers who happened to knock on our door.

I had had fantasies of humorous situations that might occur if Lucy made sexual advances to me, but I was never sexually aroused by her, even when she would see me nude and try to mouth my penis. People often internalise conflict in ways that make them sick or depressed. Not so with Lucy. I wondered what would happen if her biologically determined desires clashed with seductive social forces – what would she do if I pressed her for sexual closeness when she was most anxious to avoid me. Whenever I tested this question by trying to kiss Lucy during her fertile periods, I always got the same results. She would try to get away and if I held her she would scream in terror, lips turned back, teeth exposed. I often feel grateful that I had Lucy rather than a human daughter, because I can enjoy hugging and kissing her without fear that I am being too seductive.

Once when Lucy was in oestrus I got an idea for an experiment. Remembering how I felt about looking at a nude woman when I was a teenager, I bought Lucy a copy of *Playgirl*, which has photos of nude men. She stared at each picture of a male nude and made low, guttural sounds like those she utters when she sees something delicious. She stroked the penis with her forefinger, cautiously at first and then more rapidly. She did not caress or scratch any other part of the photographs. When she came to the centrefold she carefully unfolded it, studied it for a moment or two and then got off the sofa and spread the large picture of an aspiring actor on the floor. She stood on two legs over the photograph and positioned herself precisely to lower her bright pink genitals onto the penis. She rubbed her vulva back and forth for 15 to 20 seconds, maintaining contact with the picture. Then she changed her movements and started bouncing up and down.

But after ten years the initial experiment to test Lucy's innate maternal skills was abandoned. Janis Carter, who later became the most important human in Lucy's life, believes the Temerlins did not know what they were getting into and if they had they would never have adopted Lucy. Although she had integrated into the Temerlin family and believed herself to be a human, she couldn't have suppressed her ape side even if she had wanted to. Maurice Temerlin may have found the burgeoning sexuality of his chimpanzee daughter fascinating, but strength and attitude came along with that sex drive. Lucy grew up to be much stronger, much more territorial and potentially aggressive than the Temerlins had bargained for. Lucy was very strong but compared to other chimps she wasn't particularly well built – Carter describes her as being 'tall and slim with piano fingers' – but adult chimps are very different from malleable baby chimps.

The Temerlins began to discourage visitors and live a reclusive life. They had to fortify their house; windows had to be reinforced, doors locked and for part of the day Lucy had to be kept in a cage when as a baby she had once enjoyed complete freedom of the house.

Janis Carter was a twenty-five-year-old Ph.D. student at Oklahoma University in 1976, researching the cognitive abilities of primates. Carter needed extra cash to help her through her studies and Roger Fouts suggested she work for the Temerlins, who would pay her for helping to care for Lucy.

Carter had heard that previous helpers had given up the job and was rather nervous about meeting Lucy. Carter recalls, 'At first I didn't teach her ASL, I just fed her and cleaned up. I never went in with her. The only contact was through the wire of her cage. But we hit it off. Sometimes she would grab the hose and with a really nasty look on her face she'd hose me down. I wouldn't run, I'd stand there and take it. I let her get her feelings out and that way there was room for us to bond. Our relationship developed into a deep friendship. Maurice Temerlin could see we had developed a sisterly type friendship and one day he opened her cage door and Lucy bounded out. I was scared shitless, but she just hugged me. Later I wanted my boyfriend to meet her. We drove her in the car out to the Temerlins' ranch where she could physically play vertically and horizontally. My boyfriend wanted to take some pictures of me and Lucy. He said something like "Don't move", something negative and controlling. He wasn't intending to command me, but Lucy took this the wrong way. She defended me and attacked him, biting his hand very badly. I knew then Lucy couldn't be with just anyone – it was me. We'd bonded, we'd just clicked.'

Lucy was very particular and would, for instance, always carefully screw lids back on to things rather than forcing the lid on in the way most chimps would. In language tests Lucy's intent was clearly communicated. Her ASL signs were easy to read because of her slim fingers and perfectionist nature. About the time Janis Carter started to work with Lucy the Temerlins had acquired a four-year-old chimp called Marianne as a companion for Lucy and also to see if Lucy would teach Marianne ASL. Eventually she did.

Marianne adored Lucy, but at first Lucy didn't like Marianne at all. Lucy was an arrogant chimp, and considered Marianne to be a very ignorant ape. Lucy saw herself as a human, not as a chimp. In tests where she was asked to choose between two species categories – 'human' indicated by photographs of people and 'ape' indicated by photographs of chimps – Lucy always placed a photograph of herself in the human category. Lucy was born captive and treated like a human. As far as the ape was concerned she was an all-American girl, if a little more hirsute than most.

But now Lucy was hard to control. It had become obvious that Lucy and Marianne could not live out the rest of their lives with the Temerlins, but no one wanted to see Marianne and Lucy end their lives in a laboratory. Carter's boyfriend had a home and land in the mountains of Tennessee and suggested to Carter that she live with him and the two of them build a large enclosure for Lucy there. But Carter was unsure whether she wanted to be his wife and did not feel able to commit to the plan.

The Temerlins had heard of a young woman called Stella Brewer (now Stella Marsden) who was trying to rehabilitate a small group of 'orphaned' chimps in the Niokolo-Koba National Park in Senegal in West Africa. A mountain in the park called Mount Assirik has a small isolated colony of wild chimps.

There followed weeks of bureaucratic confusion in Washington before the correct permits were drawn up to allow the chimps to emigrate to Africa. Conveniently, American laboratories do not experience such bureaucracy when hundreds of African chimps are forced to immigrate to the United States. The flight out of Oklahoma airport was booked to depart very early one August morning in 1977. Carter remembers the journey from start to finish as being horrendous. The Temerlins joined Lucy, Marianne and Janis Carter on the flight to Dakar in Senegal via New York.

As the captive-born Lucy and Marianne arrived in West Africa, hundreds of other wild-born chimps were leaving for America, destined for laboratories and private collections via the illegal but highly lucrative ape trade. Stella Brewer met the Temerlins' group at Dakar airport and drove them to Abuko. Brewer and her father had in 1968 established the Abuko Nature Reserve 12 miles from Banjul, the capital of Gambia; consisting of the last remaining 180-acre patch of rainforest left in Gambia.

After two weeks the Temerlins returned to America but Carter, as Lucy's best friend, stayed on. The Temerlins had put Stella Brewer under pressure to accept Lucy and Marianne into her rehabilitation project at Mount Assirik. But ultimately it had not been possible as Brewer's project was experiencing difficulties. The wild chimps that Brewer hoped would 'adopt' her youngsters were becoming very aggressive towards her chimps, making visits to the camp with the sole purpose of attacking them. In 1977 Jane Goodall had yet to discover the truly xenophobic nature of territorial chimps. Stella Brewer was unaware that she was wasting her time. Carter remained alone in Banjul and tried to help Lucy and Marianne acclimatise to the reserve while Brewer returned to her band of seven chimps at Mount Assirik. Neither woman knew what to do for the best and a year passed by.

Initially Carter had thought she was leaving the United States for three weeks, so she never said goodbyes and made no provision for her belongings, firmly expecting to return to her boyfriend and to finish her Ph.D. As the weeks turned into months and she became more desperate, she counted the mornings and the afternoons on her calendar to the date of her return flight. But rehabilitating Lucy took a long time and Carter had to keep extending the ticket. Twenty-three years later she is still there. She had only taken a tiny suitcase with her with a few clothes. She never saw any of her possessions she'd left behind in Oklahoma again; she has no idea what happened to them. Carter visited the US for a couple of weeks after her first year in Africa. After that first trip home it was more than three years before she briefly visited her parents again.

Carter says, 'Many, many times I regretted taking on such an enormous responsibility. I said to myself, "You're Lucy's best friend, she needs you." I'm very responsible when I have feelings of love. At first I wasn't happy – I had all of the responsibility but no control. I'd grown up in Hawaii; I was a beach person; I'd never camped. I didn't know anything; I didn't know how to put up a tent and light a kerosene lamp. I kept fainting, and I didn't know why. I thought, "Jesus! I'm such a weakling." I didn't realise I was allergic to chloroquine [taken to protect against malaria]. None of the shops sold American goods. I loved milk – still do – and I really missed pasteurised supermarket milk. I could only get full-cream powdered milk in Gambia. Lucy missed her wholemeal bread and fruit yoghurt. She refused to eat and lost a lot of her hair. I gave her a big sun hat to protect her skin from the sun.

'I thought Lucy's release had been arranged, but it was left to me. I started as just a passenger, but I became the instigator. It was so hard trying to encourage Lucy to eat a wild chimp diet; she hated leaves and wild fruit. It was so bad that we both lost our period for six months. We started cycling again together on a chimp thirty-five-day cycle instead of me having a human twenty-eight-day cycle. For the seven years I was with Lucy on the island I cycled like a chimp.'

After a few months and at the suggestion of Stella Brewer Carter decided to explore the three Baboon Islands in the River Gambia National Park one hundred and seventy miles from Abuko, to see if they were a suitable home for the chimps. The largest island offers over 1,000 acres of natural high forest, scrub and swamp. The islands were already inhabited with many species of bird, baboons, red colobus monkeys, green vervet monkeys, hyena, warthogs, monitor lizard, snakes, crocodiles and hippopotamus. Some of the island's edible plants were the same as those found at Mount Assirik, so Brewer's chimps would manage there and the women could teach them about other available forms of edible vegetation.

When the chimps first went to the Baboon Islands in February 1979 the women saw this as a temporary measure and continued to search Africa for a suitable habitat for them. Brewer was worried that life on the island would eventually become too difficult for the chimps. There would be no escape from a forest fire, no room for a major population increase and no opportunity for escape from prolonged hierarchical or territorial fights.

After a systematic ten-year search Brewer and Carter have found no other suitable site in the whole of the African continent. Their chimps cannot settle in areas already inhabited with chimps; they cannot go anywhere where bush-meat or any other form of poaching is practised; they cannot go to a war zone or anywhere where deforestation is rampant. Brewer and Carter now have fifty-five chimps in their care and there is no place for these rehabilitated apes anywhere else in Africa.

Since the chimps have settled into life on the island there have been 31 births and 16 deaths and three youngsters have disappeared, possibly stolen by animal traders. Carter and Brewer now have more second-generation chimps under their care than first-generation rehabilitants.

Chimpanzees had probably never inhabited the islands in the River Gambia, though they had lived on the mainland, so the women were effectively reintroducing chimps to Gambia after the last remaining wild chimps became extinct there in the 1950s. While at the Abuko Nature Reserve Carter had been presented with a couple of new young chimps confiscated from the local ape trade to Gambia from Guinea, including a baby male she named Dash. With additional orphans, Carter and her chimps now numbered ten. They stayed at one end of the island and Brewer, with her six chimps and ten new orphans, stayed at the other end. The women existed independently of each other and rarely met up. They didn't ignore each other but they didn't want the two chimp groups to meet and fight and because the women were bonded independently to their own chimps they lived separate, if parallel, lives. (As the years passed, the chimps matured and the two groups began to interact, and Carter and Brewer began to collaborate more fully.)

Stella Brewer is British. She has led an expatriate lifestyle in Africa; her father was Gambia's wildlife officer. At Mount Assirik Brewer had never been totally alone. She always had African helpers René Bonang and Bruno Bubang and young European or expatriate women to help her, and this pattern continued on the island. For the first few years Janis Carter had no help, either in debriefing Lucy of everything the Temerlins had taught the ape or in making decisions on the running of her island camp. People would come over to the island by motor boat and deliver food and other human necessities to Carter to keep her going, and occasionally Carter or Brewer might take a boat and sail down to see one another for a cup of coffee and a chat. But as the two groups of chimps were specifically bonded to Carter and Brewer and it wasn't safe for strangers to pop by, effectively Carter lived alone with ten chimps for seven years of her life.

Carter's American boyfriend visited her on and off over the years, but their relationship couldn't compete with Carter's bond to the chimps. She told me, 'My boyfriend visited me many times – we'd stay in Banjul – and one Christmas I went back to the States. He was very supportive about what I was trying to do and wanted me back, but I was changing more and more. Each time he saw me I was more chimp-like in my behaviour and it took weeks for me to come out of that and be a human again. I needed time, but too soon he'd leave for the States again. He'd put a cup on the table for me and like a chimp I'd mess with it. I couldn't leave the ape world and become human again in a few weeks. I should have prolonged the transition; it was a traumatic experience coming back to the human world. My sister visited me in my first year here, and thought what I was doing was great. But we lost touch for years and years. She continued living in America, married, had three kids and a career. We've only recently bonded again.'

Janis Carter had to make a hard decision: whether to return to America and to her boyfriend and end up having a life like her sister's; or, to turn her back on all that a young American woman is supposed to want – the security of marriage, children and a career with private health care and a pension plan. Although Carter loved her boyfriend, she had found *complicité* with Lucy. Carter was torn in two by her ultimate decision to stay with Lucy and devote herself to chimpanzees.

'When I'd return to the island and see the chimps again after a break, my heart would skip a beat, because they were so big. But after a day back in the jungle with them I'd be a chimp again. I travelled with them as they moved around, I slept with them, I *was* one of them. I became very intimidated by articulate people. From the age of twenty-six to thirty-five I was alone for most of the time and around chimps. These are the years in which humans discover their philosophical views on life, but I never had any discussions about such matters during those years. I'd try to share my experience with the chimps with people, but it is such an alien experience that it was easier not to bother. It was like going to the moon for an astronaut; most of us cannot begin to imagine what that would be like.'

All the apes seemed to like their new habitat on the island and settled in well, except for Lucy. She didn't like the other chimps at all and used to sign to Carter that they were 'dirty animals' and refuse to go near them. Lucy signed to Carter all the time; when she saw Carter walking past she would ask, 'Where you going?' Lucy would often ask for magazines to read and when the monsoon rains came she commented, 'Too much to drink.' Lucy wanted Carter to take care of her and would sign to Carter to 'climb' and 'get fruit' when she saw fruit in a tall tree, but she couldn't be bothered to get it for herself. Marianne mimicked Lucy's ASL and the chimps signed to each other. Lucy would sign 'hug me', and Marianne would hug her.

Marianne made the transition through rehabilitation and into wild living seem easy, but she was younger. Lucy was older than the rest of the chimps and was scared of them. They sensed this and were often hostile to her. On the tip of the island Carter built a cage with a door and a lock that offered her a safe place away from the playfully destructive chimps. Here she kept her few possessions – a tent, food provisions, medicine, a torch and a camping cooker. This was Janis Carter's home. When Lucy signed Carter would turn away; she no longer wanted to engage with Lucy on that proto-human level. But Lucy would take hold of Carter's arm and force her to look at her, then sign again, so Carter would get up and walk into the swamp to escape. Carter had to wean Lucy from signing, but sensed that Lucy felt betrayed by this and was very hurt. In turn this made Carter feel horribly guilty.

Dash was growing up and he was about to become the alpha male of Carter's group. Carter told me, 'To begin with I was dominant and Lucy and Marianne were my sidekicks, but as Dash matured his hormones started pumping. I knew it was just a matter of time before he'd take over. I've always had long hair and if the chimps were naughty they would yank my hair, but when Dash turned on me it was different. Dash had tried to push me around before, but Lucy and Marianne always defended me and I them, but one day all the chimps started screaming and climbed into the trees, including Lucy. I turned to see Dash staring at me – he was standing bipedal, his hair all on end, swaying from side to side. He looked enormous. I knew this was it. When a male chimp displays like this they completely lose it; they are on a big high and in some ways do not know what they are doing. He charged me. Everything went into slow motion. He grabbed my ankle and pulled me along the ground a long way. He threw me around; it was a great display of dominance to the other chimps. Luckily for me, he tried to pull me through a thicket with plenty of long thorns and my body kept being caught.

'I managed to crawl on my hands and knees through the thicket down to the river and rolled into the water. Luckily the river was flowing in the direction of my cage. I managed to swim, somehow, to the cage. I was shaking so hard with fear that enormous convulsions took over my body and I could hardly open the lock. Dash didn't know where I had gone and was looking for me. Later they made their way back to the cage and it took all my courage to go out with them for a walk that evening. I stayed with them till after dark and watched them make their night nests and then walked back up the path to the cage. I heard footsteps behind me and turned around to see Dash coming after me. I thought, "Holy shit, he's going to attack me at night too!" He came right up and put his arm around my neck; I thought he might break my neck. Then he climbed up me, like he used to as a baby, and sat on my hip and hugged me. He was now very heavy and I was too scared to look him in the eye. He looked the other way also. We stayed like that in silence for a bit and then he gently got down and walked back to where the others were nesting. None of the other chimps saw him do this. I think he was trying to say "I've got to be dominant in front of the others; it's the other me, I can't help it." He'd been my baby son; I'd brought him up. I was heartbroken to have him turn on me like that, but I took his warning seriously and started to make plans to leave the island.'

The attack from Dash came when Lucy was almost ready to be on her own with her own kind. Self-taught, Janis Carter had shown Lucy how to live like a wild chimp. The two urbanites from Oklahoma had come a long way together. Lucy was now finding her chimpanzee confidence but the love between them made separation difficult.

Various people commented it would be cruel to take Lucy to Africa; apart from the culture shock, how would the chimp survive when she was used to eating hamburgers and drinking gin? The rehabilitation of Lucy became, in effect, just one more experiment to be negotiated. For the first half of her life, Lucy lived as a proto-human in America's heartland, and for the second half of her life she lived as a culturally sophisticated chimp shipwrecked on a desert island. I wonder how often Lucy thought about Maurice Temerlin, her human father, her kitten or *Playgirl*? Or how often she fancied the taste of gin or a Big Mac?

No one before had ever released a captive-born American chimp in the wilds of Africa. It is an achievement to be proud of. Janis Carter passionately believed Lucy should live a 'natural' life, far way from human society, but she has received criticism for taking Lucy to Africa. Jane Goodall has been especially critical.

'Jane Goodall and I have a disagreement over this. Jane has said that it was unfair and selfish of the Temerlins to send Lucy to Africa, that it was comparable to taking a three- to four-year-old Western child and dumping them in a market in India and expecting them to survive. But that's not what happened to Lucy. Lucy was not abandoned; I held her hand every step of the way. Jane says Lucy had a lifestyle she was used to and should have been allowed to live out her life that way, but it wasn't possible. The Temerlins couldn't cope. It's like saying that people who are okay in institutions should stay there rather than being rehabilitated and given a better quality of life. You can always make a captive life better, but nothing can replace the freedom that chimps have evolved to live.

Lucy was an extreme case. Rehabilitation is expensive on time, but not financially. Lucy required me to devote my life to her, but younger apes can go into groups and need you less. Technically, Lucy proved it was possible. I didn't cost Jane Goodall or the Jane Goodall Institute any money; it cost me my life and that's my choice. I was trapped but I felt a commitment. Jane's comments have damaged rehabilitation possibilities. I can't understand her motives. She and I will always disagree.' The chimps in Goodall's sanctuaries have plenty of space, but the forest within the sanctuary cannot sustain the chimps and the animals need to be provisioned. The Baboon Islands, on the other hand, are self-sustaining ecosystems that can support the chimp population. Carter and Brewer's chimps are not reliant on hand-outs.

I asked Janis Carter if she would ever rehabilitate another Lucy. 'I'm too old to do it now. It spans years. It's the type of thing you do once. I removed myself from my own society for an extensive period. I can look back and see my experience with Lucy has cemented the work I do today, but if I keep moving forwards I can help the lives of lots of chimps instead of just one.'

To everyone's delight Marianne had a baby called Wury. Lucy was very interested in Wury and liked to touch him; Dash was also pleased with the baby and tickled him under the chin. But because Marianne had never seen how a mother chimp should cradle her baby to nurse it, she was a proud but incompetent mother. Chimps have to learn by example, just like young women. Carter led Marianne and Wury into the cage so they wouldn't be disturbed and, cradling a toy baby chimp, Carter showed Marianne how to breastfeed while supporting a babe in arms. Marianne had expected Wury to just hang on and feed, but this was too hard for him. He became malnourished and could have died, but following Carter's 'lesson' Wury started putting on weight. Younger female chimps in the group watched Marianne's improved maternal skills and learned how to breastfeed their own babies when the time came.

In more recent years, when wild-born daughters of the ex-captive female chimps of the two groups on the Baboon Islands have reached sexual maturity, the incest taboo has led them to transfer to the other group, just as ordinary wild female chimps do. Dash, Marianne and Wury are today living a wild, autonomous life of freedom.

After Carter left the island in 1986 she moved 176 miles to an isolated piece of coastline near Banjul but she continued to make frequent visits to the island to check on everyone's progress. From '86 onwards Carter built herself a house, her first real home in years, spent time fund-raising and fighting the ape trade and building a well, as her new home had no water or electricity to begin with. One night in 1987 Janis Carter had a nightmare about Lucy: Lucy was alone, walking bipedally up the steps of an aeroplane. Carter awoke with a jolt. The next day it was reported that Lucy was missing. (By now Stella Brewer's assistants, René Bonang and Bruno Bubang, were also helping Janis Carter monitor her chimps and Carter and Brewer had become joint directors of the chimp rehabilitation project.) No one knows exactly how Lucy died. She had been seen two days before and seemed to be in good health.

It was decided that Carter would entice Dash into the cage and Bonang and Bubang would search for Lucy's body. Carter kept in touch with the men by radio and had to warn them to hurry because Dash knew they were on the island and was displaying in the cage and breaking his way out. The men found Lucy's remains in a place where Carter and Lucy used to spend time together. As they radioed Carter to tell her they had found Lucy's body, Carter ordered them to leave the island immediately as Dash was breaking free and would surely kill the men. Hastily Bonang and Bubang collected up Lucy's remains and fled the island.

Carter examined Lucy's remains but found it hard to ascertain exactly what had happened. Lucy's hands, feet and skin were missing; they could have been eaten by wild pigs, or she could have been killed and skinned by a fisherman. Black hair is needed for the talisman that locals wear. Or she could have contracted a fatal, quick-acting virus. But she had not fallen from a tree and broken her neck as some have suggested because her spine was intact. Carter buried Lucy's remains on the island.

Today, in her late forties, Janis Carter is a very busy woman. Recently married to a Guinean, she runs a chimpanzee rehabilitation, education and conservation programme in Guinea. Carter initially started exploring Guinea because she wanted to see where Dash had been taken from. Carter believes that Guinea has the highest population of Central African chimpanzees of any West African country. At the same time more chimps are poached from Guinea for the illegal but still buoyant ape trade in laboratory chimps.

Because Guinea was communist and closed to outside influences until 1984, the country's development lags far behind that of other African countries, which is bad for its human population but good for its wildlife. Carter received a two-year grant from the European Commission to organise a survey of wild chimpanzee populations in Guinea and is rehabilitating a group of orphaned chimps in Guinea's Parc Haut Niger, where the Mafou and Niger rivers meet. But after the recent chaos within the EEC Carter's funds have been abruptly stopped.

Carter told me a bit about her work: 'Rehabilitation is just treating the symptoms, we have to save the indigenous wild population, and local education programmes are the best way to do that. Bushmeat is a big problem in West Africa; some people like to eat chimpanzee meat. If someone was caught selling a baby chimp they were fined £50, which is no deterrent as they could sell the chimp for £300. The law had to be changed and I managed to do that. I've been worn to a frazzle. I can't find safe locations for the chimps and I don't have any money: it all causes sleepless nights.'

Rebecca Ham, who has worked with Janis Carter on her project understands the depth of Carter's empathy for chimpanzees: 'Some orphaned chimps had been put on an island off the coast of Guinea. Some had died; no one had seen the animals for years. Janis wanted to visit the island. When we got there two adult females came running over to us. It was totally amazing – Janis turned into a chimpanzee! She knew how to appropriately greet these animals we'd never met before. She held the right posture; she made the right sounds and gestures. I stayed back and copied her. I don't know what would have happened to me if I'd been on my own. I wouldn't have known what to do – it's quite intimidating to have two fully grown chimps run towards you. This experience taught me a side to Janis I'd never seen before. Janis is amazing. Living on the island with Lucy affected her. She's easy to misunderstand. When she drives along over rough terrain, instead of saying "Hold on!" to passengers she gives involuntary chimp calls! She's so funny. Maybe she dreams in chimp.'

Stella Brewer's initiation to primate conservation is not as extraordinary as her colleague Janis Carter's personal story because she grew up in Africa. As a child Brewer had 'mothered' all sorts of orphaned African baby animals, but in 1969, when she was eighteen, a Guinean animal trader knocked on the front door of the family home, just outside Banjul in Gambia. The man placed a small box on their veranda and untied it. Inside was an emaciated baby chimp, later named William. This chimpanzee would influence Stella Brewer just as much as Lucy changed the course of Carter's life.

  Distressed by the sight of the chimp, Brewer did what some would say is the worst possible thing – she rescued it from its torture. She paid the man, who returned on the three-week bus journey back to Guinea, no doubt to enact the whole thing again. He would shoot dead another female chimp, steal her baby by pulling it off its dying mother's body, tie it up inside a small box, perhaps sell the mother's body for meat and then make another long journey from the animals' forest habitat to the home of another expatriate white family. If this new baby chimp survived the experience, another family somewhere else would probably take pity on the animal and offer it a home. If, on the other hand, everyone whom the Guinean approached turned down the offer of William as a pet, this short life would have been sacrificed to save all the other wild chimps that this particular animal trader would be encouraged to go on to capture.

But how could Stella Brewer let the baby chimp die? Brewer untied the flex that was wrapped around William's body, cleaned him of excrement and bathed his sores with antiseptic. After she had fed him porridge he was placed in a crate lined with clean straw. Only as he fell asleep did the chimp's frozen grin of fear fade from his face. William would become the most important thing in Stella Brewer's life. Although she was close to her parents and her sisters, and even though she had been in love with her future husband, David Marsden, a nineteen-year-old VSO volunteer in Gambia, since she was thirteen, William's needs came first in Brewer's heart.

While caring for William, Brewer was made aware of an increasing number of 'orphaned' chimps turning up in the vicinity. Some of these chimps she found in the market in Banjul, others were being kept as pets by local Europeans. Eighteen months after William's arrival, Brewer was 'mother' to eight baby chimps, which were all confiscated and given sanctuary at the Abuko Nature Reserve. But Brewer wanted them to be free. She had just been given *In the Shadow of Man* by Jane Goodall and was amazed to discover there was another British woman in Africa living with chimpanzees.

One day Stella Brewer was taking her brood for a walk in the reserve when they disappeared into thick foliage. She could hear them hooting and screaming with excitement. After pushing her way into the undergrowth, Brewer stared in horror at her babies. Tina, who was approximately six years old and out of all the chimps had spent the longest time as a wild animal, was holding a baby green vervet in her hand. From the behaviour of the chimps Brewer could tell they meant the squeaking baby monkey harm. Brewer held out her hand to Tina, but the chimp tightened her grip around the monkey and climbed higher into the trees. From her vantage point Tina looked down at Brewer. Holding eye contact and without hesitation, Tina bit the baby monkey's head in two.

Sickened and disgusted, Brewer left the chimps to their bloody feast. She felt appalled at what she had seen – how could her babies behave this way? She thought this apparently abnormal behaviour was a product of the psychological damage the animals had suffered during their initial capture. But as Brewer read through Goodall's book she realised she should be pleased. It meant that her chimps were behaving like the wild Gombe chimps. Hunting monkeys for food is normal behaviour for wild chimpanzees.

Stella Brewer's chimps were born free and deserved to regain their dignity. She decided she wanted to rehabilitate her chimps, taking on the role of a surrogate chimp mother and teaching them how to survive in the way their blood mother would have done. In 1970 after hearing Goodall on the BBC World Service discussing her chimps Brewer wrote immediately for advice from Goodall, who replied with an invitation for Brewer to come and stay with her and Hugo Van Lawick at Gombe. Brewer would be able to observe the behaviour of wild chimp mothers with their infants; although habituated to humans, the behaviour of Jane's chimps was of the natural order. Goodall's chimpanzee Flo exhibited the best mothering skills that Goodall had ever observed and would be a great role model for Brewer. At this stage in her career Goodall was supportive towards attempts to rehabilitate chimps, admiring Brewer's commitment to the orphaned chimps and believing

Brewer had the 'patience, the dedication and the courage to make the experiment work'.

When Stella Brewer arrived at Gombe, she started work straight away. Among other chimps, Goodall asked her to observe a female chimp called Pallas, who had just lost her three-year-old son to pneumonia. Pallas carried the body of her dead youngster for days before Goodall's students could take it away for an autopsy. Brewer would watch Pallas as she sat purposefully alone staring down into her lap at her upturned hands. She was very depressed and seemed to be locked into memories of holding her baby in her now empty hands. In addition, Brewer recognised similarities between Figan's behaviour and that of her own special chimp, William. Both young males exhibited ambition at an early age. In Figan's case it would take him to the top. Brewer particularly wanted to familiarise herself with the behaviour of the mother chimps. She wanted to emulate them and teach her orphaned chimps how to survive in the same way a mother chimp guides her infant.

Goodall and Van Lawick liked Stella Brewer and wanted to help her. Goodall suggested Brewer write a book to help raise funds for her rehabilitation project and Van Lawick suggested that he film Brewer and her chimps to help publicise her work. Goodall introduced Brewer to her publisher, Collins, and it was agreed that Brewer would write an account of her life with chimps. It was called *The Forest Dwellers*. Hugo Van Lawick raised money for a film entitled *Stella and the Chimps of Mount Assirik*, 1997. Narrated by James Mason, it was broadcast by the BBC, and shows twenty-three-year-old Stella Brewer, blonde, very pretty and 'natural' in front of the camera, carrying baby chimps and holding hands with William as she walks through the trees teaching the animals how to eat bark and how to make the correct warning calls when they meet a scorpion. Van Lawick obviously saw Brewer as the new beguiling ape lady for his next project.

Meeting Jane Goodall had made everything possible for Stella Brewer. She had learned much from the maternally attentive female Gombe chimps on ways of raising youngsters, as well as discovering the types of leaves, flowers, fruits, nuts and bark chimps can safely eat and the appropriate chimpanzee calls of delight when these foods are seen, and noting the appropriate calls for danger when a chimp stumbles upon a snake. She had also received expert guidance from Goodall and Van Lawick in fund-raising and publicising her project.

Stella Brewer returned home to set her plans in motion, buying herself a Land-Rover and a tent. The chimps were put into crates for their 400-mile, two-day journey from the Gambian Abuko Reserve to Mount Assirik in the Niokolo-Koba National Park in Senegal. The World Wildlife Fund gave her a small grant and her publishers gave her an advance, and these funds kept her going for the first two years. January 1974 saw the establishment of twenty-three-year-old Brewer's rehabilitation project.

Late one night at Mount Assirik, Stella Brewer came out of her tent and to her surprise found William sitting on the ground gazing up at the full moon. On her way to the bushes Brewer whispered to William, asking him what he was doing out of his bed so late. Chimps are not nocturnal; they make their nests in trees at nightfall and should be fast asleep by the time it is dark. On her way back to her tent, Brewer decided to sit down next to William for a while. The moon was bright and the night sky was festooned with stars. As Stella settled down beside William he looked up and romantically put his arm around her. They leant their heads together and, bathed in moonlight, remained in mutual peace and harmony for another hour. Brewer told me that, out of all the chimps she has known, William was the most special.

William had never been trained to do 'tricks'; he was just a great human behaviourist. He watched people as carefully as they watched him. Once, when Brewer returned to camp after taking some of the younger chimps for a walk, she found William helping himself to a cup of coffee. William unscrewed the coffee jar and put a teaspoon of coffee and then three teaspoonfuls of sugar in a mug. The kettle had been left half over the stove and the water was still very hot; William, careful to hold the kettle away from his body and without spilling a drop, gently filled his mug with hot water. Carrying the mug William walked over to a tree and sat down to drink. The coffee was hot so he blew on it, but it was not cooling quickly enough, so William found some stones and put them into the mug. This helped to cool the coffee down enough for him to suck it up.

Although Brewer was always impressed with the ease that her chimps adopted human behaviour through their own volition, she did not want them to do so. She hoped they would unlearn all the human baggage they'd acquired so she could place them back in nature as innocents.

On one occasion André Dupuy, who was then the Senegalese Director of National Parks, visited Stella Brewer's camp at Mount Assirik with guests. Everyone seemed impressed with Brewer's progress as she showed them around and told them of her intentions. While the group left camp for a short excursion into the woods, William seized his opportunity. When they returned Brewer was embarrassed to find her chimp sitting on Dupuy's Land-Rover, a lit cigarette in his mouth, a Zippo lighter and a packet of Marlboros in his hand. William was fascinated with fire, and was not afraid of it. Defiantly William smoked the cigarette and when Brewer asked him to return the Marlboro packet and the lighter he ran away with them. Dupuy left without his things and only much later did Brewer manage to get them back from the chimp. Brewer says that if William had been human he would have been a gangster.

In June 1975 a twenty-seven-year-old Italian woman, Raphaella Savinelli, turned up unannounced at Mount Assirik accompanied by her three-year-old pet chimp, a male called Bobo. Savinelli was born into a wealthy family, growing up in castles on Sicily and yachts on the Mediterranean and flying in private jets to the family's smart apartments in Milan and Rome. Raphaella was a model and an artist accustomed to getting everything she wanted. In Milan Bobo had become a handful and Savinelli had written to Jane Goodall for some advice. Goodall had written back telling Savinelli about Brewer's rehabilitation project.

Without bothering to contact Brewer to ask if she would accept another chimp, Savinelli set off with Bobo for Africa. When they landed in Dakar it was late at night. Bobo was released from his crate and leaped into Savinelli's arms. Savinelli pretended to the manager of a hotel close to the airport that Bobo was her baby; cuddling the bundle of chimp concealed in a blanket, she was led up to her room. That night, as in Milan, Savinelli and Bobo slept in each other's arms. In the morning, when they left, there was no disguise – Savinelli and Bobo walked nonchalantly, hand in hand, through the hotel's reception, guests and hotel staff staring in utter amazement. In those days Savinelli enjoyed real-life drama, pushing the boundaries and shocking those around her for attention.

The next day she bought a jeep and a map and simply drove to Mount Assirik. Brewer was surprised when Savinelli and Bobo turned up, but in Africa anything can happen. The two women became friends almost immediately, but William was hostile to Bobo. As the eldest male in the group and with no one to challenge him, William was dominant and he demanded respect. Bobo had, just like his mistress, become very used to getting his own way. Bobo tried to stand up to William, but quickly thought better of it and made a submissive gesture by putting his head down and his rear up. William accepted this appeasement. Brewer and Savinelli were relieved that William accepted Bobo – it meant the two newcomers could stay. 'Mal d'Africa', as the Italians say, was seducing Savinelli with the promise of a great adventure.

The major difference between captive chimps and wild chimps that makes captive chimps potentially lethal is that captive chimps take liberties with humans and in doing so learn of their own power. Wild chimps do not know their own strength. Stella Brewer was never afraid of William and knew he would never bite her. Once, Brewer stopped William from stealing food. In his frustration he knocked Brewer off her feet and put her foot in his mouth. As William squeezed her toes between his powerful jaws, out loud Brewer dared him to do his worst; he thought better of it. Half an hour later William was sorry he'd bullied his 'mum' and came over and tried to make friends. Brewer says, 'I was never scared of Willy – he was my boy, my baby, I'd just hit him or bite him back!' They had the sort of understanding that chimp mothers or human mothers have with their sons, and were bonded to one another at a primal level.

But William was not bonded to Savinelli. One day William was once again throwing his weight around and he bit Savinelli's thigh. A stitch was needed, which Savinelli did for herself without anaesthetic. Savinelli didn't like William much; she didn't like the way he would bully Bobo and didn't see why she should now leave camp for her own safety because William was threatening her. Although Savinelli wanted to rehabilitate the chimps, she didn't understand the true nature of the animals. She idealised them, especially Bobo.

Brewer, who knew William better than anyone, felt he was going through a 'bad patch'. William was eleven years old, in wild terms a youthful age to be leader, a teenage tearaway not yet mature enough to have the responsibility of complete power. But without competition from older, stronger, wiser males to keep him in check, he was able to push everyone around like an unruly hooligan.

Around the time that the wild chips were becoming increasingly hostile and William was disrupting life at camp, Stella Brewer needed to leave camp for a couple of days to see the bank manager in Banjul about her diminishing funds. She left Savinelli in charge with Savinelli's lover Patrice Marti, a young Frenchman who ran a nearby hotel that offered holiday-makers the chance to see the wildlife of the Niokolo-Koba National Park. When Brewer returned, Savinelli told her that William had disappeared. He'd gone for a walk and had not returned.

With the alpha male gone, the politics of the group was changing, making room for Bobo to exert his will. Brewer was afraid for William. For two weeks, day and night, she walked alone through the hills calling out his name – but nothing.

A year after William went missing Savinelli left Mount Assirik. She'd had enough, and without warning she disappeared, leaving Bobo and Marti behind and Brewer to carry on the work alone. It was 1977; Janis Carter had arrived in Gambia with Lucy and Marianne. After a total of three years at Mount Assirik, Brewer decided to give up and transfer her remaining chimps to the Baboon Islands. After moving there, Bobo would be found drowned in the River Gambia. Bobo was thirteen years old when he died, and no one knows what happened. Apes cannot swim – if they fall from a tree into deep water they cannot survive.

Today Stella Brewer lives for most of the year in Gloucester with her husband and two sons and visits the chimps at least twice a year. René Bonang and Bruno Bubang run the Baboon Islands project in her absence. A few years ago Raphaella Savinelli resumed contact with her and visited her Gloucestershire cottage for a few days. It was then Brewer learned what had actually happened to her favourite chimp, William. With both Bobo and William gone, Savinelli felt able to tell Brewer the truth.

Brewer had kept a candle of hope burning for William. Brewer wanted to believe that somehow William had managed to join the wild chimps and was out there, somewhere, living free. For years guilt had eaten away at Savinelli, but she couldn't keep it inside any longer – she confessed to William's murder. Savinelli's boyfriend, Marti, had thought it was a good idea to execute William to prevent him from repeating the earlier episodes of biting or bullying her. Agreeing that this was the best option, Savinelli and Marti shot William in the head and threw his weighted-down body into the river. With William out of the way Savinelli had no need to leave camp for fear of his temper tantrums, and Bobo was able to grow up to be more powerful within the chimpanzee hierarchy. After hearing this news Brewer felt shocked for days. She had never guessed the truth.

This account of William's death is reminiscent of common Indian and Middle Eastern polygynous myths where the jealous second queen murders the first queen's son so that her own son can become king and inherit the kingdom. It is possible that Savinelli's love for Bobo and her desire to stay with him motivated her to murder William, but then guilt forced her to leave without a word.

Brewer comments: 'When she told this to me I couldn't take it in, I couldn't believe it. I just wanted her to leave. I didn't confront Raf over it. She'd had a terrible time after leaving Mount Assirik. She'd travelled the world, had a daughter she'd called Stella, lost custody of her daughter and sunk into heroin addiction. She was very low when she admitted to me what she had done to William. Later, when I mentioned it to Bill McGrew, who was studying the wild chimps while we were there and had known both Raf and Willy, Bill said he'd always suspected foul play, but I hadn't. Finally knowing what happened to Willy after all these years has left me feeling very sad.'

Motivated by love, Goodall, Galdikas, Carter and Brewer all meta-morphosed into apes for periods of their lives. Women who spend time with apes seem to be not only able but willing to regress to this prelapsarian condition, whereas men cannot or will not.

Many religions and myths, such as the Old Testament book of Genesis and the story of Pandora's Box, have blamed women for man's descent from grace. Could this be why women are so willing to return to a time before the fall, a time before language and self-knowledge? Is this why Carter wanted to take Lucy away from the twentieth-century sophistication she'd been born into and allow her a cultural rebirth back into paradise? Biruté Galdikas believes a woman is empowered by the purity of her convictions. There is an evangelical idealism common to these women that draws them towards a life of innocence that in practical terms is hard to achieve and perhaps only really exists in the perfection they yearn for inside their heads. For apes are not children or aliens, nor are they proto-humans waiting or wanting to evolve into beings like us. But they are our relatives. Loving them and idealising them, for whatever reason, is no crime.

1. Above: *Jo Thompson with captive juvenile bonobo at Twycross Zoo.*

2. Above right: *Gertrude Lintz at home with one of her chimpanzees.*

3. Above: *Founder of the IPPL, Shirley McGrill, with 40-year-old Igor, an ex-laboratory gibbon rescued by McGrill when he was 26 years old.*

4. Left: *Mary and Richard Leakey compare their hominid skulls in Nairobi.*

5. Above: *Jane Goodall gives a banana to famous wild adult male chimpanzee, David Greybeard, in the early days at Gombe.*

6. Left: *Louis Leakey with his protégé Jane Goodall.*

7. *Dian Fossey with the adult Digit. She had known the mountain gorilla from his infancy and their relaxed close proximity here reveals the extent of their trust in one another.*

8. *Louis Leakey, Jane Goodall and her son 'Grub'.*

9. Left: *Biruté Galdikas in the early days at Camp Leakey, Borneo with infant orang-utans Sugito and Sobiarso.*

10. Above right: *Galdikas receives a hug from delinquent male, Gundul.*

11. Below: *Barbara Smuts and her Kenyan baboons find some shade.*

12. *Umeyo Mori and a wild adult male Ethiopian baboon take a rest.*

13. *Mori's mother Satsue Mito with the famous Japanese Koshima Macaques.*

14. *Mariko Hiraiwa-Hasegawa in the Mahali Mountains on a follow with the chimpanzees she grew to hate.*

15. Above left: *Stella Brewer with young William, the most intelligent chimpanzee she has ever known.*
16. Above right: *Raphaella Savinelli and Brewer at Mount Assirick, before everything went wrong.*

17. Left: *Janis Carter re-visits her rehabilitated chimpanzee sister Lucy. Lucy never stopped using sign language.*

18. Above left: *Gombe's Alpha female Flo plays tenderly with Flint – her youngest and most tragic child.*
19. Above right: *Dian Fossey with the decapitated corpse of Digit.*

20. Above: *Fossey lectures on gorilla poaching at the Leakey Foundation's Man and Ape Symposium.*

21. Left: *Liz Williamson, current director of Karisoke, with her trackers.*

22. Above left: *The star of* Gorillas In the Mist, *Sigourney Weaver, poses with an infant chimpanzee in gorilla costume.*
23. Above right: *Maureen O'Sullivan as 'Jane' and Jiggs as 'Cheeta'. They hated each other.*

24. Below: *Publicity still from* King Kong.

25. Above left: *Neat role reversal in* Planet of the Apes: *Kim Hunter's primatologist chimpanzee character falls in love with an 'intelligent human', a.k.a Charlton Heston.*

26. Above right: *A detail from Peter Blake's unfinished painting* The Tarzan Family at the Roxy Cinema New York.

27. Below: *Ailsa Berk as Kala, baby Tarzan's chimpanzee mother, on the set of* Greystoke, The Legend of Tarzan, Lord of the Apes. *Berk is bonding with the baby Tarzan and an assistant director gives a helping hand.*

28. Above left: *Sally Boysen with rescued infant Bob.*
29. Above right: *Amy Parish with captive adolescent bonobo.*

30. *Sue Savage-Rumbaugh with adult bonobo Kanzi. He is laughing.*

31. *Sarah Hrdy carefully observes the Indian langur monkey.*

32. *Decades on, Stella Brewer and Janis Carter in the Gambia.*

33. *Sarah Hrdy* (right) *discusses primate behaviour with her protégé Amy Parish.*

34. *Supporter Gordon Getty with Dian Fossey, Jane Goodall, Biruté Galdikas and discoverer of 'Lucy', Don Johanson.*

35. Above left: *A pensive Debbie Martyr: 'Nothing is more important than discovering the orangpendek.'*
36. Above right: *Now director of The Orangutan Foundation International and entering her 30th year in the field, Biruté Galdikas is still saving orphaned orang-utans.*

37. Above left: *Liza Gadsby takes a dip with an orphaned infant drill baboon on her shoulder and an orphaned chimp on her hip.*
38. Above right: *Retired chimp actor Jiggs relaxes at home.*

39. Left: *Female solidarity the bonobo way – two females stroll bipedally together.*

40. Right: *Two females share lunch.*

41. Below: *Two females mutually engage in GG rubbing to help reinforce their matriarchy.*

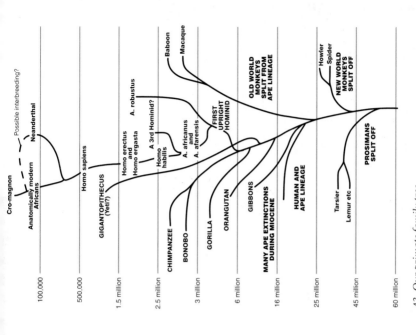

43. Our primate family tree.

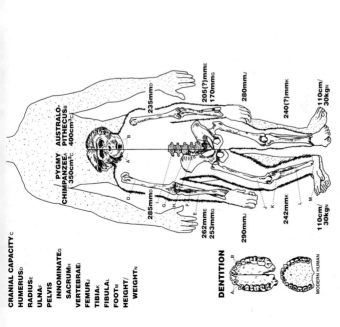

42. What did Lucy look like?

Adrienne Zihlman's 1982 conception of the relationship between a bonobo (pygmy chimpanzee) – the australopithecine fossil named 'Lucy' – and a modern human. This drawing exercise was used in an educational publication to show that the living species most like our human ancestor is the bonobo.

44. *Mary Leakey with her greatest discovery – the* Laetoli *fossilised footprints of our apemen ancestors.*

45. Below: *It's now 40 years since she first went to Gombe, and Jane Goodall, who has inspired so many, is still putting chimpanzees first.*

6
The Woman Who Loved Apes Too Much

Dian Fossey's horrific death made headlines around the world, publicising the systematic extermination of the shy mountain gorilla she had died trying to save. Her death was a PR exercise par excellence. Since Fossey's murder people have commented on her life without many of those ever being able to imagine themselves in the same dilemma. It seems certain at times she was hostile to people, she also felt bitterness towards government officials, conservation charities, bewildered students and ex-lovers. But in spite of all that and her failing health it was her love for the gorillas, her loyal staff members and the mountains themselves that drove Fossey on.

Richard Leakey remembers, 'Dian Fossey had a sharp tongue and a very short temper. A lot of people had good reason to be disturbed by things she said to them, by ways in which she had influenced their careers and ways she had influenced their earnings as employees. Like a lot of people who go out and make their name she had character attributes that don't always win you friends.'

The fact that we cannot be sure who murdered her helps fuel the

Fossey myth. Fossey was hated, but like all fascinating women she was neither solely a despicable villainess nor a consummate heroic victim.

Hunters need to employ various strategies when capturing infants from different ape species. For instance, when poachers approach a solitary mother orang-utan, the hunter locates her form in the tree-tops, shoots her dead and tries to catch her baby as they both fall from the tree. When chimps are hunted, dogs are often needed to locate the chimps because as a group they can move quickly across the ground and through the trees. The best way to capture an infant chimp is to kill its mother first thing in the morning as she gets out of her night nest; afterwards, if you make enough noise, the other chimps will probably run away in fear, allowing you to capture the orphan.

But life isn't always that straightforward for chimpanzee hunters. British primatologist and conservationist Ian Redmond told me of a story reported to have occurred in Cameroon. After a day in the forest hunters returned to their village with their booty, an infant chimp. They planned to keep the animal in a makeshift cage until it could be sold. Unbeknown to the hunters they were followed. For two days the terrified villagers were besieged by angry chimps that smashed up huts trying to get their baby back. When the hunters finally released the infant it ran from the village and into the trees to join its community and the chimps left together en masse. They could be heard screaming and calling to each other long after they had disappeared into the forest.

Capturing an infant gorilla is different again. A gorilla family unit consisting of a silverback, several wives and children at different ages of development will often fight hunters to the death en masse rather than give up a baby. It is not just the mother gorilla who dies; poachers may have to kill eighteen individuals before they get their prize. George Schaller completed a population census in 1960 and estimated that in that year there were somewhere between 5,000 and 15,000 lowland gorillas living in 15,000 square miles surrounding the Virungas and only 400 mountain gorillas living up in the volcanoes. Gorillas, especially mountain gorillas, were already very rare and poaching of infants was a disaster.

When Dian Fossey was told in 1969 that two baby mountain gorillas had been 'officially' captured by park guards to be sold to Cologne Zoo she could easily imagine the carnage. In the 1960s gorillas were already an internationally protected species, but the Rwandan government wanted to respond graciously to the German zoo's request for two gorillas. It needed a strong ally in Europe. Both the Rwandan and German governments were content to ignore the international ban on the trade in apes.

Enraged, Dian Fossey insisted the baby gorillas were returned to her. The animals were close to death and would not survive the journey to Cologne. Rwandan officials handed them over to Fossey to see if she could help them to stay alive. Back in Karisoke, using her instincts, she managed to successfully nurse the infants. Fossey took on the necessary role of surrogate mother, even sleeping with the animals. She grew to love the black, fluffy infants, which she named Coco and Pucker, and after they had returned to good health started to make plans for their rehabilitation back into the forest. But the mayor of Cologne still wanted his gorillas and the Rwandan government didn't want to disappoint.

In the long term Dian Fossey could not protect the two little gorillas. It was a political decision to take back the animals from her, two months later, crate them up and fly them to Germany. There was nothing Fossey could do about it. She contacted numerous conservation charities asking for their assistance, and in turn they wrote to the zoo and the governments involved, but nothing worked. Fossey knew the baby gorillas would have a miserable existence and die prisoners in a concrete cell. The two gorillas both lived for nine years at the Cologne Zoo. Just as they approached adulthood one became ill and died and the other pined to death a month later. Today there are no mountain gorillas in captivity. Only lowland gorillas seem to be able to adapt to a captive life and go on to breed.[1]

Endangered animals cannot be protected by conservation charities unless governments uphold the law. Fossey, a lone woman, had done her best, but she was psychologically scarred by this experience. From that time on she would bitterly refer to environmental charity work as 'theoretical conservation'. Dian Fossey believed that field conservation work was the only worthwhile approach because charities and international laws could not protect wildlife. She decided to take the law into her own hands. Living up on that mountain meant she could do anything. Fossey knew local people thought she was a witch and, aware how the powerful psychological forces of *sumu*, or black magic, governed their lives, bought herself Hallowe'en masks. Like in a gothic melodrama Fossey started to frighten poachers.

Jane Goodall spoke about Dian Fossey's behaviour: 'Dian was the sort of person who made enemies quite quickly. She was a very violent sort of person. She had a personality about as different from mine as you possibly could get. I mean I probably wouldn't have been brave enough to go running after poachers wearing Hallowe'en masks, but she did. And someone who is going to behave like that, a sort of Diana goddess out in the forest, is bound to attract rumours and jealousies. She was just that sort of person.'

Fossey was very hard on herself for not being able to protect the mountain gorillas and she was just as tough on her African staff, not trusting them and talking to them only if she had to. Fossey loathed some of the people around her, yet she refused to leave the place. Dian Fossey was a truculent and determined woman when *National Geographic* photographer and cameraman Bob Campbell was sent by Richard and Louis Leakey to film her in the summer of 1969. It is no wonder they did not get on at first. Fossey was in no mood to bond with Campbell, nor did she feel relaxed enough to have him point his camera at her. She had expected her old friend Alan Root and knew nothing of Bob Campbell's reputation as a cameraman. Fossey's only concern at that time was that the men who had invaded her mountain and used weapons to capture Coco and Pucker from the wild were punished, though they never would be.

Bob Campbell was already married and lived in Nairobi. His wife, Heather, was used to him being away from home on long projects. On and off he spent three years with Dian Fossey at Karisoke.

The images of Goodall and the chimpanzees created by Hugo Van Lawick's *National Geographic* photographs were evocative of a fairy tale. National Geographic now wanted this *Beauty and the Beast* representation from Fossey and her gorillas. They desired colour photographs of woman and ape as one in nature to adorn their magazine cover and the centrefold. Somehow Campbell had to win Fossey's trust.

At first Fossey had not relished Bob Campbell's presence in camp and she sank into paranoid solitude, making her own dinner and eating alone in her cabin. A quiet man of few words, Campbell chose to do likewise, eating alone in his tent, which he had pitched as far across the meadow from her hut as possible. They hardly spoke to each other for the first few weeks.

When in the forest tracking gorillas they never spoke. Campbell, at first, kept back, because he knew his presence disturbed Fossey and made her feel self-conscious. She had not trained in methods of data collection and ironically Campbell was more experienced than the scientist he'd come to film. Fossey sensed this and was terrified of his criticism. Though Dian Fossey was the world expert on the mountain gorilla during the 1970s and early 1980s, her general knowledge of biology was very poor and she would often make embarrassing errors in public.

Campbell told me emphatically that it was he and not Fossey who habituated the gorillas first. He said he needed close-up photographs of Fossey with the gorillas, but Dian was still a long way from making physical contact with them, partially through fear. Campbell said he made friends with the famous male gorilla, Digit, before Dian did and that he could habituate any wild group of gorillas, perhaps in a couple of hours, and certainly within forty-eight hours. Once he had won the animals' trust, apparently Campbell would gently introduce a terrified Fossey to the group. Campbell's claim seems extraordinary, and it is not mentioned by Fossey in her autobiography, *Gorillas in the Mist*, but he was experienced at living in the African bush and photographing wildlife.

Dian Fossey was utterly dependent on her National Geographic funds and the small intermittent sums Louis Leakey could send to her. She had no money of her own; she had spent it all staying on in Africa. Often National Geographic was late with its grants and Fossey would go without food in order to continue paying a salary to her staff. She never complained and never asked for favours, but internalised all of her feelings. For this Campbell respected her. Karisoke was no place to make your home. Humans were not adapted to the climate in the same way as gorillas. Living at camp with Fossey made Campbell aware of her troubles and of what a remarkable woman she was.

Internally Dian Fossey was desperately lonely. She badly needed love and support, but her external cynicism portrayed a woman who needed no one. Campbell and Fossey began to eat dinner together and talk to one another and Fossey slowly let her guard down. Late one night, long after Campbell had said goodnight and walked out of her hut and across the meadow in the dark to fall asleep alone in his tent, Fossey woke him up. She wanted to make love with him, but Campbell resisted her ardour and she quickly retreated. The next morning Fossey tried to brush the incident aside by saying she had only entered his tent because she wanted to thank him once again for all his help.

Fossey felt humiliated by Campbell's rejection of her. She was also physically run down and worried that she had contracted tuberculosis. Feeling vulnerable, Fossey decided to leave Campbell in camp and travel to Nairobi to visit Louis Leakey for a week. She knew Leakey would look after her and arrange for her to see a doctor. Dian Fossey had not seen Leakey since she had been kidnapped and raped two years before. He never actually visited any of the trimates' field sites. Goodall asked him to come and see her 'darlings' many times, but he was always too ill or busy with other projects to make the journey. Karisoke was too inaccessible for him and Biruté Galdikas's Camp Leakey in Borneo much too remote. But that did not mean Leakey was not devoted to his ape ladies. In Nairobi he organised Fossey to see a specialist and emphysema was diagnosed. It was obvious she needed a holiday, so Leakey took her on a safari. He had flirted with Fossey before; certainly she would have known that Leakey had a soft spot for women. While on safari and on the rebound from Bob Campbell, she fell into Louis's arms. Both Fossey and Leakey were lonely and for a few days they became lovers. Just because Campbell didn't want Fossey in his bed it didn't mean no man would want her. At the end of the week Fossey had to return to Karisoke and Leakey had to visit the United States for a lecture tour, but knowing she had secured the love of one man had given Dian Fossey confidence to go back to Karisoke and face Bob Campbell.

Back in camp Fossey received a string of love letters from Leakey.

Dearest love, in such a short time our love was born and grew and matured into something deep and secure – however much you try to pretend its [sic] only temporary. Which it is NOT, Not, Not. Last night you made me so utterly happy as you suddenly

accepted that you too loved me. And last night with
you peacefully asleep in my arms . . . was bliss . . . What
a heavenly week we had with so many lovely memories
to share while we are apart and for all the future. Take
care of yourself my dearest one. I love you so. Louis.

Fossey did not write back to Leakey. Two months later in Karisoke
she sent her standard formal field report to Leakey, as she always
had, but she made no reference to his passionate prose. Blinded by
his love for Fossey and encouraged by her cold, official field report,
Leakey continued to send her adoring letters. Fossey wrote in her
diary at this time, 'Don't know what to do about L. God what a
mess.'

But Leakey's love for Fossey cushioned her insecurities and one
night she returned to Campbell's tent. This second attempt to
become his lover was a success, but their relationship would never
be equal. Fossey hoped for more; she wanted Campbell to talk
sweetly to her the way Leakey had, but he would not. He could see
Fossey wanted him and while in camp he pragmatically played the
part as best he could, but Campbell always intended to return to his
wife when the filming was over.

A love triangle had developed. Leakey was in love with Fossey,
who in turn was in love with Campbell. And Campbell was a professional wildlife film-maker who just wanted to do a good job,
satisfy the Leakeys with his work and go on to film Richard Leakey's
next dig at Koobi Fora. Campbell's commitment was more to himself and to the Leakeys than to Dian Fossey.

Louis Leakey had arranged for Dian Fossey to go to Cambridge
to take her Ph.D. with Robert Hinde in January 1970. But Fossey
didn't want to go. She couldn't bear to leave the gorillas, feeling sure
they would suffer in her absence. The chaos the poachers left in
their wake made her feel indispensable to the animals, and neither
did she want to leave the man she loved; but to be taken seriously

on the international scientific circuit she needed to write up her dissertation with Hinde's guidance. When Fossey left the rainforest for Britain's equally wet weather she felt torn in two. Six months after her relationship with Bob Campbell began, Dian Fossey was in Cambridge in England, arranging an abortion.

This must have been one of the hardest decisions she ever made. To terminate the life of an unborn baby you have accidentally (or not) created with the man you love is an anguish many women know all too well. But Fossey knew it couldn't be. If she took on her biological role as a mother she would have to relinquish her moral role as gorilla warrior. Campbell knew nothing of this pregnancy; he certainly would have found joint parenthood with Dian an inconvenience. This experience hardened her resolve. Ultimately she was alone; her commitment to the gorillas intensified.

No doubt she did the right thing, but had Fossey become a single mother and university lecturer, only occasionally visiting Karisoke, she would probably still be alive today. But would the mountain gorilla?

Louis Leakey and Dian Fossey spent time in different British hospitals at about the same time during January and February 1970. Leakey had been feeling ill before he left Africa, but wrote to Dian in Cambridge and told her he would call her as soon as he arrived in London so they could spend a quiet evening together. The ailing Louis drove alone 300 miles from Olduvai to Nairobi non-stop to catch his plane to Heathrow. Before leaving Olduvai he had felt some chest pains that he chose to ignore, but as the plane touched down at Heathrow Leakey felt he was about to collapse. An ambulance was called but he insisted on being taken to Vanne Goodall's flat and not to hospital. Vanne called the doctor, who immediately sent him to hospital, where he had a second serious heart attack.

Dian Fossey wrote to Leakey when she heard of his heart attack. Cambridge is a short journey from London, but she couldn't face the trip. Vanne Goodall told me she thought it odd that Dian didn't visit Leakey straight away and certainly Louis would have wanted Dian to visit him, but at the time no one knew her pain; Fossey had kept her abortion a secret. A month later she felt able to see him, by which time they were both recovering and in denial over their emotional isolation from each other and from others close to them. They were both lonely but Dian now obviously regretted her love affair with Louis and he was left needing her much more than she ever needed him.

In London together, Leakey and Fossey found a sort of mutual comfort by using Richard Leakey as a scapegoat for their pain. Richard was running the Leakey Empire in his profligate father's absence and decided to go against two of Louis's wishes, firstly to continue funding the Tigoni Primate Centre, and secondly to send Fossey a new assistant for Karisoke. Richard saw more important ways to spend money than the gorilla project – such as his own palaeontology expeditions. Louis and Richard would argue over money many times. Ironically Richard met his second wife, Maeve Leakey, when she was one of Louis's many female researchers at the Tigoni Centre.

At the zoology department in Madingley, Cambridge, Fossey wrote her thesis. It was hard going, but in the end 'The Behaviour of the Mountain Gorilla' turned out to be the most comprehensive account of the animal's nature ever written.

Fossey was especially interested in the relationships that adult male gorillas have with each other and how they behave towards their own group, and she was also fascinated by mother–infant relationships. In addition she had studied the animal's ranging patterns and their individual vocalisations. Schaller's observations were superficial by comparison. Fossey had outdone Schaller, as Mary Leakey had cynically prophesied when Fossey first became Leakey's gorilla girl. But Dian Fossey did not care about her academic success now. All she wanted was to return to Karisoke and

Bob Campbell, who had become the love of her life. While Fossey had been in Britain Campbell had spent most of his time at Karisoke continuing to habituate the gorillas and to film them.

Back in Africa Fossey let her self-sufficiency slip and started to rely on Campbell. But when they were together she tried not to cramp him too much. She wanted him to feel they were naturally compatible. She hoped Campbell would become her soulmate and share Karisoke with her. She wanted a husband-and-wife bond, like that of Joan and Alan Root, whom she had met on her first trip to Africa. Fossey believed her relationship with Campbell was special enough for him to relinquish his wife and his life in Nairobi and stay with her, but verbally Campbell never gave Fossey such hope, unless sleeping with her regularly was tantamount to proclaiming his lifelong commitment. Even before the filming project was officially completed in 1972 he left, as he had always intended.

Dian Fossey's emotional isolation lies at the heart of her tragedy. Unlike Jane Goodall or Biruté Galdikas she was destined to be alone. At their initial urges to live with apes, both Goodall and Galdikas received support from their families, if with some trepidation, whereas Fossey dryly used to joke that her family refused ever to talk to her again when she left for Africa. Certainly her mother and stepfather were not supportive of her mission. Neither did Goodall or Galdikas enter the field alone. Galdikas was already married to Rod Brindamour when she met Louis, and husband and wife took on the immense task of studying the orang-utan in Kalimantan together. At different times Jane Goodall had either her mother, her sister Judy or Hugo Van Lawick for company. When Goodall and Van Lawick's marriage broke down, Goodall was not alone for too long before Derek Bryceson, the Kenyan Minister for Parks, fell in love with her. They married and Goodall found another kindred spirit with whom to share her love of the African wilderness. Tragically Derek Bryceson died of cancer in 1980. Goodall had nursed him through his illness and it was returning to the chimps at Gombe after his death that helped pull her through her sadness. Unlike Goodall or Galdikas, Fossey never had a husband or a baby. She only had the gorillas to live for, except for six months or so at Karisoke, when Dian Fossey found love and happiness with another human being.

In the later half of 1971 Dian Fossey and Bob Campbell achieved some of their best collaborative work and were probably very close and tender with each other. Photographing Fossey sitting among a group of wild gorillas exploded the myth that gorillas are a raging force of danger. White hunters had perpetuated such stories to flatter themselves when posing with the dead trophy of a silverback, but amazingly Dian Fossey, a mere woman, was posing with live gorillas! From her close proximity to the animals it was evident that, when unthreatened, these giant, leaf-munching apes are tolerant of people. Many photographs that seem to show aggression in gorillas are in fact photos of the animal yawning, exposing their formidable canines.

Never knowing when a gorilla would be in sight and a photo opportunity would come along, Bob Campbell always had his camera at the ready. He had to make Fossey look attractive in the pictures at short notice, but this wasn't easy. The high altitude had taken a toll on her health and her appearance. She was out of breath and wheezing most of the time, looking exhausted and bedraggled. Climbs were especially hard on Fossey. Her lungs were in bad shape and, unable to take in enough oxygen, she gasped for breath constantly, her face contorted in agony. She also loathed her own image and was a reluctant model.

Climbing up the rock face at 10,000 feet was hard for Bob Campbell too. He was in good shape when he arrived at Karisoke but he had not anticipated the physically gruelling nature of the work. Neither of them had enough to eat and Campbell lost weight quickly. It took time for him to become accustomed to the hot sweats, induced by the gruelling climbs, followed by freezing chills that enveloped you as you crouched motionless in the icy rain to watch the gorillas. This sudden change in body temperature made chest infections all the more likely. At these physically tough times Campbell's unwavering respect for Fossey was cemented.

Louis Leakey and Robert Hinde both sent students to Karisoke to help. Sporadically, enthusiastic students would arrive, such as Leakey's recruit Michael Burkhart, but these young men and women would often then crack under the workload. Robin Dunbar was supposed to be Dian Fossey's first Ph.D. student, but just before leaving for Karisoke he decided that gorillas were too frightening and opted for baboons instead.

The arrival of students never seemed to ease Dian Fossey's anguish. Ian Redmond was perhaps the exception. Blond and bearded, with a warm smile, Redmond now runs the conservation group, Ape Alliance, fighting the bush-meat trade. Back in 1976, when he first joined Fossey, he was one of the very few scientists she could tolerate. Ian Redmond is a conservationist from the heart and Fossey would have soon realised he wasn't just interested in getting a Ph.D.; he genuinely cared about the gorillas.

Most of Fossey's students were not as committed to the fate of the gorillas. Some students could not stand the cold and wet and left within a few days of arriving without even telling her. Others would get lost in the jungle, or stoned on the local marijuana. Fossey felt angry and let down by these students. A familiar theme in all the long-term sites is that the women feel they are the only ones who truly care about the work.

It can certainly come as a shock for people to find themselves geographically isolated at a research station like Karisoke, but often the greatest adjustment is that of returning home. Such disorientation is common among field biologists and VSO workers. Many students recount feelings of alienation on going back to university to write up their theses. It is common for these people to find life in the West 'hollow' and 'selfish' and to be physically shocked by the noise and filth of Western cities after being alone in an exotic location for months on end. They find it hard to share their adventure with others, because it is difficult to convey the intensity of the trip to other people whose life experience has remained so far from their own.

In 1979 Robert Hinde wrote a report on culture shock. Hinde had become aware that many students felt disorientated after leaving their research sites. Other stories suggested long-term researchers were losing a grip on reality. After conducting interviews, Hinde documented that scientists experience their field work positively and find the personal challenge in completing their research stimulating. The researchers appreciate the beauty of the country and the attraction of a simple lifestyle. But in the early months they also feel homesick and inadequate in coping with their new environment. Married people whose spouse remains at home can suffer jealousy arising from paranoid fantasies of love affairs being acted out in the marital bed in their absence. Researchers grow accustomed to the lack of physical contact; when a new researcher or a visitor arrives it can become excruciating to greet them. Simply shaking hands can be horrendously difficult.

When questioned on culture shock, Jane Goodall has dryly commented, 'When I return, I haven't found living in the West any harder than I found it before I went.' On returning to civilisation Dian Fossey said, 'It takes about a month to remember to flush the loo!'

Fossey had a well-developed sense of humour and often cracked impromptu jokes, but in camp she certainly found it hard to talk to people. Fossey found the student Michael Burkhart unbearable. He had difficulties with the nature of the research he was supposed to be helping with. He couldn't negotiate his way through the forest, kept getting lost and found it very hard to estimate how many gorillas he had encountered. In 1970 Campbell helped Fossey to sack Burkhart and after he packed his things and left camp Fossey started to trust Campbell more. Campbell was proving himself to be reliable, supportive – and he was also tall!

While at camp Bob Campbell would watch people arrive uninvited; whether they were tourists or guests of local Rwandan dignitaries, they would expect to be taken into the forest to see the gorillas doing something fascinating and they would also expect Dian to feed them. These impromptu visits put Fossey under much stress. She never had enough time for everything that needed to be done. Neither did she have enough money to share out. Campbell knew Fossey felt that any distraction from her gorilla project was a distraction she couldn't afford.

Dan Topolski, Oxford rowing coach and travel writer, told me that he visited Rwanda early in 1972 as a 'naive' twenty-seven-year-old. With two other friends, he arrived in Ruhengeri, the nearest town to Karisoke. When they asked locals how to find Dian Fossey, people pointed up at the mist-covered green mountain. Topolski and his mates trekked for five hours up the mountain to visit her, but she was not in camp when they finally arrived, soaked and covered in mud. Bob Campbell, accompanied by Fossey's golden Labrador, briskly strode towards them as Topolski approached the three green-painted huts on the grassy slope that was Karisoke.

Campbell, looking like a colonial explorer in khaki shorts, called out to Topolski, 'You there, what do you want?' He informed Topolski in no uncertain terms that Dian Fossey didn't like uninvited visitors. 'Dian was out with the gorillas and we would have to leave.' This was just weeks before Campbell left her for good.

Topolski wanted to spend the night and meet the enigmatic Fossey; he wished to photograph and interview her. But Campbell, her official photographer and camp bouncer, was protective and only allowed the young men to remain there for an hour or so before they were sent back down the mountain. Before Topolski left Rwanda he learned from locals that Dian Fossey was considered to be crazy. Topolski told me, 'She was the butt of crude jokes. They said she must be fucking the gorillas – why else would she be up there?'

Fossey's favourite gorilla, known as Digit, had bonded with Fossey and Campbell during filming. They often located him alone, on the periphery of the group, a blackback, mature but not yet a senior male. Digit was a friendly gorilla. Fossey would lie back in the long wet grass and he would lie down with her; sometimes they would even fall asleep together and Campbell would snap way. These were the intimate photos National Geographic was paying him for, and it would be this type of physical contact, encouraged by the camera, that Fossey would be criticised for.

Soon Fossey was pregnant again. She did not bother Campbell with the news, but took a trip with him to Ruhengeri under some pretext and then went for an abortion alone. Saying nothing afterwards, she made the journey back to Karisoke. Some time later, in great pain, she started to haemorrhage. The operation had been far more crude than her first termination in Britain, and this time she bled for days. Ignorant to the problem, although he really should have guessed, Campbell helped her back down the mountain to Ruhengeri hospital. Still unsure what was wrong with Fossey, he decided to go on to Nairobi to spend Christmas of 1971 with his wife and to meet up with Richard Leakey to discuss the filming of Richard's Koobi Fora trip. This was the end of their love affair.

Campbell returned for a month or so in early 1972 and then left for good. He'd turned Dian Fossey into an icon, but it was just a job. Richard Leakey needed him now. Fossey didn't try to make him stay. She had her gorillas; they needed her and would never reject her love and leave her.

Fossey continued at Karisoke, and the intermittent arrival of students of varied abilities continued. Finding ways of protecting the gorillas dominated her days and nights. While destroying poachers' snares and guarding the gorillas Fossey continued to learn a great deal about the animals' social life and their similarities to man.

Right up until his death in October 1972 Leakey was frequently required to smooth things out for Fossey with disgruntled Rwandan officials. As an African and a natural politician, he knew how to win friends and influence people. Fossey did not. Because Leakey's efforts on her behalf were usually covert, after his death Fossey became much more isolated than most people realised. Without Leakey and Campbell, there was no one to protect her. From 1973 onwards she had to rely on her wits and her instincts, but she was outnumbered.

The Parc National des Volcans is 125 km^2, the Karisoke study area covers one-fifth of that. The gorillas roam freely and researchers can only observe groups that pass through the study area. There are thirty-two gorilla groups, six of them habituated to human observers. Of those six groups two gorilla groups live part-time over the border in the Democratic Republic of the Congo. Digit was approximately two and a half years old when Fossey first encountered him in 1967. Born into Group 4, his parents were both aging when he arrived. Not long after Fossey first habituated Group 4, Digit's mother disappeared, presumed dead. After that Digit shared night nests with his father, but a year later his father, the group's silverback, died of pneumonia, leaving Digit an orphan. But because gorilla family groups are just that, Digit was surrounded by relatives – aunts, uncles and siblings – and thus when very young retained a place in the bosom of the group. But as he grew older he had no one to help him fight for a social position. Orphaned children and apes are more likely to die before they reach their fifteenth birthday than those cared for by their biological parents, another parallel being that orphaned children and orphaned non-human primates have the lowest social status without a mother to fight for them and must work very hard to assert themselves.

In the gorilla social hierarchy, the sons of the dominant female and the dominant silverback will probably grow up to take over the group. There will be other males in the group, who are second and third in command, as well as younger adolescent males, who may stay on the periphery of the group or may leave to join another group or start their own. For maturing males who are sons of males who have left the group or are of lower status than the existing silverback, the best reproductive strategy is to leave and form a new family, where they can become the boss. Young, strong and confident males achieve this by herding or kidnapping unrelated young females away from other groups, so they can mate without competition. Recently matured females never wander freely for long; a fertile female is a prize that mature silverbacks will fight over. The tougher and more experienced the silverback, the larger the number of wives he can keep. Peripheral male gorillas often remain on the edge of their group and act as guards. Referred to as 'watch dogs' by Fossey these male gorillas are expendable.

Digit's role in adult life was that of a watch dog. As a young juvenile Digit had four half-sisters to play with and much of his day was spent racing through the foliage and climbing and swinging through the trees with them. Adult gorillas are rarely seen in the trees. The males especially are too heavy, though if the tree is strong enough adults will make night nests in the lower branches.

As the juveniles grew older and the females were reaching sexual maturity, Group 5's silverback, Beethoven, took one of Digit's sisters, Bravado, to be his wife, and Group 8's silverback took another two of the sisters. After the loss of his playmates, Digit became isolated from the group and Fossey watched his personality change for good. He spent more and more time alone, an appendage to his group, and this partial isolation made him vulnerable. On one occasion he was attacked by an unknown male and left with a canine puncture wound to his neck. It persisted in oozing pus for five years, and as a result a weakened Digit stooped for much of his young adult life. Digit seemed to be lonely and scarred by life's cruel blows; it was easy for Fossey to empathise with him.

Because of encroachment into the Parc des Virungas from cattle farmers, woodcutters, honey-gatherers and poachers with dogs setting snares for antelope – let alone gorilla hunters – the gorillas existed within a reduced 19 square kilometre area. This overcrowding meant the study groups' ranges were constantly overlapping with each other, causing tension. On one occasion Fossey observed Group 5 interact with Group 4. Digit backed up his group's silverback, the inexperienced Uncle Bert, as they both directed defensive behaviour towards Beethoven, who was more mature and much more experienced than both Digit and Uncle Bert and did not retaliate. For two days Fossey observed both groups peacefully residing within 150 feet of each other.

During this time Bravado returned to her natal group to play with Digit, and the two adolescents were delighted to be with one another again. They mutually groomed and enthusiastically played tag for hours. Other youngsters from Bravado's new group joined her activities with Group 4. For a while everyone got along, but at the end of the second day Beethoven walked confidently into the middle of Group 4 and herded Bravado back. He didn't want her to start getting ideas about leaving him. This was the last time Digit and Bravado saw each other and the last time Fossey saw Digit play happily with another gorilla.

Fossey's life with the gorillas was satisfying and it offered her a redemption denied by humans. But Fossey's war with the poachers was escalating and her reputation preceded her. Robert Fritts was then the US Ambassador in Rwanda and he wanted to improve her image with the government, who regarded her as a cranky Western woman and not as a scientist or conservationist. Between 1974 and 1976 Fritts would invite Fossey down to government dinners where she could impress the Rwandan officials with her gorilla lore and introduce a screening of the National Geographic film that Campbell had shot a few years before.

On such occasions Fossey would arrive at the embassy in her scruffy working clothes and boots, scuttling into a bathroom with her bag of tricks and sometime later emerge a fashionable, elegant woman, wearing a full-length cocktail dress and silk scarf, make-up, and jewellery (until her jewellery was stolen from her cabin while she was out with the gorillas). It was hard to believe it was the same woman. The lights would dim and the Rwandan establishment would sit and watch her film, the light from the screen reflecting across their faces. Watching herself with Digit and other gorillas, Fossey would have been able to see herself the way Campbell had seen her. At times she must have wondered why he had left – hadn't it all looked so good?

Dian Fossey wasn't a political person – the duplicity sickened her – but Fritts knew it was important that she did not isolate herself in a foreign land. Fossey understood that his advice was sound and when her film ended would bite her lip and force herself to work the room, chatting and laughing animatedly with all present. Appreciating Fritts's help, in return she gave him an open invitation to come at any time with his family and see the gorillas. In 1975 Fritts took Fossey up on her offer and soon afterwards arrived at camp with his wife and two small blonde daughters.

Fritts expected that his family would travel to see the gorillas as a single group, but Fossey would not allow it. She told Fritts that he would become aggressively protective towards his daughters when he saw the size of the gorillas and the animals would sense his defensiveness; the ensuing tension would escalate and could possibly become dangerous. Fossey had seen it before. She insisted that the girls had to go with her and that Fritts and his wife went with a student. The two separate groups left camp and the Frittses did not see their girls again until they met back at camp after nightfall.

On their way back to Kigali Fritts's daughters told him of the extraordinary experience Dian Fossey had given them. The girls had both found Fossey to be kind and gentle. The climate was hard for these girls and they needed to stop regularly. After caring for youngsters with disabilities Fossey had an affinity with children; during her life at Karisoke she always tried to show tenderness to local African children.

When Fossey located Digit, who, as usual, was alone, she whispered to the girls to sit quietly, look straight ahead and not to scream, no matter what. The two little blonde sisters were about to have a close encounter.

Digit approached them from behind, and reached out and touched their long, golden locks. The gorilla rubbed the little girls' hair gently through his fingers; stepping closer he lifted their fine, silken hair to his nose to smell it. Digit had never seen a blonde before and the two girls entranced him.

It's not only humans who have an appreciation for the aesthetic. A male chimp at Gombe was observed to make daily visits to the waterfall, where he would sit alone in tranquil surroundings soaking up the calming atmosphere. The animal neither drank from the water nor ate any vegetation; his trip to the waterfall seemed purely for pleasure. Fossey believed her gorillas took pleasure in beauty. She felt sure that Digit found pleasure in the little girls' soft, blonde hair. He certainly had never reached out to an image of the human female primate like that before.

But escalating poaching and gun-running in the forest were making the gorillas increasingly nervous. Fossey once recalled an incident with Digit that showed how nervous the gorillas had become. 'In November 1976 my tracker and I were searching for Group 4 when we suddenly saw them some 40 meters away off to our left, day nesting in heavy drizzle. I was debating as to whether or not to bother them when out of the brush to our right ran Digit to inadvertently meet the tracker at a distance of 8 meters. Digit immediately stood upright and gave two prolonged screams exposing his upper gums and all canines. Alarmed, Group 4, led by Uncle Bert, took to their heels and knuckles and fled away into the trees. At that moment Digit seemed undecided whether to charge or to flee until I stepped into his view and pushed the tracker down behind me. At once he dropped to all fours and fled towards his group who had rapidly covered over 100 feet. Digit caught up with them, leaving a strong, pungent, fear odour that remained in the air for over 15 minutes.'

At this point Digit was eleven years old and had just become a silverback. He spent more time out of social contact with his group and remained about 150 metres from the group at most times. Both Digit and Fossey had become loners. As the years rolled by they would take solace in each other's company. Gorillas have a sense of humour. They chuckle, especially when tickled, and smile and 'purr' when contented. Digit would find Fossey and they would sit together in the long grass, tickling, purring and mutually grooming one another. There are many photos of Fossey with Digit. Theirs was a relationship of mutual reciprocity. Fossey publicised the gorilla's fragility and for her he symbolised all that was innately pure and good. Digit shared sticks of wild celery with Fossey and they spent many hours in each other's company, the two of them exchanging mutually comforting 'Oooo-aahh-mmpss' – gorilla speak for 'This is the life!' The 7 million years of evolution between our two species made no difference to Fossey and Digit.

After Ian Redmond had been living at Karisoke alongside Dian Fossey for fourteen months, he was feeling homesick and Fossey had cheered him up on New Year's Eve 1977 by sharing her bottle of whisky with him and the African staff. Fossey had yelled out, 'Happy New Year!' as she burst into his hut while Redmond struggled by candle light to type the monthly report. Redmond told me, 'I found Dian hard going at times. The biggest conflict between us was my endemic slowness of typing. It was my duty to type up field notes to send to National Geographic as the monthly report.

She'd get very cross with me if it was late. She was always worried she'd lose their funding. I respected Dian, she was self-made and well aware of her failings. She tried twice as hard as everyone else.'

Late on New Year's Day a tracker returned to camp to inform Fossey that Group 4, Digit's group, could not be located and that he had found poachers' tracks from the evening before. On 2 January Fossey told Redmond they must go out and see what had happened.

That day Ian Redmond found Digit's body. He had been dead two days. It wasn't easy for Redmond to recognise him at first. He could see it was the body of a young male by the silver hair on the back, but as Redmond turned the body over to look at the face he realised the head was missing. Digit had been decapitated and his head taken away. Digit was so named because the third finger of his right hand had been permanently damaged in a poacher's trap when he was an infant, but, as Redmond looked at Digit's hands for clues to his identity, he realised that both hands were also missing. But as the body size and shape matched Digit's – the back was slightly stooped as Digit's had been after the bite to his neck – Redmond realised there was no doubt.

There were plenty of tracks around Digit's corpse. Redmond studied tracks left by the gorillas, the tracks of a passing herd of forest elephants, the poachers' tracks and the tracks of their dogs. According to Redmond the elephants, while moving away from the approaching group of men and dogs, disturbed Group 4, who in trying to avoid the herd of elephants ended up walking towards the poachers. The men, led by their dogs, came across Digit first; he was about 65 metres from his group when the poachers struck.

In defence of his group Digit charged at the hunters. He would have let out alarm calls to Group 4, who took this chance to escape, leaving Digit to fight the five dogs and the six poachers. Surrounded, he would not escape alive, but he did put up a great fight that must have lasted at least ten minutes, thus giving his kin time to leave the vicinity. Although his whole body was covered in machete slashes, Digit died from six fatal spear wounds to arteries in his neck and chest. When Redmond found him he lay on his side at the edge of an area of flattened, blood-stained vegetation approximately 50 feet in diameter. It had been some battle.

After finding Digit's corpse, Redmond walked back to locate Fossey. He didn't want her to stumble across Digit's remains the way he had. Redmond was crying, in shock from what he had seen. He had studied Digit for little over a year but was now feeling as if a close friend had been murdered. Redmond knew Fossey would take the news much harder. She had studied Digit for over ten years.

Many years later Redmond wrote an essay entitled 'The Death of Digit' in which he recounts the experience of finding Fossey to break the news. 'At 12.15 we came across Dian's footprints heading off the Border Trail on elephant tracks at 10,200 ft. The elephants had gone southeast, then curved back towards the border above where we had crossed earlier, then over Border Stream and into Rwanda. It was 1.45 p.m. before we caught up with Dian and Kanyarugano; as we walked we occasionally whistled to try to attract their attention. Whistling is a good means of long distance communication in the forest, but that means it is used by poachers too, and so until I rounded the corner in the trail which had hidden my approach, Dian was unsure of just who was coming. Which is why, as I came striding round the bend, I suddenly found myself looking into the muzzle of Dian's automatic pistol. She was kneeling in the middle of the path, holding a gun in a two-handed grip ready to challenge us had we turned out to be poachers. On seeing it was me, a mischievous grin crept across her face and she kept the pistol levelled at my chest, saying, "Halt! Who goes there?" I smiled weakly and slowly pushed the gun aside and down. With what I had to say I didn't fancy being at the wrong end of that. She caught my expression and said, "Poachers?" I nodded, still searching for words and her face fell. "Group 4?" she asked quietly. "Digit," I replied.'

Ian Redmond said that when he told Fossey that Digit was dead, 'It was as if a steel shutter had dropped behind her eyes.' In Dian Fossey's autobiography *Gorillas in the Mist* she speaks of hearing about Digit's death: 'There are times when one cannot accept facts for fear of shattering one's being. As I listened to Ian's news, all of Digit's life, since my first meeting with him as a playful little ball of black fluff ten years earlier, passed through my mind. From that moment on, I came to live within an isolated part of myself.'

Kin selection and altruism was a decisive step in the evolution of early human societies and a sample of this type of behaviour can be recognised in Digit's last moments. He altruistically chose to protect his kin from poachers, allowing Uncle Bert to lead the group away. Digit didn't just save aunts and cousins, but sacrificed his life for his then unborn daughter. Without anyone realising, Digit and Simba had successfully mated. A hormone released in a male primate after he has mated lasts for the length of gestation so the infant born to the female mate is generally accepted by the male to be his. Digit would have anticipated Simba's baby.

When a silverback is killed from poaching or dies from natural causes or is defeated in a fight, the harem disperses and females join other groups. The silverback of the new group will not offer protection to another male's infant, as that baby does not share his genes, so any females carrying a suckling infant will lose that baby. The silverback kills the infant by fatally shaking it, throwing it against trees and biting it.

Shortly after Digit died, Simba gave birth to their baby daughter, Mwelu. A few months later poachers struck again, and Uncle Bert and Macho were both killed. Without a silverback to lead them, Group 4 split up. Simba joined Nunkie's group and, when Mwelu was only eight months old, Redmond found Simba alone with very swollen breasts and the corpse of her baby daughter lying on the ground (Fossey was away from camp when this happened). After performing an autopsy on the tiny body Redmond concluded that Nunkie had used his canine teeth to puncture the infant's skull and had stamped on her chest, crushing her rib cage. Shortly afterwards Simba came back into oestrus and mated with Nunkie and a new baby was born. Simba still lives on the mountain and when last seen was alive and well, despite everything life had thrown at her, including the Rwandan civil war.

Fossey feared Digit had been killed because he was her favourite gorilla and Group 4 was her favourite group. Certainly, over the years Group 4 suffered more than its fair share of poaching, but Redmond thinks the theory of a vendetta against Fossey's favourite animals is unlikely. Poachers would not be able to recognise one individual animal from another in the way a researcher could. Later, when one of the poachers was caught and interviewed, still wearing clothes splattered with Digit's arterial blood, Redmond and Fossey learned that the poachers had been promised the equivalent of £20 if they came back with the trophy head and 'ash-tray' hands of a silverback gorilla. It was at this time fashionable to use the dried hands of a gorilla as novelty ash-trays in private homes. In addition, Digit had killed one of the poachers' dogs in the fight.

Fossey and Redmond decided they must try to use Digit's death positively. Redmond had a Bolex movie camera with him and together they re-enacted Digit's last moments on film. Redmond used the camera to create both Digit's point of view of the poacher's approach and the poacher's point of view of Digit, crashing through the undergrowth to simulate the fight. Fossey suggested that Redmond should recreate Digit's death as he fell into vegetation, so Redmond placed a spear back into one of Digit's open wounds and filmed the body. He took endless photographs of the corpse and of Fossey sitting with what was left of her friend. Most of the footage remains unused in National Geographic's vault and has never been seen.

Redmond wanted to use previous film of Digit to make a documentary of Digit's life and death, thus attracting a major fund-raising campaign for the gorillas. But Fossey had reservations. She knew many Rwandan officials were corrupt and, if foreign revenue for the park passed through their hands, they were free to siphon off money and might be encouraged to kill off another gorilla to keep the cash flowing in. Fossey was also very much against the idea of eco-tourism. She was very resistant to the gorillas' and her own privacy being invaded by 'idle rubbernecks,' as she called them. But ultimately Fossey agreed with Redmond: the grisly photographs of the headless Digit should be published around the world and revenue should go to funding a more efficient anti-poaching patrol. Digit was buried alongside Fossey's cabin. No poachers were sent to prison in connection with Digit's death.

Two other well-known students of Dian Fossey were Kelly Stewart, daughter of the Hollywood actor James Stewart, and Briton Alexander ('Sandy') Harcourt. Stewart would have grown up immersed in the privileges of Hollywood but she chose instead to visit Karisoke in 1973 and toil away as an anonymous primatologist. Stewart and Harcourt were both young when they went to Karisoke and there they met Dian Fossey and each other for the first time. At camp, some months later, Stewart and Harcourt fell in love, much to Fossey's disgust.

While at Karisoke Harcourt and Fossey had a profound difference of opinion. Harcourt was of a younger generation and had trained at university in the finer details of method collecting. He argued with Fossey over her outdated methods and he wanted to be left alone to get on with his research, but she insisted that all students, before they got on with their own research, helped the anti-poaching patrols to destroy the previous night's poachers' snares. Many young researchers resented this chore.

When a gorilla died Fossey thought it was because she wasn't doing enough to protect the animals. She blamed herself and decided she had to keep on stepping up the reprisals. Because poachers had decimated populations of bushbuck elsewhere in the forest, the few bushbuck left migrated nearer to Karisoke for protection. Sometimes poachers would set their traps foolishly close to camp and every so often a poacher would be caught. Fossey made sure that any poacher brought to camp suffered a humiliating experience. Often their hands would be tied together with wire from their own snare and their stripped body would be smeared in gorilla dung and she would threaten to mutilate their genitals. Wearing a Hallowe'en mask, she would yell abuse at them in various languages, fire guns over their heads and confiscate the poachers' precious *sumu* pouch. (The amulet would contain samples of an individual's hair and other items thought to inflict black magic curses on that person and at the same time endow the owner of the pouch with great power.)

Dian Fossey was accused of torturing poachers, but this is not true. She waved weapons at the captured poachers, but she never touched the men and always insisted she was never left alone with them, just in case she lost her temper completely and did attack them. She also wanted to make sure witnesses were always present to prevent lies being told about her. Her revenge against the poachers was psychological rather physical.

Once Fossey had a poacher named Hatageka brought to her by her team. Fossey took his *sumu* pouch from him. Infuriated, he struggled, unsuccessfully, to take it back. She also forced him to give her a long list of names of people and locations involved in the smuggling of gold out of Rwanda – gun-, gold- and drug-running took place across the border with Zaire unabated. Fossey knew corrupt government officials were involved and she wanted them prosecuted, but no one was ever charged. Fossey only succeeded in making lifelong enemies.

Harcourt and Stewart were scandalised by Fossey's brutal treatment of poachers. They felt her ways of humiliating and terrifying these men were not going to stop the hunting of the gorillas. Harcourt organised eco-tourism in 1979. Fossey contemptuously referred to the tourists and guides as 'The Car Park Gang', after one was built for the tourists' jeeps. Harcourt hoped that when the preservation of the gorillas became of financial benefit to the Rwandan authorities and to the locals, the hunting would stop.

But in the long term eco-tourism in this part of the world has done little to help save the mountain gorilla. In fact it has been the gorillas habituated to tourists that have been shot dead, such as the tolerant silverback Rugabo and his mate Salama, killed in 1998. Poachers from Zaire wanted Salama's infant; after its parents were killed the infant took refuge up a tree. Something startled the poachers and they left without the baby. Staff of the Dian Fossey Gorilla Fund, DFGF, found the two-year-old infant and the bodies of his young parents. They have managed to reintroduce him to his group, where the youngster will be looked after by his older half-siblings and aunts. Whether he will survive to adulthood is debatable. Young gorillas usually remain in almost constant physical contact with their mothers until at least four years of age.

Primatologist Liz Williamson runs Karisoke today. When I asked her about the benefits of eco-tourism she told me that it puts these shy creatures, the very last of their kind, under too much pressure. The few hundred mountain gorillas can't be treated like ambassadors representing their species, like groups of chimps can because they are all that is left of the species. Dian Fossey knew this was true. She felt the park should remain a human-free sanctuary where gorillas would not be disturbed by tourists trampling through vegetation, clicking cameras and possibly sneezing potentially fatal bacteria over the animals.

But it wasn't just Fossey's methods and her resistance to eco-tourism that came between her and Harcourt. He owed Fossey his life. He was gored by a buffalo, which penetrated his groin with its horns. Luckily Fossey heard his cries and fired her gun to frighten the bull away. Fossey then carried Harcourt back to camp. In 1983 Harcourt decided to leave Karisoke for the safety of America, although he has retained ties with the place. Today Kelly Stewart and Sandy Harcourt are married and live in California. They both continue to study gorillas and both lecture in primatology at Davis University; Harcourt specialises in primate reproductive anatomy.

Joan Travis remembers Fossey well. Joan and husband Arnold, supporters of palaeoanthropology, have been alfalfa farmers in Los Angeles for fifty years. Whenever the Leakeys and the trimates were in LA they would stay on the Travis' farm. (To this day Joan remains a governor on the board of the Leakey Foundation and Don Johanson's Institute of Human Origins.) Joan remembers, 'Dian was very sensitive to the needs of others, but she wouldn't talk about her problems, she was a very private, emotional, earthy and direct person. She was very loyal to people she loved and to the gorillas, Dian had the character of someone who would die for their cause. She liked to call herself the 'Old Lady of the Mountains'. One day a small package covered in beautiful stamps arrived from Rwanda for me. Dian had sent to me an exquisite gemstone box. Inside the box, on a miniature cushion of red velvet, there was the first bowel movement of a new-born mountain gorilla. Dian thought such a thing far more precious than jewels. I still have it.'

In 1981 Dian Fossey, Jane Goodall and Biruté Galdikas met in LA for the Leakey Foundation's Man and Ape Symposium, hosted by palaeontologist Don Johanson. At this point in Fossey's life she had suffered all of her major losses – Coco, Pucker, Digit and Campbell – as well as hostilities from students and threats on her life from poachers.

Fossey spoke about her feelings towards poachers: 'A *substitut* [sheriff] began working with me and putting these poachers in prison. There are some of them still in there; they are eighteen-year terms and Ruhengeri prison has got a 60 per cent mortality rate every year – we see to that.' There was an embarrassed silence when Fossey made this remark, the audience shuffling uncomfortably in their seats and just one or two hardcore conservationists whooping. Fossey wouldn't have cared; she said what she felt. She was not interested in the poacher's reasons for killing the gorillas. The fact that his children were going hungry was of no concern to her. Back in her home state of Kentucky, first-degree murder is punishable by the death penalty and to Fossey the poaching and genocide of the mountain gorilla was premeditated murder. Digit, aged fourteen, had died two years before the Man and Ape Symposium, and Fossey did not bother to pay lip service to the politically correct way of addressing this type of tragedy. Poachers were bastards who deserved to die, and that was that.

Dian Fossey lacked confidence throughout her life and even though at this time she was the world expert on gorillas she was very insecure when speaking publicly. In her halting Southern lilt she said: 'The cohesiveness of the gorilla group and the tightness of kinship bonds are the most striking aspects of this animal's social structure and they are unique among all non-human primates and I would like to think once upon a time the early hominids also lived in these tightly, kinship, genealogically linked, groups and I see my gorillas as just carrying on something we've all inherited. They have more closeness, more cohesiveness than humans do in their defence and protectiveness of each other.'

The Rwandan government was very keen on establishing tourism to the Parc National des Volcans, where Karisoke was located. Dian Fossey's work, publicised by the National Geographic, had put the tiny country of Rwanda and its wildlife on the international map. The Rwandan government were astounded to discover that this crazy American woman and the black, furry gorillas were a tourist attraction! At the height of eco-tourism, before the latest war ravaged the country, the Rwandan government could expect to earn $5 million per annum from tourism. When the country erupted into civil war in October 1990 after years of sporadic fighting, the Tutsi, with aid from the Ugandan government, destabilised the Hutus' government and one of their first strategies was to destroy eco-tourism. Gorillas and white people were murdered and landmines were planted in the park.

The Interahamwe are led by Hutu Rwandan government extremists. In 1994 these men sanctioned the torture and murder of more than one million Tutsis and Hutu moderates. Kabila's Tutsi supporters are left feeling betrayed by him and today the fanatical Interahamwe reside mostly just over the Rwandan border in the Democratic Republic of the Congo. They continue to fight their own covert and barbaric war in the mountains. It was 150 armed Interahamwe soldiers who marched into Uganda's Bwindi Impenetrable Forest and kidnapped and killed eco-tourists on 1 March 1999.

The holidaymakers were wildlife enthusiasts and had visited the 331-square km Bwindi National Park especially to observe mountain gorillas. There are approximately 300 gorillas that live in the forest. It is speculated that they may in fact be a separate subspecies of mountain gorilla from the 330 that live in the Virunga volcanoes. The Bwindi gorillas' vet, Gladys Kalema, was safely in Uganda's capital, Kampala, at the time of the attack, but fourteen white tourists and tour leaders were captured at the campsite in Buhoma, at the gateway to the park. They were dragged from their tents, beaten and then marched up into the forest. Surprisingly, six were later released but the four remaining men were murdered and four remaining women were reportedly raped before being murdered. The Interahamwe want to publicise their cause in order to retake control of Rwanda and destabilise Uganda. This tragedy stopped all eco-tourism, but with military escorts it has recently bounced back.

Laurent Habiyaremye was the country's director of tourism in the early 1980s. He did not want Fossey blocking his plans to start gorilla tourism in any way. Whenever he sent messages that he would be arriving with a large group of VIPs to see the gorillas, Fossey would hide and forbid her team of trackers to take anyone to see the gorillas, which infuriated and humiliated Habiyaremye. In punishment he never issued Fossey with a park permit and visa for any longer than two months at a time. He was well aware that Karisoke was her home and her whole life and that she was very ill with emphysema; the gruelling 2,000-foot climb down the mountain every two months to renew her visa and permit caused her to spit blood. Then she had to travel to Kigali, 60 miles away, a four-hour journey, partially across rough terrain that limited the speed to 10 miles per hour.

Understandably Fossey hated these journeys. She said of Habiyaremye, 'He's trying to kill me before my time.' Then amazingly, in the summer of 1985 Fossey was issued with a visa for two years and a permit for six months. There was now no need for her to leave camp for some time; she could relax.

Wayne McGuire was a heavily bearded Ph.D. student from Oklahoma University. He arrived in Karisoke on 1 August 1985 after writing to Dian Fossey for four years, hoping that she would accept him as a student. He wanted to establish how loving and protective male gorillas are to their own infants. When Fossey finally agreed to him coming, McGuire was pleased but ambivalent about meeting her. Dian Fossey was infamous around the world. Among academics and conservationists, stories of Fossey 'going bushy' being 'crazy', 'becoming a raving lush', 'fucking around with gorillas and torturing poachers' were two a penny.

Fossey told Wayne McGuire soon after his arrival: 'I know you've heard a lot of bad stories about me, but ignore them and concentrate on the gorillas.' This was sound advice. Fossey also told him to 'keep your gear and your passport together. If you ever hear guns or firing in this camp, grab your gear and follow the stream down. Don't worry about me or anybody else, just save your own skin.'

Dian Fossey was expecting retribution. She'd set herself up for a fall, and it wasn't paranoia that made her sleep with a pistol under her pillow. She knew it was only a matter of time. She had plenty of enemies, from Habiyaremye to Sebahutu, the 'Godfather of Poachers', as Fossey referred to him. In 1985 Sebahutu was serving a five-year prison sentence in Ruhengeri jail, and Fossey had been the woman behind his conviction. Prisoners are not fed at all, but starve unless their relatives visit them daily, bringing food. Fossey heard that Sebahutu would die in jail because he had incurable tuberculosis. Sebahutu, who held political power in the style of a gangster would have hated Fossey a great deal. He may well have given the command that she had to die before him. Fossey told McGuire about Habiyaremye's desire to take over her camp, saying that she was stockpiling kerosene. She would burn the place to the ground before she let Habiyaremye have it.

Dian Fossey was murdered in the early hours between midnight and 5 a.m. on 27 December 1985. She was fifty-three years old.

Wayne McGuire was the last person to see her alive. On the evening of 26 December 1985 he was returning from the forest after spending the day with Group 5. He saw Dian and he says she seemed distant and tired. Wayne asked her if she was okay; she told him she was simply exhausted. With that they both went to their respective huts.

Christmas 1985 at Karisoke had been a little flat and depressing. Fossey, although melancholy, had decorated a Christmas tree and had wrapped presents for Wayne McGuire, only for him to feel embarrassed as he had nothing for her in return. Fossey wouldn't have minded; she was well known for her generosity with gifts.

That Christmas Fossey was an old lady, disintegrating. Her body and her mind couldn't take much more. It was time to die or leave and she wouldn't do the latter. She told Wayne, 'When I die there will be no one to save the gorillas and Karisoke will become a tourist attraction and cease to exist as a research centre.'

Wayne McGuire says he slept through the night and did not hear anything unusual. He was fast asleep in his bed when at 6 a.m. Kanyaragana and other Karisoke staff rushed into his hut shouting 'Dian kufa, kufa!' – *kufa* means 'dead' in Swahili.

The Africans led McGuire to Fossey's hut. 'I was shocked. The house had been ransacked, only the Christmas tree had been spared. In Dian's bedroom the table had been overturned and Dian's mattress had been pulled part way off the bed. Drawing closer I saw Dian's legs. She was lying on the floor with a pistol and a cartridge beside her. In her right fist, she held a clump of what the authorities later claimed was the hair of a white person. A brutal gash ran diagonally across her forehead, over the top of her nose, and down her cheek. Her eyes were wide open. I remember her expression – the ultimate look of horror, permanently frozen. A boy began screaming. Even though I knew she must be dead, I held her wrist and checked for a pulse. I found myself whispering "Oh Dian"; she was cold. (Philippe Bertrand, a French doctor who examined her body, told me that death hadn't come instantaneously, but that she had died in minutes.)

'I told the chief of the park guards to radio for the police. It took about twenty minutes to get through, because our equipment was so antiquated. I remember thinking that someone with a wound like that should have bled more profusely. Even so, there was a lot of blood on Dian's face and in her hair, and on the rug under her head. Others have said there was blood on the walls as well. Bertrand determined she died from a fractured skull and bleeding. Whoever killed her must have come in through the wall. Everyone knew Dian had pistols. Since she kept the kitchen door, the front door and her bedroom door locked, someone trying to get in through the door would have had to knock down two of them. Had someone ripped off the wall and got to her before she could grab her gun? Had her food been drugged? Or had she taken a sleeping pill with a beer? After all she hadn't been sleeping well. (Bertrand didn't check for poisons or drugs, so we'll never know.)'

Whoever murdered Dian Fossey was fortunate that she slept alone and soundly that night, because she must have remained unconscious through much of the time it took them to rip out part of the corrugated iron wall to her bedroom. By the time she awoke, they must have been halfway inside and there would have been no time for her to load her pistol. Perhaps an insider had taken the bullets out of the gun? Many of her trackers and anti-poaching guards were relatives of the poachers. Dian Fossey couldn't really trust anyone – other than her gorillas.

For a while Habiyaremye danced on Fossey's grave. He visited the scene of her death and tapped with his stick the metal wall where it is presumed the murderer made his entrance. Just before her death Universal Studios had bought the rights to her book, and after her death a screenwriter arrived to conduct his research. Habiyaremye told him that Universal could only film in the park if the film did not mention Fossey's death and that he, Habiyaremye, could play himself in the movie. His requests were ignored by Universal Studios.

Police searched her hut for clues, although they never took any fingerprints from the blood-stained machete that lay by Fossey's corpse. Ian Redmond flew out to sort though Fossey's things. Disturbed, he read a carbon copy of a letter Fossey had typed and sent to him, but which had never arrived. Redmond learned she had a list of names of poachers and corrupt officials she wanted to see prosecuted. Fossey kept this list with poacher Hatageka's *sumu* pouch. Redmond found the pouch but the incriminating list was gone.

Somewhat surprisingly Wayne McGuire chose to stay on at Karisoke for another six months after Fossey's murder, until he was forced to leave. He says he wanted to complete his research, even with a mad machete-man on the loose. McGuire says that over the previous months he and Dian Fossey had become good friends and that he had no reason to kill her, but the Rwandan government thought otherwise and, six months after her murder, he was sentenced to death by firing squad. There is no extradition treaty between the U.S. and Rwanda. Luckily for McGuire the Rwandan government 'allowed' him to sneak out of the country and to return home.

Eventually Rwelekana, Dian Fossey's most loyal tracker, was arrested for her murder and sent to Ruhengeri jail. A few months later he was 'found' hanged in his cell. Local police claim he committed 'suicide'. Had someone confided in Rwelekana? Perhaps he knew too much.

After Fossey's murder, Biruté Galdikas said, 'I knew Dian would be killed, I knew this was her destiny. I last saw her in New York City in 1983. She said she would be returning to Rwanda and told me not to expect her back.'

Although Fossey had said she wanted to be cremated and have her ashes scattered in the park, she had not written this in her will. Consequently it was decided by her family and colleagues that she should be buried next to Digit in the gorilla graveyard. Fossey had buried dead gorillas in a tiny graveyard next to her hut. Fossey was a lapsed Catholic, so a priest conducted the service. Bob Campbell was not present, but a friend Rosamond Carr who lived at the bottom of the mountain and admired Fossey greatly attended.[2] At seventy-four she struggled up the mountain to pay her last respects. But there were no other close friends present to mourn Dian Fossey's passing.

Wayne McGuire today helps primatologist/evolutionary psychologist Jane Lancaster on her long-term research on the social behaviour of a selection of individual men who live in Albuquerque. The project is known as the Albuquerque Men Study.

*

At Karisoke things are now run very differently. The DFGF pays a salary to British primatologist Liz Williamson, who in January 1996 became director of the Karisoke research centre, but she hasn't set foot in the park since May 1997 and neither have her African trackers, who are habituated to the gorillas. The whole region is a war zone and military authorisation is needed to enter the park. Between 1995 and 1997 the Tutsi ethnic cleansing resulted in the murder of 1 million Hutus. There are perpetual revenge attacks; every day people are being shot dead in the park and bodies are left to rot.

Between 1997 and 1999 twelve gorillas were killed mostly by cross fire between soldiers, and the social structure to the various study groups has been in turmoil. Although the gorillas are habituated to the trackers, the animals have become very nervous and stressed by the war that has encircled them. In 1997 a normally relaxed and habituated silverback held down one of the African trackers for half an hour, showing the man his massive canines, but thankfully the gorilla resisted harming the researcher. Liz Williamson was with the tracker when this happened. She took a photograph of the terrified man being intimidated by a massive but equally terrified gorilla.

Early in 1997 Liz Williamson had been based in camp but as the war escalated and the Hutus were laying landmines and running guns right past the camp it became too dangerous for her to remain. First Williamson moved to the town of Ruhengeri, but after three Spaniards were murdered it soon became obvious that she should be based in the capital city of Kigali, closer to the embassy, for security. She is still there. Williamson has suspended all gorilla conservation efforts in the park and remains on stand-by until it is safe enough for her and her African staff – those who haven't been arrested, that is – to return.

In 1997 five Italian aid workers were killed when their car was ambushed and robbed en route to the park.

The Rwandan genocidal civil war is an uncontrollable evil. Some 3,000 people have been murdered at the park entrance and 1 million refugees are reportedly living in the park. Because of the war and the continuing hostilities spreading throughout the Congo region of Africa the park guards and anti-poaching patrol teams have not been able to do their jobs. Consequently there has been a rush of illegal activity in the park, with people grazing cows there and setting animal traps. Several of Liz Williamson's skilled trackers, including Nemeye Alphonse, have been arrested and are still in prison waiting for their cases to come to trial. Long-term employee Nshogoza Fidèle was murdered by Interahamwe at his home. Williamson's remaining trackers face severe economic hardship as food prices have tripled. Many staff members have been robbed of all their personal belongings and some have abandoned their fields along with their homes.

Every few years the mountain gorilla research has been abruptly halted. The previous time the Karisoke research team was forced to leave the gorillas was back in 1993. Working in a politically unstable region results in gaps of time passing when the animals are not observed or protected.

Ian Redmond believes we should not habituate animals that we cannot protect. Knowing that people have to hunt animals to feed themselves and that more and more refugees are looking for food, Redmond would like to see the introduction of feral pigs at a number of primate sites where hunting continues. He hopes this extra meat source would help take the pressure off the primates and other wild species that are being hunted into extinction.

Because poachers were black-skinned Africans, Dian Fossey did not want the gorillas to become used to her African trackers. 'I have purposely not habituated gorillas to the Africans who so loyally work at camp and without whose help the study could never have succeeded. The reasons for this policy should be apparent: that split second that it takes a gorilla to try and identify and recognise the African who approaches him as observer or poacher is just the time needed to cost the gorilla his life from a bow or spear.'

I questioned Liz Williamson about Dian Fossey's policy. 'I very much doubt that the gorillas could mistake a poacher for a researcher. They recognise everyone who works with them and often taunt newcomers. I have seen the gorillas' reactions to strangers on the trail, or to military patrols – it is silent flight. Gorillas are smart. Dian's policy was understandable, but we could not operate like that now. Not only would it be unacceptable, but the Rwandan trackers are invaluable – they are the constant human presence at Karisoke, with up to thirty years' knowledge and experience. Through the writing of daily reports of their observations since 1993, the trackers have come to know and appreciate the gorillas more. This greater involvement has increased their commitment to the gorillas' protection. Last year, trackers were taking great personal risks in their attempts to monitor the gorillas. Expatriates had no access to the gorillas for two months before the trackers were forced to suspend their daily monitoring of the gorilla groups by a deteriorating security situation. Today all trackers and researchers must maintain a distance of 5 metres between themselves and the gorillas to minimise the risk of transmitting human diseases to the gorillas. Gone are the days of lying in a gorilla's lap!'

I asked Williamson about the gorillas' welfare. 'The mountain gorillas in the Virungas [Rwanda, DRC and Uganda] are, I believe, more threatened now than ever before. They risk being accidentally shot in crossfire or stepping on landmines in military actions or being deliberately targeted. They are exposed to human faecal and other waste, as well as decomposing corpses, increasing the risk that gorillas will be infected by human diseases. But probably the greatest danger comes from poachers, whose antelope snares can easily trap a gorilla at a time when monitoring and anti-poaching activities are suspended, rendering the gorillas particularly vulnerable. In 1997, 673 illegal snares were removed by Karisoke anti-poaching teams.' However, there has not been a reported incident of direct poaching of gorillas for body parts since 1982, five years after Digit's death. At least Fossey succeeded in stopping that in her lifetime.

Williamson continues: 'We had no detailed information about the gorillas for a year, other than news that groups are occasionally encountered by military patrols. These patrols do not enable us to determine whether the groups are intact or if individuals exhibit conditions which require veterinary attention. Many more snares have been set since all patrols were suspended in June 1997, thus the likelihood of gorillas becoming snare victims is high. As we are unable to monitor and intervene, accidental snaring is probably the biggest threat to the gorillas in the current situation.'

Liz Williamson has a long personal history of conservation work in Africa. In 1982 before her Ph.D. Williamson undertook a chimpanzee survey in Liberia's Sapo Forest with Janis Carter to investigate opportunities to release ex-laboratory chimps. She did her 1983–1988 Ph.D. on the behavioural ecology of Western Lowland gorillas in the Lopé Reserve, Gabon. Bill McGrew was her supervisor, giving her Dian Fossey's just published book *Gorillas in the Mist* as a present. During this time Liz Williamson and Primatologist Caroline Tutin became very good friends. Tutin worked with Bill McGrew at Mount Assirik when Stella Brewer was there. Together Williamson and Tutin have tried to habituate the lowland gorillas living in the Lopé Reserve in a bid to study the apes' social behaviour and life histories, instead of just their ecology. For some inexplicable reason the lowland gorillas have not been as amenable to humans as the mountain gorillas, yet in captivity it is the lowland gorilla who survives better than the mountain gorilla.

In 1991 Williamson moved to Zaire to perform a survey of the Eastern Lowland gorilla population. Over the years she has concentrated on gorillas more than other primates. I asked Williamson if they were her favourites. 'Absolutely, and I cannot explain this. Until I came to Karisoke, I rarely saw a gorilla. For instance, during gorilla surveys in Cameroon and Zaire, I was motivated by a passion for gorillas, which I *never* saw. I didn't really choose to become a primatologist. As a child I wanted to come to Africa and protect wildlife from poachers – George Adamson's book *Bwana Game*, David Attenborough on television sitting amongst the mountain gorillas and Fossey's book all encouraged me.' I asked Williamson if she had a favourite gorilla. She told me she hated to admit it, but yes, she does have a favourite – a very gentle and very beautiful blackback called Ugenda.

Williamson told me she has never seen her work in Africa as a career. She keeps thinking each project will be the last and she will return to 'normal' life in Britain. But at the end of each project another position makes itself available and leaving Africa has proved hard. When she was offered the position of director at Karisoke her dreams came true.

I asked Williamson if being a field primatologist made it hard to fall in love. 'All I will say is, yes, it is difficult to sustain a relationship if you are a field biologist or conservationist. I was in a ten-year relationship which ended when I decided to return to Africa again . . .' I asked Williamson how much longer she will go on with the work at Karisoke with war and destruction all around her. 'For as long as I am physically capable of climbing up the mountain, or until I make some political faux-pas. The current situation is far too depressing, yet I wouldn't still be here if I wasn't an optimist.' Williamson's trackers are finally back at work after receiving paramilitary training. Military staff of ORPTN, (the Office of Rwandan Parks Tourism National), now patrol the Virungas, making the area much safer for Williamson and other residents, eco-tourists and the gorillas.

At the time of Digit's death, when asked about mountain gorilla conservation, Dian Fossey said: 'We may have discovered and seen the extinction of a species of gorilla all within the same century.' But the species outlived Fossey and has seen the dawning of the 21st century. Yet for how many more years we will have these magnificent creatures, no one knows.

The last time Liz Williamson could get up to Karisoke she visited Dian Fossey's grave. The miniature graveyard had not been desecrated. Fossey's tombstone stands, her epitaph reading:

Dian Fossey
1932–1985
No one loved the gorillas more
Rest in peace, dear friend
Eternally protected
In this sacred ground
For you are home
Where you belong

Across the top of the headstone, in capitals, reads the Africans' nickname for Dian, a term she welcomed – NYIRMACHABELLI, meaning 'the woman who lives alone on the mountain'.

7

Aping for the Audience

Primatology has developed a particular relationship with popular culture. Field primatologists can find fame through their science because the nature of the work lends itself easily to a culture that wants to capture it and romanticise it. Natural history documentaries filmed on location and starring women and primates, especially the great apes, are very popular. These animals are highly photogenic, and it is easy for an ordinary person to correctly interpret much of the animals' behaviour. The films do not need a persistent commentary; the actions of the apes speak for themselves. We immediately see shadows of ourselves and we are delighted and perplexed by the obvious similarities. The women and their long-term commitment to the animals also provide entertainment. Field biology is full of risks and often lonely. No matter how beautiful the landscape it is evident that the scientist has not had a much deserved cooked dinner, long relaxing bath and comfortable night's sleep for many months. We feel inspired by the primatologist's personal sacrifice and delighted by the animal's behaviour; consequently both the women and the apes become stars.

But it is not just the world of the documentary that embraces primatology and images of apes. Painters have depicted them in their paintings; for centuries artists have pondered over the seductive bond and void between ourselves and our nearest relatives. The movies also relish the celebration of apes. One chimpanzee, named Jiggs, became a highly successful Hollywood actor playing the part of Cheeta in the Tarzan films. In recent years, the real thing has been replaced by specialist actors wearing animated ape costumes. Hollywood will probably never see the likes of Jiggs again; cinematically and politically it is no longer acceptable to make endangered animals work publicly in show business even if they are still employed in medical laboratories. But the most obvious synthesis between primatology and the creative arts has been the movie, *Gorillas in the Mist*, the film that tells the life story of Dian Fossey.

Looking at the documentaries first, a number of persuasive documentaries depicting the women scientists with their apes have thrilled the public. The first article on Goodall's work in the *National Geographic* magazine was published in 1963 and the first National Geographic TV special on Goodall, entitled *Miss Goodall and the Wild Chimpanzees*, narrated by Orson Welles, was aired in 1965. The documentaries prove that it's not just the animals that are photogenic; the women that study them are just as gorgeous and have been purposefully filmed and photographed in ways to exploit their sex appeal. Jane Goodall was beautiful, and her safari shorts revealed her shapely legs. In the 1960s the image of a lone, white woman in darkest Africa conjured up romantic notions of conquering and claiming natural treasures. Today we would say that latent racist messages of colonialism can be found in those National Geographic films too. Goodall became a reluctant pin-up, but she agreed to the filming because it would help raise funds for her project. As a result Jane Goodall and her chimps, such as the tool-maker David Greybeard, Flo and her son Flint, all became

stars. In the 1960s and 1970s she received sackloads of fan mail
from both men and women who had fallen in love with her
through her media exposure. She received proposals of marriage
from men she had never met. The films inspired generations of
other primatologists such as Barbara Smuts and Rebecca Ham.
Many women primatologists and conservationists currently working in the field have told me that the films of Jane Goodall out
there, alone and at one with nature, clinched their dreams when
they were little girls. Goodall herself says that when she was a child
the original Tarzan movies and the story of Dr Doolittle talking to
the animals motivated her passion for chimps and for Africa.

Women such as Fossey and Galdikas may have dreamed about
living at one with nature away from Western materialism, but from
time to time it followed them out there. By popular demand, many
documentary film crews have manfully struggled with their
modern technology through a primal landscape in order to capture
the women and their beasts on celluloid. Though many people
harbour the yearning to leave for the jungles of Africa, very few are
able to tear themselves away from family and friends. The next best
thing is to turn on your television set and watch a National
Geographic special about Jane Goodall and the chimps.

Most people are far too social to want to be alone with members
of another species. These women can be fascinating because in
choosing to live alone with apes they are rejecting us. The reality of
their primitive existence living in forests with these almost human
beasts becomes our vicarious thrill.

All three of Leakey's trimates either married or had love affairs
with the men who filmed and photographed them with the apes;
no doubt the romance enhanced the photography. Bob Campbell
has said he soon realised that having a love affair with Dian Fossey
was integral to gaining access to those magic images of woman
and ape. After she fell in love with Campbell her hardened exterior
melted and she was willing to be manipulated by him.

Donna Haraway, Professor of History of Consciousness at the University of California, is particularly interested in the media's depiction of primatology. 'National Geographic had a whole set of storytelling practices that were a mixed blessing. Look at the way the Jane Goodall stories were told, both in print and on TV in terms of first-contact narratives. They are almost like science fiction narratives, where the explorer visits the alien culture and enters into this narrative of achieving contact and perfect communication. You get the blonde stranger, the exotic animal, the heroic quest. The blonde woman raises her child like Flo the chimpanzee raises her child. The study of war and the great tragedy, the movement from girl guide to elder wise woman. You have an extraordinary density of storytelling conventions and that's just the Goodall stories; look at Fossey, look at Galdikas. It's full of historically specific, racially loaded, sex-gender-loaded stuff. I'm not saying it's a bad thing, but it's a very special cut. I always loved the National Geographic reports and read them avidly, but National Geographic's fascination with Jane Goodall, Biruté Galdikas and Dian Fossey had a huge impact on the general public and I believe on the science as well.'

The films certainly proved to have mass appeal. Goodall became a star overnight and from that moment on anonymity would be denied her. Thirty-five years later, in the United States, Jane Goodall is still mobbed by fans. Her famous golden ponytail is now a white one but, at times, she is still forced to wear a brunette wig and dark glasses to escape from persistent fans. Attracting the attention of others was not what originally motivated Goodall; for her and probably all other field primatologists who continue to do field work, attention from others is an embarrassing anathema.

Coincidentally, many women primatologists are blondes and many of them are as beautiful and glamorous as movie stars. The striking appearance of these women makes us wonder why they decide to live away from their own society where they would be considered good breeders and valued for their looks. Instead they choose to hide away in impenetrable forests. Many of these women have found this persistent interest in their appearance 'embarrassing, patronising and a pernicious distraction from their work'. When they are back in the West they give lectures on their study animals to raise revenue to support their research. Afterwards members of the audience approach them to ask questions, and at this time the women primatologists are often told how attractive they are and how impressive it is that a beautiful woman has chosen such a difficult career. Male primatologists are never objectified in this way. A male field worker would never be told that it is amazing that a man so handsome would choose to live so primitively and at the same time be so clever and knowledgeable. Good-looking women are seen as prizes that should be married off to rich men rather than being out there alone, tempting fate.

Initially reticent, Goodall quickly realised the necessity of publicising her work. The extra power it gave her in the world of conservation was at the time worth more to her than the loss of her personal autonomy. Jane Goodall became a dominant leading female through her study of the dominant female chimp Flo. Just like Flo, Goodall found herself surrounded by subordinate females wanting to achieve her status, bathe in her light and share in her glory.

Goodall's fans follow her progress with fascination and awe. Many make long journeys to listen to her lectures and have her sign their copies of her books. Jane Goodall today possess a regal quality about her. She is now in her mid-sixties, but the passage of time has not drained her spirit or energy; in many ways Goodall is as young and as determined as she was when she first went to Africa. Her character is a mixture of stillness and strength. She has absolutely no self-doubt when it comes to chimpanzees and she freely argues her point of view on chimpanzee behaviour, but she exudes tranquillity, quiet confidence and inner calm. When she walks she seems to float and her voice is gentle but strong. Sitting and watching ever so quietly and intensively for so many years has shaped Jane's personality. Had Goodall remained a secretary from Bournemouth no doubt today she would be quite different.

Like thousands of others, Dian Fossey adored the National Geographic films and the photographs of Goodall, which seduced Fossey into creating a similar role for herself. In turn, after Fossey's work had been broadcast around the world, Galdikas saw a role for herself emerging. Two years later this was a tangible reality. The early photographs that Rod Brindamour took of beautiful and sexy Biruté Galdikas embraced by orphaned orang-utans have a sublime purity. The young Galdikas and the baby orang-utans shared the same large, brown, watery eyes and a still, inscrutable mouth. What has angered many women scientists since is that the fascination with those three women has continued and the equally extraordinary stories of other women in the same field have been somewhat overlooked. Could some of this resentment stem from jealousy? I have come across a dismissive attitude towards the trimates' celebrity among many men and women primatologists. I have, for example, been told the women are not proper scientists but more like glorified PR agents for the conservation lobby.

The images of Digit and Fossey, sometimes fast asleep together in the long wet grass, were the intimate photos that National Geographic hoped for. Close-up pictures of Goodall, Fossey and Galdikas with their respective animals were fed to a receptive public, enthralled by the women's close proximity to the animals. The scientific establishment, on the other hand, had quite a different response to these images, criticising this type of physical contact.

Touching, grooming and cuddling supposedly wild animals that are part of a scientific study is bad science. How can you be objective when your study group is more like your extended family? When we look at these pictures of Fossey with her arms around the gorillas, we do not think about the persuasion required to achieve such closeness to the animals. Although she felt exhilarated by the contact, Fossey was simply trying to please her lover and her funding body. Intimate photographs like this have since gone out of fashion, because the general public understands much more of the animals' vulnerability to disease through contact with humans.

Jane Goodall received similar criticism over feeding bananas to her chimps. Van Lawick and Goodall knew that National Geographic wanted vivid images of woman and nature; close-ups of the apes were a must. Goodall was terrified that the money would dry up if National Geographic did not get the desired pictures. At this time, before the friendly tool-using David Greybeard had made himself at home in camp, the best way to photograph the nervous chimps was to feed them bananas. The chimps would come and snatch at bunches and run away with them, as would baboons, and Van Lawick would start rolling. On one occasion, when Richard Leakey visited Gombe to check up on things for his father, he saw the provisioning of bananas, which he regarded as bad science and very damaging to Goodall's study. He returned to Nairobi and informed Louis of what he had seen.

Louis Leakey was furious with Goodall. Feeding the animals caused tensions between individual animals and between the baboons and the chimps that ordinarily would not occur. As a result Louis felt he had to dismiss many of Jane's observations; her reports of the mother–infant relationship and the male hierarchy seemed contorted to him. It was some years later, not long before his death, that he conceded that the banana-feeding probably made little difference to the chimps' behaviour. Today, after receiving much criticism from the scientific community, Goodall says, 'I regret the banana-feeding very much' but she's philosophical: 'I was afraid the money would run out, due to the funding agreement, that was the way it was.'

In 1996 I met Biruté Galdikas in the rainforest of Borneo where

I was filming an interview with her for British television. At the same time, Galdikas's charity, the Orang-Utan Foundation, was also filming Galdikas, and coincidentally a large Japanese film crew was there too. Biruté Galdikas's every move in that wild place was observed by three film crews, all of us loaded up with state-of-the-art video equipment. The three documentary teams followed her and recorded her spiritual utterances as she trudged through rain-forest mud; it became farcical.

The Japanese crew had most members so we took up the rear and over their shoulders filmed whatever their director asked Galdikas to do. He was very keen to have Galdikas sharing a cup of tea with two female orang-utans and their babies, a scene reminiscent of the chimpanzee tea parties popular at London Zoo some twenty years ago. But that is just the point – in the West today we do not want to see great apes behaving as surrogate humans. It seems undignified. But Japan has its tradition of endowing apes and monkeys with human-like status, so there is still an appetite for this sort of imagery and Galdikas was keen to give them what they wanted.

Biruté Galdikas wants eco-tourists to come and see her and she wants film crews to make a record of her work because, like a preacher, she wants to facilitate 'conversion experiences'. She believes that when people meet an orang-utan face to face or even when just watching films of orang-utans, a conversion experience can take place. People exposed to the gentle soul of a female orang-utan with her baby will automatically take the animals' threatened extinction to heart. Galdikas is always looking to expand the orang-utan's fan club, so she wanted to project images that a Japanese audience would want to see, such as those almost human creatures drinking cups of tea. Japan imports more rainforest wood than any other nation and in South-East Asia every tree logged is one less to be inhabited by an orang-utan. Galdikas fights a war against the very rich timber companies who are destroying the Bornean paradise. She is outnumbered, but her voice is loud and irritating to the wood merchants; she has had a number of death threats, but refuses to go quietly.

So we filmed Biruté and two female orang-utans, Princess and Siswi, with their babies drinking very sweet tea. No doubt a spoonful of sugar helped the film go down very well in Japan, but in the United Kingdom the image of Biruté Galdikas indulging half-tame orangs that had a passion for cups of white sugar, with a little tea to taste, caused acute indigestion for more censorious viewers. Galdikas takes the criticism of being too emotionally involved with the animals on the chin, willingly sacrificing her scientific reputation if it means she becomes one step closer to saving the rainforest. Direct action is Biruté Galdikas's style and conservation of her apes will always come before anything else.

The magic of the National Geographic documentaries of the 1960s and '70s has not dated; the behaviour of the apes and the quiet meditation of the women scientists still delights. But it has been the fictionalised account of Dian Fossey's life in the 1988 Hollywood movie *Gorillas in the Mist* that has presented the general public with their most enduring exposure to the world of primatology. When asked about primate studies most people cite this movie as having enlightened them about the subject.

Sigourney Weaver had read about Dian Fossey in the *National Geographic* magazine while at university. When the screenplay for *Gorillas in the Mist* came along she made it known she wanted to play Dian Fossey. She spent four months filming on location in Rwanda with real gorillas and chimps in gorilla costumes. At the time Weaver said that 'playing that part has changed me enormously'.

On her return to the United States Weaver became a patron for the Dian Fossey Gorilla Fund (DFGF), other patrons including Michael Crichton, author of *Congo*, which was later adapted for the screen in the movie *Congo* released in 1995. Its spectacularly stupid plot concerns a primatologist who takes his female gorilla back to Africa in the hope that she will teach sign language to wild, killer gorillas, enabling them to 'speak' English and reveal the location of King Solomon's diamonds. For $15,000 the fund allowed the producers to use on screen their radar image of the Parc National des Volcans taken by the space shuttle *Endeavour* in April 1994; the satellite pictures flash up for only a moment. The Los Angeles and London premières for *Congo* were hosted by Paramount, with the $130,000 profits being donated to the DFGF.

Science fiction writer Arthur C. Clarke is also a patron of the DFGF. He wrote the screenplay for the film *2001: A Space Odyssey* (1968) directed by Stanley Kubrick. Its opening sequence, called 'The Dawn of Man', depicts a group of pre-hominids browsing for food and quarrelling with a rival tribe over access to a waterhole. A mysterious monolithic slab appears in their territory. Violence erupts and one large male beats another to death with a bone – the first murder is committed. The murderous ape throws his weapon into the air; Kubrick cuts to a spaceship in flight and we are instantaneously transported forwards in human evolution by 5 million years. Kubrick consulted Richard Leakey on the staging of the apemen for this sequence. On set, both Louis and Richard met Clarke and after Louis told him about Dian Fossey's work Clarke became a lifelong fan of Fossey.

After completing *Gorillas in the Mist* Sigourney Weaver commented on the filming, 'I remember thinking, Dian wanted this movie made, I'm not going to fuck it up, she will protect me. And I remember lying back with four babies jumping and urinating on me and pulling my hair and it felt like heaven. It's a big reason I wanted to have a child myself. There were certainly times in my baby's life when I thought, okay, this is a gorilla . . .'

A woman sacrificing herself for the thing she loves most is a familiar theme in western culture. Madame Butterfly, Camille and Sappho all choose the path of female martyrdom – but dying for an animal is a different matter. Unlike the original *King Kong*, where Kong, the huge and hirsute leading man, dies for his object of love, the screaming Fay Wray, *Gorillas in the Mist* tapped into something new by reversing the self-destructive role of King Kong with that of the damsel. In these days of conservation awareness, the beast's value had appreciated to become greater than hers. (There are many more women in the world than mountain gorillas.)

Gorillas in the Mist pays homage to *King Kong* in a number of other ways. Both screenplays, for example, have the film within the film device. In a sort of *ménage à trois* the leading ladies are icono-graphed with the beast, with the hero voyeuristically filming them together. Both films also represent the personification of the beauty and the beast mythology. Fay Wray commented that being in Kong's grasp was 'a kind of pleasurable ordeal'. Before she began filming, the RKO executives were secretive about their new movie: 'The only thing they would tell me was that I was going to have the tallest leading man in Hollywood. I thought of Clark Gable.' Fay Wray was an experienced actor, directed by Eric von Stroheim in *The Wedding March* and also writing short stories and plays. But we shall always remember her first for screaming and fainting at the sight of Kong. Jessica Lange made a name for herself appearing with the beast in the forgettable cheesy 1976 remake, but cleverly managed to move on to other things.

Tales of women falling for beasts are ancient. In 1757 Madame Leprince de Beaumont was the first woman attracted enough by one such myth to write down the Beauty and the Beast story. In Jean Cocteau's beautiful film, *Le Belle et la Bête* (1944), Belle gets her man when the beast's spell is broken by her love and he reverts to his human form. In *King Kong*, although Fay Wray's fear does turn to love, it cannot save the beast, because hairy black African male primates shouldn't try to love white women. In *Gorillas in the Mist* the depiction of Dian Fossey's tragedy also reveals that when her fear turns to love it is not enough to save either of them.

If *Gorillas in the Mist* is analysed, the plot is revealed to be a fetishistic appraisal of woman at one with nature. The spectacle of a white woman falling for a black African is not a new theme. Colonialism has left the West with a legacy of right and wrong that goes to the heart of primatological iconography. White women falling in love with black men seems erotic in a way that white on white or black on black is not, and the frisson of women loving apes is a logical extension of that racism. National Geographic films showing a lone woman walking through the forest looking for apes reinforced a particularly sexy image. Donna Haraway described their racial impact: 'We have had racially inflected, nationally inflected series of accounts of white First World people studying Third World animals in black Third World countries in post-colonial Africa. There is a colonial history to the *National Geographic* magazine and its whole cultural apparatus that has influenced the primate stories a lot. In the early material on Goodall it is especially visible. You see the blonde stranger washing her blonde hair in the pure stream; there are an amazing number of racial codes constructed into those early reports.'

Primatology is perceived to be very much a white woman's profession. There are plenty of Asian and South American primatologists but the subject does not appear to attract black people of either sex. Yet at the long-term study sites in Africa, such as Gombe, many local black African men are employed as research assistants. They are trained in methods of data-collecting and contribute an immeasurable amount of factual observation on chimp behaviour. Jeanne Altmann's site, Amboseli in Kenya, would not have produced so much observational data on the baboons without her African staff. These men work full-time and usually live locally, running the site year round while Western scientists are coming and going. Some of these men are ex-poachers; their experience and skill in animal tracking help them to find and observe and protect the animals, as well as catch other poachers. Their vital contributions, however, are not widely acknowledged by the media. They are lucky to have cameo roles in films.

With the release of the movie *Gorillas in the Mist* in 1988, three years after her death, Dian Fossey's life became fully mythologised. There is no question that Dian Fossey loved the young adult male gorilla she called Digit and her passionate hatred for the poachers that hacked him to death probably incited her own similar fate. But in the movie Fossey's character was softened by the scriptwriter, making her less complicated and more obviously sympathetic than she really was. Fossey's love affair with Campbell only lasted for two of her eighteen years at Karisoke but the film allows the romance to dominate the story. The screenwriter would argue that 100 minutes only allows for limited character development, while the producers would say they were making a commercial movie, not a documentary, and the truth should never get in the way of a good story. But what the public perceives as truth becomes so. *Gorillas in the Mist* was a big hit, with the titillating subtexts of Tarzan and Jane and King Kong seen for 'real'.

In the closing sequence of *Gorillas in the Mist* Dian Fossey's grave is shown next to her beloved Digit's grave. The actor playing Fossey's faithful tracker gently places stones around the two graves, uniting their two souls for ever. This is a local African custom; if husbands and wives are buried next to each other, relatives encircle their two graves with a line of stones. But in *Gorillas in the Mist* director Michael Apted uses this marital bond in death to unite the souls of a woman and a gorilla. In real life Fossey and Digit's graves are not united by a ring of stones, although they are next to each other in a very beautiful and normally quiet spot on the hillside.

To coincide with the release of the film, the original cover of Dian Fossey's science book *Gorillas in the Mist*, which had shown Fossey with a gorilla, was replaced with a still from the movie depicting Sigourney Weaver and a chimpanzee in gorilla costume. When these multiple replications on the front of a scientific book are considered, it reveals how our understanding of primatology has been blurred by popular culture.

Early on in their careers both Dian Fossey and Jane Goodall became past masters at making their own story palatable to a Western audience. In *In the Shadow of Man*, the story of her life at Gombe, Goodall tells how she coped in the rainy season. 'I think I spent some of the coldest hours of my life in those mountains, sitting in clammy clothes in an icy wind watching chimpanzees. There was even a time when I dreaded the early morning climb to the Peak: I left my warm bed in the darkness, had my slice of bread and cup of coffee by the cosy glow of a hurricane lamp, and then had to steel myself for the plunge into that icy, water-drenched grass. After a while, though, I took to bundling my clothes into a polythene bag and carrying them: there was no one to see my ascent and it was dark anyway. Then, when I knew there were dry clothes to put on when I reached my destination, the shock of the cold grass against my naked skin was a sensual pleasure. For the first few days my body was criss-crossed by scratches from the tooth-edged grass, but after that my skin hardened.' On the surface this account reveals Goodall's oneness with nature, but this anecdote is also touched with erotic and masochistic overtones.

Before the trimates went to work, the popular image of field research was male-dominated. Male scientists studied aggressive male-dominated animals, and family life and the behaviour of infants were not the issues. Cultural images of women field researchers now have a powerful effect. In *The Lost World*, Steven Spielberg's sequel to *Jurassic Park*, it seems utterly right and proper that the zoologist studying the behaviour of a stegosaurus family at dangerously close quarters is a woman. We want her character to risk life and limb to touch the baby – and she does. We, the public, have come to expect primatologists to be women, and we expect them to die for their animals. In the 1998 Disney movie *Mighty Joe Young*, directed by Ron Underwood, a remake of the RKO 1949 movie of the same name, a woman primatologist dies trying to save an infant gorilla of King Kong potential from poachers. With her dying breath she makes her young daughter, Jill, promise to protect Joe from these male killers. Joe, played by John Alexander in animated costume, grows up spectacularly; because of a mutant gene he reaches 15 feet in height and weighs over 2,000 pounds. He is a gentle and intelligent giant, who will fight to keep poachers off his African mountain. Jill, played by Charlize Theron, also grows up spectacularly, developing into a 6-foot-tall blonde beauty. One way or another she manages to keep her promise to her mother to protect the beast from harm. In one scene Joe, now in California, enters a fairground hall of mirrors and takes pleasure in his distorted reflection. Through such representations, today's children are absorbing a cultural diet of women and apes mutually adoring one another.

So many women have studied the intelligence of apes that it is no coincidence that in a neat role reversal the only female ape character in the 1967 movie *Planet of the Apes* is a primatologist who studies the intelligence of dumb humans. Charlton Heston is the astronaut who returns to Earth at some point in the future to find that apes rule the planet and humans have lost their ability to speak. Chimps are sinister intellectuals, gorillas are aggressive law enforcers and orang-utans are conspiratorial elders. Ziva, played by actress Kim Hunter, is an anthropologist/psychologist chimp who studies the behaviour of the degenerate humans. How intelligent are they? Can they be taught to speak? She becomes fascinated by Heston's astronaut, who can read and write but not talk because of an injury. At this time an American primatologist, Beatrice Gardner, was undertaking a much publicised project to teach a chimp named Washoe to talk. Science was inspiring the media with its psychological experiments on ape intelligence. Now the media inspires young women to be scientists by casting actresses as primatologists. But let's not forget the apes themselves.

*

Movies with ape characters will never go out of vogue, but today few real-life apes are cast to play their fictional counterparts. Nowadays actors are paid handsomely if they can convincingly ape it up in front of a camera; Peter Elliot and Ailsa Berk are two actors who have made successful careers for themselves from portraying apes. Peter Elliot appeared in Jean-Jacques Annaud's 1981 *Quest for Fire*. Set 80,000 years ago, it represents pseudo-communities of *Homo erectus*, *Homo sapiens*, and *Homo sapiens neanderthalensis*, who fight each other for the secret of fire. At times the film seems closer in tone to an elongated 'Monty Python' sketch. The *Homo sapiens* heroine teaches the neanderthalers' hero not only how to rub two sticks together to make fire but also how to rub two bodies together to make missionary sex. In fact, we know chimpanzees and orang-utans to a lesser and bonobos to a greater degree have had sex in the missionary position for millions of years. The ape-human gestures in the film were created by Desmond Morris and the primitive language by Anthony Burgess.

Peter Elliot and Ailsa Berk first aped about together in the 1984 film *Greystoke: The Legend of Tarzan Lord of the Apes*, directed by Hugh Hudson, who tried to keep the picture faithful to the original 1913 novel by Edgar Rice Burroughs. Berk played Kala, Tarzan's ape mother and Elliot played Silver Beard, Tarzan's ape father. Both performances are utterly convincing.

Berk and Elliot received a tremendous amount of guidance on the emotional lives of chimpanzees from Roger Fouts, the behaviourist now based at Washington University. Berk made a point of watching all of Jane Goodall's National Geographic Films, and Flo the chimpanzee became her role model. In *Greystoke* the adult chimps were all portrayed by actors wearing animatronic body costumes. The ape masks and costumes for *Greystoke* were made under strict security. After the movie was finished they were kept in storage for some time and finally destroyed to prevent other productions appropriating the design. The shape of the head of a contemporary ape or an early hominid is different from that of modern man. Our jaws are much smaller and our cranium is larger, so the masks are quite alien in shape for the actors to wear. The ape actors could see through the eyeholes and wore scale lenses in their eyes to give the impression of an ape's iris, but in most of the ape and ape-men costumes that Berk and Elliot wear they cannot see through the mask, except, perhaps, through the nostrils. Usually the creature's eyes, nose and mouth are radio-controlled by animators.

Adult chimpanzees are never used in dramas or commercials because they are too independently minded and too strong to control, but infant chimps are happy to please their trainers and their bite isn't as nasty. For example, the chimps used in the British PG Tips tea commercials – grey wigs and dungarees apart – are all infants and juveniles. The chimps in the early PG Tips commercials all came from Molly Badham and Natalie Evans's collection. Both childless, Badham and Evans have lived together for forty years. Evans bought Badham a woolly monkey as a pet in the late 1950s, but it died. After this both women started visiting zoos, pet shops and animal dealers, acquiring more and more apes and monkeys. Soon their interest became known in primate circles and sick animals were brought to them to be nursed back to health. In 1963 the women opened Twycross Zoo. Self-taught, they have managed to breed chimps, gorillas and gibbons and they also have six bonobos, the only ones in Britain. At Twycross most of the primate keepers are women and the zoo has a camera-friendly policy. A good deal of filming still goes on there, though today 'a refreshing cup of PG' is drunk by chimps from a primate collection in Germany.

Playing Kala in *Greystoke* meant that Ailsa Berk had to bond with a real baby chimp named Lucy, who was going to play her baby in the film. Lucy is owned by the Chipperfield circus family. Before filming started, Berk worked with Lucy at the Chipperfield's farm. They trained every day for three months, hoping to encourage Lucy to react to a costumed Berk as though she was her real chimpanzee mother. Ailsa Berk bonded with the young chimp and after filming was over continued to visit the animal for another three years. Breaking the bond was difficult.[1]

Peter Elliot went on to play Digit in *Gorillas in the Mist* and Ailsa Berk played Max in the 1986 film *Max Mon Amour*, directed by Nagisa Oshima. In the movie Charlotte Rampling falls in love with a chimp, Max, while her husband, played by Antony Higgins, tries hard to deal with his jealousy as his wife shares a bed with an ape. When asked if it was different acting opposite an animal rather than a man, Charlotte Rampling said, 'No, no, the emotions are the same. In a way it was like playing opposite Paul Newman. The chimp reacted differently, that's all.'

In dramatised sequences from a documentary entitled *In Search of Our Ancestors* Ailsa Berk portrays the earliest known hominid, the 3-million-year-old *Australopithecus afarensis*, Lucy. Don Johanson, who discovered the 40 per cent complete fossil in Hadar in Ethiopia, proudly presents the film in which Berk, clad in a nylon body suit with yak hair and foam head attached, brings Lucy back to life.

Peter Elliot has more recently played the gorilla in *Buddy*, the film that tells the life of Gertrude Cunningham, as well as another gorilla in *Instinct* (1999), where Anthony Hopkins plays a prima-tologist who loves his mountain gorillas so much he 'goes feral' and lives with them in communal harmony until the poachers arrive.

Orang-utans are even more popular than gorillas in Hollywood, but orang-utans tend to play themselves rather than having actors impersonate them. Steven Spielberg paid homage to an orang-utan in his movie *E.T.* When the children first encounter the alien they hide him from their mother and try to figure out what he is: 'Maybe he's some animal that's not supposed to live . . . could be a monkey or an orang-utan.'

Clint Eastwood has found himself playing opposite orang-utans in *Every Which Way But Loose* (1978) and its sequel *Any Which Way You Can* (1980). A real-life female orang-utan impersonated a male named Clyde, who had a taste for beer, helped Clint in fights and car chases and nipped into the local zoo for a quick bit of sex. Naturally, the orang-utan stole the picture from the star. But, digging deeper, how did orang-utans make their way to North American from South-East Asia when they are supposed to be a protected species and the sale and import of them is illegal? Endangered animals are trafficked as easily as illegal drugs are.

In *Every Which Way But Loose* the orang-utan actors were owned by a private American animal trainer who used his apes in a cabaret show when they weren't on hire for the movies. Adult male orang-utans cannot be used in the entertainment industry because they cannot easily be controlled owing to their enormous strength and size. Adult males also have large leathery cheek pads, which body-obsessed Hollywood might find unsightly. The juvenile or fully grown female orang-utans are much prettier and far more compliant, but they are also very strong and can be wilful. Most animal trainers use fear to keep their orang-utans in line.

According to the Orangutan Foundation, during the filming of *Every Which Way But Loose* the orang was relentlessly helping herself to food at the catering van. The keeper was asked to keep her under control. Later the orang stole six cakes, for which her trainer took her into their trailer and beat her around the head for misbehaving. The orang-utan slipped into unconsciousness and died of a brain haemorrhage. The studio panicked. This was bad publicity. In a cover-up an orang-utan understudy was brought in to replace its unfortunate predecessor.

Before primatology gave us entertaining documentaries and affiliated feature films as a context in which to view apes and monkeys, we had the animals themselves. Hunters brought us stuffed bodies for museums and live ones for zoos. In the early twentieth century, the mere existence of apes was enough to excite people.

During the economic depression of the 1930s one chimpanzee in particular became a household name. During his lifetime science's, the media's and the public's fascination with his species has flourished. A great deal of that interest in chimpanzee behaviour and intelligence was encouraged by this chimp's very own public profile. He was an actor, and his celebrity reached out to individual scientists, conservationists, artists, the military and thousands of ordinary people.

Apes, such as the chimp David Greybeard, the orang-utan Sugito and the gorilla Digit were wild animals that have become almost as famous as the women who befriended them, but the most famous ape in the world is probably Jiggs. We know him by his character's name, Cheeta, Tarzan's simian sidekick in the MGM movies of the 1930s and 1940s. Jane Goodall says one of her adolescent inspirations was the famed chimpanzee, who made a massive impact on the general public. For millions of people in the West it was the first time they had seen a chimpanzee, and they loved it.

What was special about these movies to the young Goodall was that the chimp played a surrogate child to the romantic leads. Jiggs was treated like a relative with equal status to humans. The Tarzan family was depicted as the ideal nuclear family, with Jane and Tarzan as Mum and Dad and Boy and Cheeta as the kids. Hollywood's Tarzan and Jane films had a biblical resonance, evocative of Adam and Eve in the Garden of Eden.

In 1934 two-year-old Jiggs acted alongside Johnny Weissmuller and Maureen O'Sullivan in *Tarzan and his Mate*. This movie was the beginning of a successful relationship between the thespians. Jiggs accepted his stage name and persona and went on to make another eighteen movies. The life of the Tarzan family, swinging vine by vine through the jungle, eating a constant diet of fresh fruit, was an idyllic fantasy world.

Born in Liberia, Jiggs was 'orphaned' at a young age when his mother was shot dead by Tony Gentry, who had been sent to Africa by MGM and RKO to bring back chimpanzees from the Third World to the First. MGM was the biggest studio in those days and Jiggs would become one of its greatest assets. Hidden under Gentry's jacket as a stowaway, he was illegally brought into the States.

Gentry had been taught how to trap and train animals by the famous trapper Frank Buck, and his talent at training apes would become legendary. Colonels in the US Air Force and Army loved watching trained chimps in the movies and were particularly impressed with Cheeta's acting abilities. The US military eventually employed Gentry to train the chimps Ham and Enos in preparation for space flight. The first chimpnaut, Ham, dressed in a pressure suit, was launched into his sub-orbital space journey in 1961. No doubt on his mission Ham was able to look down at planet Earth's Great Rift Valley and the green forests of Africa. He is now buried in the International Hall of Fame, New Mexico.

Jiggs's own story fulfils the American dream. From humble beginnings as an illegal alien, Jiggs was groomed by Gentry for stardom. The ape rose to the top, as a child star rubbing shoulders with the rich and the powerful of American society. Johnny Weissmuller said of Jiggs, 'I'd sooner work with apes than with some stars I know.' Jiggs and Johnny loved to wrestle together and when the actor died in 1984 Jiggs joined the funeral procession.

Jiggs's life has been similar to that of a circus animal. He has been taught to do many tricks, such as somersaulting, impersonating human behaviour and even writing his autograph. Because he was taken from his mother at such a young age and from then on looked after by Gentry, he could only relate to men. As Jiggs grew older he became more and more hostile to women, including co-star Maureen O'Sullivan, who loathed working with the chimp. Scripts had to be rewritten so that the ape and the actress had as few scenes together as possible because there would have to be countless retakes as the actress was bitten or covered in chimp spittle. She said of Jiggs: 'He was really rather queer, I'm afraid. He didn't like girls at all. I was never more consistently sick and miserable in all my life. I was never without an ache or a pain. I was never without a bite from one of those monkeys. I always had the same average – one fresh bite, one about half healed and one scar.' Jiggs had learned to bond with men; Gentry had become his 'mother' figure. He was jealous of Maureen O'Sullivan and any other woman he had to share his favourite men with.

Maureen O'Sullivan was not the only thespian to fall foul of the chimp. Jerry Lewis decided to ape about in front of Jiggs and for a joke pretended to eat one of his bananas. The irate animal hit him, almost knocking him out. *Bedtime for Bonzo* (1951), directed by Frederick Cordova and starring Ronald Reagan and Diana Lynne, is another classic on the chimp's CV. In the movie Reagan plays a primatologist who undertakes a behavioural study with a chimp by raising him like a child. The film was so popular that the studio produced a sequel, *Bonzo Goes to College* (1952). Naturally Jiggs starred once again and Maureen O'Sullivan was persuaded to team up once more with her simian colleague. Sadly for the ambitious Ronald Reagan, his contract was not renewed.

Jiggs's last movie was *Dr Doolittle* (1967) with Rex Harrison. The ape was getting too big and too long in the canine tooth to act; mature chimps are too concerned with status to play somebody's fool.

Gentry died in 1991 and his nephew, Dan Westfall, became Jiggs's full-time carer. Jiggs now lives in semi-retirement in Palm Springs. He is in his sixties but amazingly he is still swinging – just. His trademark back flip is more of a sideways flip nowadays, but he's fairly active and still loves to be driven down the highway to get a Big Mac. He makes the odd guest appearance, and is still invited to movie functions and receives fan mail and requests for autographs. Mostly, though, after a porridge breakfast, he relaxes by his pool, drinking his fruit cocktail through a straw and exchanging a few hoots and lip-smacks with his neighbours Kirk Douglas and Bob Hope. When he leans over the garden fence to look at Jiggs, Douglas must be reminded of his own defiant line: 'I'm not an animal!' from *Spartacus* (1960).

Jiggs's home is an air-conditioned apartment at the back of Westfall's house. He has his own refrigerator and television, and especially loves to put on a video of his old movies. He leaps about and hits the TV with excitement when he sees himself. Now it has become more effective to use state-of-the-art animated prosthetic costumes worn by actors. A shift in taste, both cultural and scientific, means that Hollywood will probably never see the likes of a chimp such as Jiggs again.

Jiggs had shared all his acting roles with his long-time chimp flatmate, Susie, also originally trained by Gentry. In their twilight years, Jiggs and Susie turned their hands to 'apestract' painting.

A lover of the kitsch and the profane, British Pop artist Peter Blake has long held the celluloid image of Cheeta and the real-life Jiggs in the highest esteem. Peter Blake was first introduced to the concept of apes painting in 1957, when he attended a show at London's Institute of Contemporary Art entitled Paintings By Chimps. This exhibition, devised by Desmond Morris, was the first to relate our primal origins to art and the aesthetic. When he heard that Jiggs was dipping his sable brushes in acrylics and coming up with something close to an abstract expressionist painting, Blake knew that an exhibition of his and the ape's 'apestract' art was essential. For Blake, the exhibition became an emotional synthesis of much of his life's work. Blake was thrilled to be professionally rubbing shoulders with the ape. The iconography of the Tarzan movies has inspired and entertained Blake for years. His primal attraction to Cheeta is well known; pop star Ian Drury has even written a song about the two of them, entitled 'Peter the Painter and Cheeta the Monkey'. When Hugh Hudson was directing *Greystoke*, he invited Blake on set, where he met Peter Elliot and the other ape-actors. Blake says he was most impressed with all the actors portraying chimps because they stayed 'in character' even when off the set and eating a banana lunch.

Peter Blake has one large painting that after thirty-three years he still cannot bear to finish and part with – *Tarzan and His Family at the Roxy Cinema, New York*. Originally Blake painted the Tarzan family as an ordinary American family, where Dad is a bodybuilder desperate to emulate his hero Johnny Weissmuller, but Peter has since reworked it. Tarzan is now portrayed as Michelangelo's David, Jane as a truly smiling Mona Lisa, Boy as Titian's Ranuccio Farnese.

Blake failed to locate a historical picture of a chimp as a reference, as chimps were only successfully imported from Africa to the West after the 1920s. Before then they had been rare in Europe as they were hard to keep alive. Instead Bake decided to contact his chimpanzee hero – or rather the aging simian's manager – for a photograph of Jiggs. Like fellow artists full of mutual appreciation, they swapped paintings: Blake sent Jiggs a screenprint of his *Madonna on Venice Beach*, which Peter inscribed with the Latin *Ars simia naturae*, and Jiggs sent Peter three of his 'apestract' paintings, signed with the ape's thumb print.

Ars simia naturae is a Latin phrase commonly used by artists from medieval times up until the eighteenth century. It means: art is the ape of nature. An image of a monkey was displayed outside artists' studios for centuries. Primates and paintings have a joint place in history, and the human instinct to focus on the relationship between man and ape precedes film-making. Seventeenth-century artist David Teniers the Younger produced a number of paintings along the theme of the replication of derivative imagery. Eighteenth-century painters Alexandre Decamps, Antoine Watteau and Jean-Baptiste-Simeon Chardin all painted monkeys as pastiche artists, and Picasso also sketched a monkey painting a nude woman, *Dans l'atelier* (1954). In Decamps's *The Monkey Painter*, held in the Prado in Madrid, a monkey artist works in his studio as a well-heeled monkey patron looks over his shoulder; his canvases of portraits of men and landscapes fill the space.

One of Jiggs's abstract paintings was predominantly green; Blake felt it represent Jiggs's origins in the Liberian jungles. Another was yellow and red, which he thought was a post-post-modern comment on his friend Roy Lichtenstein's comic book paintings on abstract expressionist brush strokes. Genuinely moved by the chimp's sensibility, Blake saw great pathos in the paintings and was impressed with Jiggs's feel for paint and composition. Blake included these paintings together with *Tarzan and His Family at the Roxy Cinema, New York* in his 1996 Associate Artist Exhibition at the National Gallery in London.

Jiggs is now the oldest chimp in captivity. Chimpanzees Susie and Jiggs lived together for most of their long lives, but Susie died in 1997 aged sixty-five. This is a great age for a captive chimp; in the wild fifty years is considered a long lifespan. Dan Westfall still cries when he speaks of Susie's last days; he loved her like a wife. Helping his uncle for years and finally taking over the day-to-day care of the apes and performing in cabaret with them has been Dan's life's work. He has never married and still lives with his mother.

Jiggs is not alone, though. His robust ten-year-old grandson Jeeta lives with him, and baboons and rhesus and spider monkeys are housed close by. Westfall has a variety show he performs with Jeeta and the monkeys.

Westfall was a great fan of Jane Goodall until she wrote about Susie in *Visions of Caliban*. Goodall has witnessed much suffering of captive chimps. Part of the Jane Goodall Institute's remit is to end this type of systematic abuse. In her book, Goodall shows a photograph of Susie posing in a tutu and ballet shoes. Westfall was devastated to see that Susie had been used in Goodall's book to make this particular point and maintains that Susie always loved to dress up in clothes and shoes and was happy to pose like that for photographs. But Goodall argues that these types of performing chimps are mistreated and often electrocuted to coerce them to behave in the way Susie did.

'Jane never met Susie. She never saw how Susie lived. Susie enjoyed dressing up. My chimps have never been mistreated – Jane said that Susie wouldn't be able to climb because her feet had been damaged by wearing shoes, but it's not true. Susie could climb, there was nothing wrong with her feet. I loved her. When she died I went bananas, we played 'If You Knew Susie' at her cremation and scattered her ashes across the cactus bed. That book certainly busted my Jane Goodall bubble. Brigitte Bardot also wrote a nasty letter to me telling me I was mistreating my animals. I love them. I'm 100 per cent against chimps in labs. They should take hardened criminals to test in research instead.' Dan Westfall has started his own ape charity, CHEETA – Committee to Help Enhance the Environment of Threatened Apes. Money donated to this charity helps to provide Jiggs and Westfall's other ex-showbiz primates with a stimulating captive life.

By any actor's standards Jiggs has had a fabulous career. He starred in over thirty movies and made countless guest appearances, spending upwards of 2,500 days on set and earning in excess of $1.25 million. Jiggs has a star on Palm Springs' Walk of Stars that says 'Cheeta the Chimp, Movie, Stage and Film'. A time capsule is buried under the star with a lock of the ape's hair, photos of him with Susie, his grandson Jeeta and Dan Westfall and a review of his life by the *National Enquirer*, who paid for the star.

Jiggs has given a great deal of pleasure to thousands of people over the years, but has Jiggs found pleasure? He has certain behavioural abnormalities that suggest early trauma and repetitive stress. And of the thousands of dollars Jiggs's movies grossed for MGM, none went towards preserving chimps in the wild.

In 1988 cartoonist Gary Larson and his actor friend Jack Lemmon took a trip to Gombe to meet Jane Goodall and her chimps. Larson was a little tentative about meeting Goodall because he had lampooned her in one of his cartoons. It depicts two Larsonesque chimps sitting on the branch of a tree. The female grooms the male and discovers a golden hair; the caption reads, 'Well, well – another blonde hair . . . Conducting a little more "research" with that Jane Goodall tramp?' Some of Goodall's acolytes were scandalised at the nerve of Larson using Goodall in a Far Side joke, but Goodall enjoys a laugh as much as anyone and to have Gary Larson poking fun at you is a great compliment.

Primatologists do need media exposure to be able to share their discoveries and insights, but many women working in the field are worried their science is being trivialised and presented as a soft option in misguiding images from popular culture. In both *Mighty Joe Young* and *Gorillas in the Mist* the actresses are seen confronting danger to save the life of a beloved gorilla. In *Mighty Joe Young* the heroine's mother was a scientist studying the fictionalised genetic make-up of a certain group of gorillas who carry a gene for gigantism; every few generations an infant grows into a massive beast. But all this is passed over very quickly in favour of a self-taught gorilla expert (Theron), crusading to save her behemoth. In *Gorillas in the Mist* the story is the same. Fossey's research is glossed over in favour of 'Fossey, the crusading conservationist'. The women are portrayed as caring for animals at all costs and most male characters in these films are murderous villains. The popular image of a primatologist is of a woman conservationist rather than a scientist using statistical analysis to research evolutionary adaptations of behaviour. Many women primatologists do find, owing to environmental pressures, that conservation is the way forward now. But they are trained scientists and popular culture tends to forget the hard graft of data collecting and testing of hypotheses, replacing that instead with the emotive fight for animal rights. In real life field research in primate studies is not solely about defending the innocent from the violent nature of mankind. Today university courses in primatology are closely tied to statistical analysis and a thorough grounding in theoretical ecology. If you want to sidestep university and work on location with endangered primates and try to protect them, you have to pack your bag independently and make your own way.

8

Talk to the Animals

Apes, especially chimps, have had their intelligence tested by humans for decades. We have them painting, we show them how to 'speak' a few husky words, we instruct them in sign language and we coach them to count. Now we are teaching them to 'talk' by using symbols that represent words and turning our attention to their ability to deceive one another. Today the emotional and Machiavellian intelligence of apes is being studied with much interest.

The human brain grew rapidly in size from two million years ago, when our language instinct to create spoken words with syntax, nouns, verbs and adjectives developed. Social animals, such as humans and apes, need to be able to communicate with all the individuals they share their lives with and the more individuals around you, the more you must communicate. The ways in which humans communicate and passion with which they do so signifies our species' intelligence. But we know that the use of the spoken word is not the only sign of intelligence within our own species. American author Helen Keller, born in 1880, lost her hearing and her eyesight through illness when a baby. Because her illness rendered her deaf and blind at only 19 months old, she had hardly any knowledge of a spoken language before losing her ability to acquire one. This meant Keller was never able to speak. But she was eventually taught sign language by her companion and carer, Anne Sullivan Macy, and went on to graduate from Radcliffe College with honours in 1904. Keller had the instinct and intelligence for communication and used these to surmount her considerable problems. But until she was taught sign language she was considered an imbecile.

Through recent behavioural and psychological studies on apes science has been able to prove that great apes have the ability to use a wide repertoire of non-verbal languages. Some of the apes' vocabulary is innate and some of it is learnt. But the use of language, non-verbal or spoken, means that individuals, ape or human, are self-aware and realise that they have a different thought from another individual (this is known as 'theory of mind'). Then comes the desire to share that thought or keep it to themselves if they want to deceive. Humans and apes consciously make these choices.

But is it possible for humans to achieve perfect communication with apes? Numerous scientists and amateurs have tried. It is self evident that we are dealing with an intelligent life form, but exactly how clever are these animals? Some psychologists suspect that great apes, especially chimpanzees and bonobos, have intelligence equivalent to that of four-year-old children. It is this belief that fuels the Great Ape Project (GAP). GAP, conceived by Peter Singer, does not have paid membership but a long list of supporters, including scientists Richard Dawkins, Jared Diamond, Robin Dunbar, Roger Fouts, Toshisada Nishida and Jane Goodall. The New Zealand parliament is the first government to take GAP seriously and is now debating whether to change the laws in New Zealand and give great apes civil rights. This would include: the right to life, the right to freedom from cruel and unnecessary medical experiments and freedom from degrading treatment. Peter Singer believes it is morally wrong to undertake medical experiments on a self-conscious creature such as a healthy chimpanzee. It is more acceptable to Singer to test instead on humans who have sunk into a permanent vegetative state, as people like this are no longer self-aware. Today's received wisdom on the abilities of the great apes, more than any other animal, has challenged our self-imposed superior position. The people behind this lobby believe great apes should be recognised as deserving equal rights to humans. This is a radical philosophy and many people are still of the opinion that it's a crazy idea.

Scientists and conservationists who have spent time with apes have realised they are dealing with creatures of sophisticated abilities and intelligence, even though the creatures cannot talk. Before Goodall's pioneering research, the vision of man the tool maker separated us from other apes. As that crumbled, a new self-image arose, that of man, the social animal with language. Primatology is now eroding this last bastion used to define humanity.

We may be the only animals to talk, but are we the only species with the cognitive abilities for language? We have to examine the biology of other apes to understand why they cannot talk. Modern man's larynx is much bigger than a chimp's and located lower in the throat than it is in apes. Human children are born with small larynxes located high in the throat, as it is in chimpanzees. This allows both chimps and human babies to breathe while they swallow food, but it also limits the ability for speech. When human children get past the breast-feeding stage at two years, the larynx expands in size and travels downwards in the neck, allowing children to talk but now preventing them from breathing and eating at the same time. This transition does not occur in apes. But if it did, could chimps talk? Some psychologists who have studied the intelligence and language capabilities in apes, especially chimps and bonobos, believe they would be talking if they had suitable physiology.

It is suspected that apes have language capacities because they are social animals. Wild chimpanzees can live in co-operative groups of approximately 100 individuals, but if the population expands much beyond that, disorder breaks out and the chimps organise themselves into two smaller groups. In time the new groups will become mutually hostile. Without spoken language chimps manage to live in hierarchical societies, where individuals occupy defined social positions, with each one perpetually competing for more status and everyone knowing everyone else.

Robin Dunbar has theorised that the evolution of spoken language has enabled humans to continue living in hierarchical, status-conscious ape societies, but they are societies of far larger numbers. In Western cities, millions of modern people live co-operatively side by side. An individual may only know a few hundred of those people and probably fewer than 100 as friends or acquaintances, but millions of modern humans are capable of inhabiting geographically chaos-free small areas. Every society has its social morals and written rules based on what the majority of its members instinctively feel to be right and to which the majority of individuals adhere. Anyone defecting from co-operative behaviour is punished. Language and the sharing of ideas have enabled people to co-operate in vast numbers and have made *Homo sapiens* the most dominant species on the planet.

Up until recently it was presumed that archaic humans developed speech of a similar complexity to ours some 40,000 years ago, a date that ties in with the earliest known Neanderthal cave paintings. But the latest research funded by the California-based L.S.B. Leakey Foundation, established by Louis Leakey in 1968, suggests that modern human speech evolved at least 400,000 years ago.

The hypoglossal canal is the key. In modern humans and apes and fossil hominids this pencil-sized, tubular bone is a canal for the motor nerve that controls the tongue. Scientists from Duke University found that the canal in Neanderthals and early humans more closely matched that of modern humans than the smaller

canals of apes and proto-humans such as *Australopithecus*.
Researchers found that the canals in humans were twice as large as
those in chimpanzees. The diameter of 2.5-million-year-old
Australopithecus africanus's hypoglossal canal was almost identical
to those found in several species of chimpanzees and gorillas.
Australopithecus clearly did not possess even the rudimentary structures essential for speech. In contrast, the 300,000-year-old
Neanderthal and 400,000-year-old archaic *Homo* canals fell within
the modern human range. Scientists Richard F. Kay, Matt Cartmill
and Michelle Balow are convinced that the size of the canal reflects
the fineness of the motor control over the tongue in humans. We
don't need a large nerve to our tongue so that we can eat – we
chew our food just as well as apes – we need it to talk.

Brain size varies considerably within any species. This is not
usually related to intelligence, but is related more to body size. As a
result, women, on average, will have smaller brains than men, and
pygmies will have smaller brains than Zulus, but as far as we know
the average intelligence of all these groups is equal. In humans it is
the wiring of the brain that counts, not brain size.

The average brain size of adult modern humans has a size of
1,350–1,400 cubic centimetres. But there is much variation, with 1
per cent of the population, known as microcephalics, having
extremely small brains, which fall as low as 600 cubic centimetres;
some of these people are mentally impaired but others are not.
Richard Leakey's 1.4-million-year-old *Homo ergaster* skeleton,
Turkana Boy, died when he was eleven years old. Had Turkana Boy
reached adulthood, he would have stood at least 6 feet tall and by
today's standards, where tall people usually have larger brains than
small people, he would be expected to have a larger than average
brain. But it is estimated that Turkana Boy's adult brain size would
have been 910 cubic centimetres, which is smaller than modern
human brain size, except for that of microcephalics. In comparison,
900 cubic centimetres is the typical brain size for a healthy modern

child at three or four years of age. Turkana Boy would have used a primal form of human speech. Chimpanzees have a brain size between 300 and 500 cubic centimetres, with an average of 400 cubic centimetres. Gorillas have an average brain size of 500 cubic centimetres, with large individuals going above 700 cubic centimetres. But again, the crucial element here is not actual brain size but brain design.

A contemporary newborn human baby is a helpless scrap of humanity, unable to defend itself in any way. An average baby weighs twice as much as a newborn ape, even though gestation is a similar length of time, because at birth its brain weighs twice as much as that of an infant ape. Unlike other animals, the primate's brain continues to grow after birth. But the growth rate in non-human primates slows down after birth, unlike the human baby's brain, which continues to grow at the foetal rate for twelve months. It could be argued that human babies are born premature and if women's birth canals were wider women would probably remain pregnant for an eye-watering twenty-one months – though if the birth canal was much wider bipedal locomotion would be decidedly more difficult for women (even more difficult than running for the bus while carrying your helpless newborn baby and the shopping). Nature has designed human birth to occur at the earliest possible moment, which explains the vulnerability of human babies when compared to other baby mammals.

By the time an infant primate reaches its first birthday the growth rate of its brain is slowing down. In adulthood an ape's brain is 2.3 times bigger than the brain in a newborn ape and an adult human's brain is 3.5 times bigger than that of a newborn human. If adult apes and humans had the same body size, the human brain would be three times as big as an ape's. Contemporary apes and monkeys take on average between five and ten years to develop, whereas human children take a leisurely fifteen to twenty years to grow up.

Primatologists who have studied the computing ability of the great apes would say that although the ape's brain is smaller than modern man's it is designed in a similar way. Richard Leakey has suggested that Australopithecines were as clever as or even a little smarter than contemporary chimps, and the intelligence of *Homo habilis*, *Homo erectus* and *Homo ergaster* lay somewhere between chimps and modern humans.

The brain size of Lucy, the 3.3-million-year-old *Australopithecus afarensis* discovered by Don Johanson in the Ethiopian Afar desert in 1976, was 400 cubic centimetres the same size as that of a chimpanzee brain. The brain size of the *Australopithecus* lineage never significantly increased over 2 million years before they became extinct. During Lucy's time she kissed, made love, sat quietly cogitating and no doubt enjoyed a good old laugh with her female friends. She would have had a complicated series of vocal calls and physical gestures with which to communicate, but she didn't utter a single world. She did not have the anatomical necessities to talk, and neither do modern apes. It was Lucy's descendants, the *Homo* lineage, that started talking. The *Homo* brain trebled in size over 3 million years as they branched further away from Lucy's direct *Australopithecus* descendants.

Why did the larynx evolve to expand in archaic humans? From the fossil record it seems the larynx of 2-million-year-old *Homo erectus* had begun its migration down the neck and was in the position of a modern eight-year-old child's larynx. *Homo erectus* was primitively talking 2 million years ago, which was also the time of a significant brain expansion that has culminated in modern man. Perhaps with the acquisition of spoken language the brain of archaic humans needed to expand to deal with the bombardment of information that language brings. As the populations of archaic humans became larger so their brains grew and their spoken language improved upon itself. This enabled large groups of early people to talk and organise themselves.

Robin Dunbar of Liverpool University and author of *Grooming, Gossip and the Evolution of Language* told me that he believes human language evolved in female hominids first. He suggests that women developed language to form bonds with other women so they could help each other care for their increasingly dependent children. The more chatty you are, the more friends you have; the more friends you have, the better your and your baby's chances for survival. Schoolgirls show an aptitude for language that boys do not possess. Modern women with children have an instinct to bond with other mothers and spend a good deal of their 'maternal time budget' offering vital reciprocal childcare. It is a credible theory.

Much ape behaviour is shared by humans and our ape relatives. All of the great apes kiss, just the way we do; bonobos use their tongues to enjoy 'French kissing' too. Kissing and the sexual act pre-date language. They do not require speech to be successful; in fact, talking is superfluous. It is thought that laughter probably evolved 6–8 million years ago. Chimps and humans share a similar slapstick sense of humour, but the difference in the vocal tract causes humans to laugh with a *ha, ha, ha* only during expiration. Chimp laughter sounds more like out-of-breath panting with *ah, ah, ah* expressed during both inspiration and expiration.

We cannot be sure what our ancestors or what the great ape's ancestors looked like. We know that 4–5 million years ago small bands of archaic humans walked out of the diminishing green canopy into the arid sunlight and over subsequent millennia strode towards the frozen north. A number of species managed to adapt to the changing habitat and our ancestors lived with them and made war with them. Legends of giants and goblins may be all we have left as a collective primordial memory of our journey. Today's great apes have not changed so much over the last few million years, whereas *Homo sapiens* has taken quite a trip.

From the dissection of contemporary ape and human brains and from the fossil record, it is evident that apes and humans are similar and that it has been a gradual and ancient separation of our two genera that have brought us here. In general terms we believe ourselves to be unique, seeing our sentience as representative of the void between us and 'them'. We speak, we create great works of art; apes do not. But could they? If their bodies evolved and adapted to certain influences like ours have, could they talk or express themselves creatively then? Do they actually have a dormant ability for language and art that has not been exposed to environmental pressures that would encourage its development? If a chimp's larynx was positioned where ours is and if a chimp's hypoglossal canal was the size of ours, could the creature tell us its thoughts? Man and ape's common ancestor did not talk and did not paint, but the structure of this joint ancestor's brain would have contained the seed for self-expression.

Primatologists have long searched for evolutionary continuity. There has been an infectious need to make real contact with these animals and close the 5-million-year gap. While talking to their somnambulistic animals, behaviourists have also offered their primates paints, brushes and canvas. Not only the Hollywood chimpanzee actor Jiggs likes to paint; many of the most loquacious apes are renowned painters too. The history of ape art and language goes back to the turn of the last century.

One of the first women to try to unpick the mind of a chimpanzee was the Russian scientist Nadjeta Kohts. Living in Moscow in 1912, she embarked upon a psychological experiment with a juvenile chimp named Joni. Kohts would tidy her long brown hair in a neat plait down her back so that Joni wouldn't play with it, then, just as one might do with a toddler, she would sit him on her lap with a pencil and a pad and let him doodle to his heart's content. Kohts worked with Joni for three years, comparing his ability to develop shape and form with her own child's artistic abilities. Nadjeta Kohts concluded that a chimpanzee's perception of colour and shape is not dissimilar to that of man. Kohts discovered that there is a crucial time when a child's scribbles diverge to become more figurative, but that a chimp never makes the intellectual leap to the creation of representational images.

Over three years Kohts observed Joni develop a style of his own. If the chimp was given lined paper to draw on he would embellish the existing pattern by drawing a series of lines over the top of the originals. Kohts went on to study the artistic talents of two capuchin monkeys and three more chimps, and observed that the apes had a superior drawing skill and a more sophisticated approach to developing a theme than monkeys. The First World War put a stop to her work.

These days Molly Badham's bonobos and gorillas at Twycross Zoo have taken to painting and to watching television. The apes at the zoo love putting the telly on at night and watching sex and violence. They've also become fans of the Teletubbies and the Spice Girls. But the apes seem to hate party political broadcasts – can we blame them? Do ape activities of painting and developing a discerning taste for television programmes represent the uniting interface between culture and nature, between art and evolutionary thought?

Anthropologist and surrealist painter Desmond Morris feels convinced that he's proved that the great apes do have an aesthetic sense. Their paintings reveal to us the origins of *Homo sapiens*'s hardwired appreciation of art evolved millions of years ago. Morris believes he has established that apes have compositional control, that is, that they know when their paintings are 'finished' and, self-satisfied, will put down their paint-brushes.

But is it art as we know it? Back in 1957, during Desmond Morris's London ICA show dedicated to ape art, some people were enraged at the suggestion that a 'monkey' could paint and therefore should be endowed with a soul – just as many linguists and ordinary people have been outraged today at the suggestion that an ape's mind is wired to speak, even though they do not.

*

Bio-psychologist Sue Savage-Rumbaugh is today considered to be the world expert on the language abilities of bonobos and chimps. Savage-Rumbaugh has been studying the cognitive capabilities of apes for the past twenty-five years, and her work has made bonobo siblings Kanzi and Panbanisha as famous as she is. But she is frustrated by outraged doubters, who accuse her of being nothing more than a circus trainer, and by linguists, who keep moving the linguistic goalposts regarding biological adaptation just as Kanzi and Panbanisha score yet another goal.

Sue Savage-Rumbaugh grew up in St Louis, the oldest of seven siblings. Helping her parents to raise her younger brothers and sisters taught her a great deal about the development of the human infant mind and of the spontaneous arrival of speech. She first saw chimps when she was eight years old and a circus had come to town. The apes were riding motor-bikes and stilt-walking. She watched the animals and wondered what they felt about their experience working as entertainers.

At high school Savage-Rumbaugh decided to become a psychologist and in 1970 attended the University of Oklahoma to take her master's degree. Here she met behaviourists Roger and Deborah Fouts and their chimp Booee. Roger Fouts asked Savage-Rumbaugh if she would help teach the three-year-old Booee American Sign Language and she agreed. She would sit with Booee inside his cage, show him objects at random and encourage him to learn the signs for them, feeding him raisins when he succeeded. At times Booee would become bored and want to play. Savage-Rumbaugh understood his needs; she was patient and played with him. No doubt spending her formative years caring for younger siblings was good grounding for this type of work. At the beginning of the 1970s, when she started collaborating on Roger Fouts's projects, she was not a naive student, like many of her contemporaries would have been. She had been married and divorced and had married for a second time to Bill Savage, a lecturer at the University of Oklahoma. Sue had a baby son, Shane, from her first marriage and the little boy bonded well with his new stepfather.

While teaching apes ASL at Oklahoma she watched two-year-old Shane as he acquired language with no effort at all. At times Savage-Rumbaugh would doubt that the chimps really did understand everything they were claimed to comprehend; she became known to the other researchers as 'the unbeliever'.

Booee was encouraged to learn ASL but he was not expected to become physical in his lessons nor praised for doing so. Once, during a break between lessons, Booee hung from the top of his cage and did a full somersault before jumping down. Savage-Rumbaugh enjoyed his antics and gestured for him to do it again, pointing with her index finger to the top of the cage and moving her finger around in a circle. Wanting to oblige, Booee jumped back up and repeated his acrobatics. He knew exactly what her gesture conveyed and effortlessly complied. That night Savage-Rumbaugh lay awake in her bed, her mind racing over what she had seen Booee do and its implications. She had always felt that she was destined to seek out and achieve something of substance with her life. She knew that night she had found her mission: she would work with apes.

Soon after this revelation a female chimp bit Savage-Rumbaugh's right index finger clean in half. She had not appreciated that chimps do not like strangers, nor that a chimp she'd never met before might be hostile to her. Her parents were very concerned and asked her why she wanted to undertake this difficult type of work. But being separated from part of her body was not enough to undermine her resolve.

When Savage-Rumbaugh first entered the world of ape language research the science was primitive in some ways, but it was also buoyant. By the 1980s, when she was ready to publish some of her most astounding data, the climate had changed. Ape language research was seen as fundamentally flawed. She would find it hard to make people listen to the fact that her apes spontaneously acquired the comprehension of spoken language in the same way that very young human children pick up language.

In the 1950s there had been a great deal of excitement in the speculation about ape intelligence. Vicki, a baby chimp, was raised by the Hayes family along with their own newly born infant. The two youngsters were given the same amount of attention and love and eventually Vicki, after seven years of tuition, managed to say a very breathy 'Mama', 'Papa', 'cup' and 'up'. This well-publicised experiment inspired the film *Bedtime for Bonzo*, one of the chimpanzee actor Jiggs's earliest roles.

But no matter how clever an ape is, it is never going to manage to say very much because anatomically it is not built for spoken language. Humans with disabilities that prevent them from speaking are taught sign language. Today we don't force people like this to 'talk' and we do not call them dumb if they can't.

Beatrice Gardner and her Jewish parents had hurriedly left Austria at the start of the Second World War and settled in Nevada. She attended Oxford University, where she studied the behaviour of sticklebacks (many scientists, including Desmond Morris, have started their careers studying this small, but passionate fish). Back in the United States she met Allen Gardner, a psychology student, in Massachusetts and in 1961 they married and later moved to Reno in Nevada. In the early 1960s the Gardners had the inspired idea of taking in a baby chimp, Washoe, and looking after her as if she was a deaf human baby, teaching her American Sign Language instead of trying to make her talk. Washoe slept in a trailer in their back garden for four years. Gardner would never have children, but she would mother Washoe and in return the chimp would manage to sign 132 different words.

What really impressed people about the research was that Washoe would spontaneously combine words. When she saw a swan for the first time she signed 'water bird'. She would sign 'tickle me', 'Washoe sorry' and 'Gimme sweets'. Washoe also taught signs to other chimps without human encouragement. After her much publicised success, people wanted to know whether Washoe's ability to communicate was unique or if all chimps could do it? Beatrice Gardner would go on to mother Pili, Tatu, Moja and Dar; all the apes would combine signs to give names to new things they encountered, such as 'listen drink' for Alka Seltzer.

Roger Fouts learned about ape language studies from Beatrice Gardner at the University of Nevada. When he left Nevada for Oklahoma University he took boisterous Washoe with him. Bill Lemmon ran the Oklahoma Primate Centre then. When Washoe arrived, Lemmon ordered Fouts to remove her toys and blanket comforter. Even though she had never seen another chimp before Lemmon insisted the four-year-old chimp be housed in a cage next to all the other chimps' cages. After being raised as a human baby, to suddenly be screamed at by twenty hostile chimps was a terrifying ordeal for her. By all accounts Lemmon was a cruel man. He employed a talented scientist like Roger Fouts because he wanted his primate centre to succeed, but he didn't like Fouts's soft approach to the animals.

In 1980 Roger Fouts and his wife Deborah moved their family of chimps to Washington State University, where Washoe and the other Gardner chimps live today. Booee wasn't so fortunate. Chimpanzees in universities are the research property of the faculty. At the University of Oklahoma Bill Lemmon decided in 1982 that nine-year-old Booee should be sold to New York Laboratory for Experimental Medicine and Surgery in Primates (LEMSIP). It meant nothing to Lemmon that Booee had personality and character and had human friends who had been influenced by their interaction with this self-reflective ape. As a research animal he had no more rights than a fruit-fly.

Sixteen years later Roger and Deborah Fouts were allowed to visit Booee, now living in a 5 × 7 foot cage. By this time he was very sick with hepatitis. As soon as Roger and Deborah Fouts both approached his cage, Booee recognised his old gossiping partners. He perked up and signed their names and remembered all the signs he had been taught by the Fouts, just as if time had stood still and he had not endured sixteen years of experimentation. When it was time for the Foutses to go Booee slumped down in his cage. He did not want to say goodbye. Utterly depressed he averted his eyes as they left.

In 1973 behaviourist Herb Terrace started Project Nim. Terrace took a young male chimp away from Oklahoma for some intensive one-on-one lessons in sign language communication, naming him Nim Chimpsky to poke fun at Noam Chomsky, the linguist and doubter of ape language research. Influential linguists such as Chomsky and, more recently, Steven Pinker have poured scorn over the validity of the ape language research. Noam Chomsky obliquely expressed his feelings to me about ape language research. 'I cannot see what it is supposed to have to do with human language any more than the study of the way people jump has to do with flying; less so, in fact, since in the jumping-flying case at least it is clear that homologous structures are involved. That's no criticism of the work, just as it's no criticism of the study of human self-organised motion to say that it's hardly an illuminating way to study eagles.' Steven Pinker, author of the book *The Language Instinct*, told me, 'Ape language research has been too anecdotal to take seriously. Apes do not have language and the research is a waste of time and money. But to study apes' natural communication in the wild would be worth while.'

By the time Nim reached four years of age he was composing some very long utterances, such as 'Give orange me give eat orange me eat orange give me eat orange give me you.' Often Nim would repeat what Herb Terrace signed and, after returning Nim to Oklahoma and analysing all his data, Terrace felt cheated. He decided Nim had been fooling him. In 1979 Terrace aligned himself with the doubting linguists and publicly announced, 'A sequence of signs, produced by Nim and by other apes, may resemble the first multiword sentences produced by children. But unless alternative explanations of an ape's combination of signs are eliminated, in particular the habit of partially imitating teacher's recent utterances, there is no reason to regard an ape's utterance as a sentence.'

But the crucial omission in this work was that Nim, Washoe and the other chimps who were taught to sign were only encouraged to 'speak'. They were not encouraged to listen. As any mother knows, her baby listens to her for at least a year before the infant attempts to utter 'Mama' and at least another eighteen months before starting to talk coherently. But a child is obviously comprehending its mother's speech long before utterance takes place. When my son, Noah, was fifteen months he was not yet speaking, but I knew he understood all sorts of things because he reacted appropriately and would bring me familiar objects if I asked him to.

From infancy Savage-Rumbaugh's apes are encouraged to listen to ordinary human speech. Spontaneous acquisition of the meaning of the words occurs, just as it does with children. Normal human children pick up different languages with ease, even if they are raised in a bilingual setting, but trying to learn a language after the age of fifteen becomes very hard. You cannot teach an old dog new tricks. As we grow older synapses in the brain shut down if they are not in regular use. The brain is a muscle that requires exercise. For this reason Savage-Rumbaugh likes to work with an ape from its infancy.

Now based at Columbia University Herb Terrace has been researching the mathematical abilities of rhesus macaques. The monkeys have a concept of numbers from one to nine. This latest revelation from research undertaken in 1998 must leave Terrace somewhat confused – perhaps Nim wasn't tricking him after all?

It has been long believed that counting is a cultural phenomenon that depends on language, but psychologists Claudia Uller and Susan Carey have show that babies as young as five months old can count. They have been testing the numerical skills of prelinguistic babies compared with non-linguistic macaques and tamarins. Tamarins are New World monkeys that are even more distantly related to humans than Old World monkeys such as macaques. But even our distant relative the tamarin can count. Uller and Carey tested a human baby's 'preferential looking time', PLT. A Mickey Mouse doll was shown to a baby then placed behind a screen, and then another doll was shown and subsequently placed behind the screen. Before lifting the screen, one of the dolls disappeared through a trapdoor. When the screen was raised and only one Mickey Mouse was in view, the babies gaped, eyes wide. Their PLT was much longer if there was only one doll when they expected two.

From these tests it seems that babies under ten months can easily add and subtract numbers from one to three; any higher and they are less successful. When aubergines were used instead of dolls, the tamarins' PLT revealed they could count just as well as the babies, but the rhesus macaques were the most numerically successful, counting up to four objects. Macaques could even add up different objects, and knew that one carrot plus one apple equalled two things. Babies only managed this when they reached twelve months. By then they were learning the words for different objects and this helped them to count individual items. It seems that basic spontaneous numeracy is common to all primates with or without language acquisition.

Sally Boysen is a senior psychologist at Ohio State University. She has studied chimps for twenty-five years, and has tested chimps' mathematical skills using what's called a 'quantity judgement test'. In this test a chimp, already ready familiar with the process of counting and arithmetic, is given a choice of two amounts of sweets, one large, one small. They are taught that the pile that they point to is not the pile they receive, but will be given away to another chimp; it is the pile they do not point to that is ultimately given to them. So if you want the big pile of sweets, you point to the small pile. When given the choice of two piles of sweets – one small and one large – chimps feel compelled to ask for the large pile even though they know in this test the pile they point to is given away. The chimps behave this way even with inedible objects like pebbles – they have an in-built trigger they cannot suppress that makes them want the larger pile. But Boysen discovered if number symbols were used in the place of amounts of things – she uses the mailbox numerals bought in American shops – the chimps would make the correct choice, pointing to the smaller number to be given the larger number.

From her research Boysen has concluded that if piles of sweets or pebbles are used in tests like this chimps exhibit the cognitive domain of a three-year-old child, but if number symbols are used chimps exhibit the cognitive domain of a four-year-old child. These same tests have been done with three-year-old children; children of this age always want the larger pile of sweets and rarely make the right intellectual connection, but twelve months later children can understand the correct thing to do to acquire the most sweets.

Boysen also lectures in chimpanzee conservation and welfare. Seven years ago she heard about the plight of Bob, a two-year-old chimp, owned privately by a man who made a living taking Bob to shopping malls and fairgrounds and getting members of the public to pay to have their photo taken with him. This type of career for a chimp is short-lived; they do not stay cute for long and are dispensed with when they become too dangerous to bring in money. Bob's owner had taken the chimp to the annual Ohio State Fair, where he was expecting to earn good money that day, but unfortunately Bob bit the child he was posing with and the Ohio Humane Society was called in to investigate. When Bob was confiscated, Sally Boysen heard about the story. Bob's owner had, by this time, decided it was time to sell him and placed an ad in *USA Today*, asking $25,000 for the chimp. Anything could have happened to Bob; a private owner could have bought him for breeding purposes. Most primate centres do not allow any more breeding of chimps as they have no room, but the entertainment industry continues to breed them. Boysen knocked the asking price down to $14,000, and her offer was accepted. But in reality she had no money!

Boysen decided to remortgage her house, and took Bob with her down to Bank One in Columbus, Ohio, to see her loans officer, Shirley Smith. 'Bob and I were lucky the loans officer was a woman.' Bob sat on her desk as Boysen asked for $14,000. Shirley looked doubtful, but on cue Bob did some cute acrobatics and her heart melted. She gave Boysen the money. Today Bob is nine years old and lives with Sally Boysen's other chimps. He is asserting himself in the male hierarchy and will probably become the alpha male. Not unexpectedly, Bob is very good at arithmetic! Boysen told me she is against all bio-medical research on apes because her cognitive research has taught her that chimps are self-aware and intellectually just a 'whisper' away from us.

Wild vervet monkeys use a proto-archaic primate language of their own. Dorothy Cheney and her husband Robert Seyfarth have studied the vervets of Amboseli Park in Kenya for years. Vervets are highly organised and socially co-operative. When they are out foraging, one sharp-eyed individual climbs high into a tree and looks out for monkey-eating eagles sweeping down from the sky and leopards and snakes creeping in over rocks. If the look-out sees one of these predators they let out different loud calls to warn the rest of the troop to take appropriate cover. If a snake alarm is called, the vervets all stand up on their hind legs looking about for the snake. If a leopard call is emitted, the vervets run for the trees. When they hear an eagle alarm call, the vervets come down from the top of the trees, where they can be picked off, and hide under rocks or deep in undergrowth.

Even more amazingly these monkeys use such alarm calls to deceive tactically. Vervets live in large groups, and the senior male is always on the look-out for competition from lone rangers who want to take control of his females. In one such situation, a resident senior male was observed realising that an uppity young male was hanging about in a nearby acacia tree, hoping to make a move on his troop and take control. The resident male used predator alarm calls to keep the challenger at bay. Every time the interloper looked as if he was coming over, the leopard alarm call was let out and the challenger fled for cover.

Eventually the interloper became suspicious. When he saw the cocksure resident male let out his alarm call while confidently striding about in the open where he was vulnerable to predation, the game was up. The interloper walked over and took control of the females, ensuring his chance for reproduction.

The disillusioned Terrace temporarily left the world of ape language research, about the same time as a conference of academics was organised in 1980 called, 'The Clever Hans Phenomenon'. Clever Hans was a music hall star, a white horse who seemed, amazingly, to be able to count when his owner asked him to. 'Hans count to ten' – Hans would tap his hoof on the ground the correct number of times, stopping when he reached ten. It was noticed that Hans would stop when his owner gasped in air and moved his head in a certain way and that his owner would exhibit this unconscious tic when Hans had reached the correct number. It seemed Hans's owner didn't realise he was giving his horse cues; he was so proud that he wanted to believe his horse was a genius.

The cueing of apes has been criticised and is a problem for ape language research. Psychologists have often been thought of as proud parents who push their little darlings to shine. Savage-Rumbaugh tries to avoid this criticism by placing a headset on an ape so that the animal responds to recorded requests that the scientist cannot hear and therefore cannot cue. Later, when the test is completed, the scientist rewinds the tape to check if the ape's behaviour was appropriate to the recorded requests.

Sue Savage (as she was then called) met Duane Rumbaugh at the 1974 International Primate Society conference held at Kyoto University in Japan. Rumbaugh was already a well-established senior figure in the world of primate research. Savage's talk expressed some of her doubts about the methods of teaching ASL to chimps, and Rumbaugh listened with interest to this young scientist's concerns. He had already progressed from using ASL a few years previously, developing the Language Analogue, known as LANA, to take the place of sign language. Wanting to help mentally impaired children acquire language, Rumbaugh had become fascinated in Beatrice Gardner's work with Washoe. He reasoned, 'If we could learn how language develops in a chimpanzee, we would surely have a leg up in learning how to cultivate language in mentally retarded children.' Together with Ernest von Glaserfield, Rumbaugh developed and, later with Sue Savage, perfected, LANA.

A LANA board consists of many abstract geometric symbols, or lexigrams, that denote individual words, including nouns, adjectives and verbs. When a lexigram is touched by the finger of a human or an ape, it lights up. There is a projector above the rows of symbols where the lexigrams touched are chronologically displayed, so a series of words can be viewed and recorded in sequence. This allows the keyboard or symbols board to be used by an ape or a disabled child without the presence of a researcher. The scientist can later discover what has been 'said' in their absence.

Duane Rumbaugh was impressed with Sue Savage and after they returned from Japan he contacted her. Savage-Rumbaugh is an ambitious scientist. She speaks with determined conviction and passion, has large, questioning brown eyes and a big smile. Rumbaugh found the combination irresistible. He offered Sue Savage a post-doctoral research post with him at Georgia State University, which is affiliated to the world famous Yerkes Regional

Primate Research Centre and Savage knew it had a high academic profile. Its colony of bonobos added to the attraction. She had become intrigued by the mysterious bonobo or pygmy chimpanzee and very much wanted to study it after she had taught signs to an ape called Pancho. He was a half-bonobo half-chimpanzee koola-kamba. Well over 100 years old 'koola-kamba' is a West African term for a wild gorilla and chimp hybrid. Today koola-kamba is used to describe any African ape hybrids. Savage-Rumbaugh describes Pancho as a 'gentleman' who liked to drink root beer.

Sue Savage took up Duane Rumbaugh's offer immediately. Though she at first insisted that she wanted to observe bonobo behaviour and not contribute to any of their language research, Rumbaugh needed her help in developing the LANA project further and consequently she was drawn back into the research. Rumbaugh is about twenty years older than Savage and was already married, but they shared a common goal and fell in love. When Sue Savage left Oklahoma for Georgia, in 1976, she decided to leave her son Shane behind with Bill Savage, his stepfather, who agreed to raise him. Duane divorced his wife, Sue divorced Bill, and Duane and Sue married. They never had any children together and remained married for over twenty years. Recently divorced, they continue to research the acquisition of language together.

In 1981 Georgia State University opened the Language Research Centre to house Savage-Rumbaugh's research. She is funded by the National Institute of Child Health and Development. Nationwide, belief in the ape language research was crumbling in the early 1980s. Savage-Rumbaugh found it hard to get papers published even though she was approaching their work from a new perspective by using LANA and had discovered some new facts. Even now it is still an uphill struggle, but Savage-Rumbaugh is fiercely committed to her work.

She told me, 'I've unintentionally gotten involved in an intellectual battle to try to change people's views on the nature of relationships between man and apes and perhaps man and all animals. I don't want this battle, but I have learned so much from the twenty years of first-hand interaction that I have had with apes that I cannot deny it. I have to be honest as a scientist. Even if people make fun of me, even if they don't believe me, even if the work doesn't have the impact it should, I have to say what I know from my heart.'

We know that modern apes mindread each other in their bid to understand reality and survive another day. The *Homo* lineage developed a spoken language in order to approach that ultimate truth, and truth is what we are all searching for. I can use my spoken language to ask for explanations until I am satisfied that I understand. I can also use language to lie and cheat to help me survive.

Captive apes have proved themselves capable of high intelligence, able to solve problems they would never encounter in the wild. Why do they have such a great mental capacity if they hardly ever use it? In the wild apes are social; even the solitary male orangutan is social from time to time. To be successful socially you need to be smart. Like us, apes are conscious of their own thoughts and of the thought processes of their nearest and dearest.

When I hear that we share 98.5 per cent of our DNA with chimps and bonobos and that they are roughly as smart as a four-year-old child, I can't help but think of my eldest son Oscar at the age of four. Anyone who knows or has known a four-year-old child or remembers being four years old will appreciate the quick and manipulative nature of small children, who can often outwit adults. Both children and women have to use a good deal of psychology: if you are physically smaller than other people in your family group, you cannot bully them to get your way.

The scheming and planning of counter-strategies to various possible scenarios that children engage in demands a great deal of brain power. Sometimes it is difficult for us to realise that others see things differently from the way we do, yet we often correctly read each other's thoughts. A nut-sized area of the brain called the amygdala controls our sense of fear and also our analysis of our first impressions of people. These observations are cross-referenced with our past experiences, and we realise a mean-looking face or a furtive glance can mean us harm. It's been observed that baby monkeys with damaged amygdalas grow up to be socially incompetent and to hold low social status.

Understanding that others see things differently and correctly interpreting another's move can win you friends, save time or save your life. It seems chimps and bonobos have this ability too. Captive apes have been observed 'lying' to each other and to their human carers in much the same way as children who lie to get what they want.

There is a struggle for power in all relationships. From about the age of four to five normal children acquire a 'theory of mind', seeking out their peers at school and indulging in tactical deception to survive. Of course they still need the protection of the family, but they have become distinctly more independent than they were just a year before. Their theory of mind means they appreciate that others do not always share in the information that they have. This is the age when children learn the arts of manipulation and pure cunning.

A compelling argument for the Great Ape Project and a 'community of equals' arises from the recent 'theory of mind' debate. Average adult humans and children over four years of age have five to six layers to the process of reasoning. This means we sift through information five to six times instantaneously before reacting to the moment. Great apes sift information three to four times, as do very young children and people with autism. The cognitive abilities of the average ape are equal to those of the average autistic person. Most autistic children do not pass Boysen's quantity judgement tests, and make the mistake of pointing at the larger pile of sweets.

Autistic people, naturally, have full human rights, but if sentience and intelligence are the ways by which we avail ourselves of certain moral and ethical standards GAP is now arguing that the time has come to give rights to great apes. This means they must not be experimented on and treated like 'animals' when they are so much like us.

Psychologists who support the community of equals movement would say great apes and humans are located on a single intellectual continuum. We cannot see each other's minds, only each other's bodies; just as there are types of body shape known as mesomorph, endomorph and ectomorph, so there are different types of brain. You may have the body of a human and the primate next to you may have the body of a chimp, but your brain may be constructed in a very similar way.

Autism is a misunderstood condition. It is very common; most people with autism are only slightly affected and will probably live a 'full' life, albeit with some eccentric disasters and probably without diagnosis. Some high-functioning autistics have a condition known as Asperger's syndrome, a less severe relative of autism, which like autism seems mostly to affect the male of our species, though women are also diagnosed. This genetically inheritable condition runs through generations of families. High-functioning people with Asperger's syndrome are naive but often of above average intelligence and can have skills in maths, music and analytical thought. These people take spoken language literally, creatures of habit; they are often awkward and clumsy, they cannot generalise or improvise. They don't like change, they like to dominate and often want to be alone. Many of them are found lurking in academia – the absent-minded professor who lacks the ability to connect to his students because he cannot attribute different mental states is typical. The nutty professor is clever but he lacks a full theory of mind; he is gullible, does not get the joke and cannot 'duck and dive' to survive. Similarly the single-minded artist, who paints through the night ignoring his wife and children and threatens suicide from time to time, may well be touched by Asperger's syndrome. People with Asperger's syndrome do not understand the basic foundations of friendship. They often see a superficial relationship as important and treat an important relationship superficially. In terms of empathy, the ability to bond, social discourse and the capacity to consciously change allegiance for selfish gains, some of these people are less equipped than the average great ape. Kissing and love-making predate language, yet people with Asperger's syndrome rely entirely on the literal meaning of words. Body language, including facial expressions, is alien to them, and forming a sexual relationship is very hard for autistic people because 65 per cent of human language is non-verbal. Consequently, an autistic mother finds it very difficult to understand the needs of her baby and subsequently may ignore the child for much of the time.

Unaided autistic people find self-reflection hard; they rarely look back at their past experiences and learn from past mistakes, whereas apes do spontaneously self-reflect. Chimps have the ability to contemplate their feelings of pleasure or distress. Language chimp Washoe had a baby, Sequoyah, that died. After the body of Sequoyah was taken from her, Washoe was left alone in her cage and suffered depression. She would sign an interrogative 'Baby?' to her human carers, the inflection used with the ASL 'baby' sign translating to 'Where's my baby?' Washoe was self-reflective in her misery. Depression in apes manifests itself similarly to the way it does in humans. The ape will withdraw from social interaction, sleep more than normal and possibly decline food.

There is a theory that socially dysfunctional people, such as autistic people, who are likely to be unpredictable and impulsive, are not freaks of nature. The more original autistic thinkers offer their communities intellectual leaps with their surprising insights; their 'disability' has been selected for. Albert Einstein and Vincent van Gogh may well have been touched by Asperger's syndrome.

Composer Béla Bartôk definitely was and, by all accounts, artist Andy Warhol probably lived under its shadow too. Talented autistic people can focus on their work as they are not distracted by the feelings of their family, because they do not feel them. Savant abilities more commonly reside with autistic people rather than within the average population. Disabilities often engender innovative creativity in the individual as a way of surmounting the obstacle.

Chimps and bonobos do not want to be alone; they are team players. They have been observed playing in similar ways to creative four-year-old children. They have been seen second-guessing each other to avoid conflict and have a highly developed sense of their rights and what seems fair in a social context. Chimps have a theory of mind. The eleven chimps and bonobos that Sue Savage-Rumbaugh has studied have revealed themselves able to read each other's minds. The apes have a sophisticated emotional intelligence, they can improvise and they are able to play imaginatively Machiavellian games with each other and on Savage-Rumbaugh. High-functioning autistics can never be successfully Machiavellian because they cannot effectively hide their intent. They are dependent on favours and reliable guidance for their day-to-day survival.

The Machiavellian intelligence hypothesis has been established as an area of psychological study by Dick Byrne and Andy Whiten from St Andrew's University in Scotland. They believe that primates are unique in the animal world because they can use their understanding of one another's personalities to predict what other individuals might do. Whiten and Byrne, who, like Robin Dunbar, supervise mostly women Ph.D. students, have studied the devious nature of primates in detail. They have identified this scheming behaviour as 'tactical deception', an aspect of a Machiavellian nature that offers political primates success.

Sue Savage-Rumbaugh's apes often lie to get their way. Kanzi was told he was allowed no more M&Ms, his favourite sweets. Earlier in the day he had seen another research assistant place a pack of

M&Ms in a cupboard in the toy room. Kanzi predicted that Savage-Rumbaugh would not give him the M&Ms if he asked for them, so instead asked her if he could go and play with his toys. She agreed, but as soon as he went into the toy room he ignored the toys and went directly to the cupboard where the sweets were. He grabbed them and ran off with them, quickly scoffing the lot before they could be retrieved from his greedy grasp. It is obvious from the intent with which these actions happened that Kanzi had schemed his way into the toy room, tricking Savage-Rumbaugh with this ulterior motive in mind all the time. The desire to eat sweets motivates children to sneak, cheat and lie successfully in just the same way.

This lying and cheating behaviour in chimps has inspired Savage-Rumbaugh to test her chimps' theory of mind and Machiavellian intelligence, and not just their language abilities with the LANA board. The rebellious behaviour of apes has been studied just as much as their good behaviour. Savage-Rumbaugh learns just as much when an ape becomes difficult and does not participate in a psychological test as when it is co-operative. Disruptive behaviour does not mean the animal is stupid; it means the animal has a personality and is making a choice.

Savage-Rumbaugh's apes are more than able to follow the narrative plots of TV dramas, especially melodrama. They also like *Sesame St*, wrestling and movies where early man is depicted, such as *Greystoke*. Recognising their own reflection in a mirror is child's play to them. Savage-Rumbaugh is often deceived when one of her chimps distracts her while the other pickpockets the keys to their enclosure. Sue either finds herself locked in with them and has to be rescued by another researcher, or, like escapologists, the apes are busily opening doors in record time.

This type of behaviour is not a trick taught by an animal trainer for an amusing scene in a movie. Savage-Rumbaugh's apes watch her every move and repeat what they want, not what she wants.

This can be a problem for her when guests arrive at her centre to observe her research. When the cameras start rolling the apes may go on strike; they don't like strangers staring at them. These apes are not performing circus animals, but individuals who have a strong bond to each other, to Sue and to the other women who help to care for them.

Chimps Austin and Sherman grew up together, Savage-Rumbaugh caring for them from 1976 when they were infants. They shared a bedroom, but their relationship was not stress-free. Often Sherman, who was bigger, would push Austin about. But Sherman was not totally fearless; he may have liked to bully Austin when the mood took him, but if any unusual banging occurred outside their enclosure after dark Sherman would scream in fright and hug Austin close to him for comfort. Savage-Rumbaugh noticed that when Sherman would start to get bad-tempered Austin would walk out of their bedroom and into the annexe next to the door, where, out of sight, Austin would bang the walls and throw some toys about before rushing back towards Sherman, feigning fright. Sherman would then panic and cling to Austin for protection. Austin wasn't as strong as Sherman, so he used his intellect to read Sherman's mind, understood Sherman's fear of the unknown and manipulated Sherman's emotions, thus saving himself from being bullied.

Austin had started life as a disturbed little bundle of nerves. At two months of age he was found to be unable to feed probably from his mother. He was taken from her and hand-reared. Like all infants who suffer the loss of or separation from their mothers at a young age, he became disturbed. He would cuddle a blanket and rock from side to side. With Sue's care and attention he managed to outgrow this dysfunctional behaviour, but even as an adult if he saw a picture in a magazine of a cute human or chimp baby he would tear it to pieces.

Wild chimps also tactically deceive. Once, at Gombe, a lower-ranking male was observed to approach a feeding site where some bananas had been placed. The lucky chimp was about to take the bananas when a higher-ranking male appeared. The newcomer would surely take the bananas for himself, so the lower-ranking individual acted casual, meandering around, avoiding the bananas and waiting for the higher-ranking male to leave. When the higher-ranking male disappeared from view, the first chimp quickly picked up the bananas, at which point the higher-ranking chimp appeared again. He had become suspicious of the lower-status male's 'casual' behaviour and had hidden to watch. The higher-ranking individual strolled over and took the bananas for himself and revealed himself to be worthy of his status. He had brains as well as brawn.

One of Sue Savage-Rumbaugh's tests placed Sherman and Austin in two separate rooms that were partitioned by a reinforced glass panel. The glass partition had a two-way opening big enough for the chimps to reach through to the other's room. Food was placed in sealed boxes in Austin's room, in front of Austin but out of sight of Sherman. A key would be needed to open one box to get sweets, money was required to open the second box to get fruit and a wrench to open another box to retrieve corn chips. Both apes had played with tools many times and knew how to open the boxes. They also knew which symbol on their keyboard represented each tool.

But unfortunately for Austin, all the necessary tools were given to Sherman. And in turn Sherman had no food in his room. The apes needed to talk to each other via their keyboard and needed to co-operate. Austin asked Sherman for all the appropriate tools, which Sherman gave to him using the two-way drawer. One by one Austin opened each box, collected the food and gave half of it to Sherman. They were a tight team who solved this problem without coaching. When the roles were reversed, things went just as smoothly. Had Austin eaten all the sweets instead of sharing them fifty-fifty, he would have lost the co-operation of Sherman in other tests.

Austin and Sherman knew that it was in their mutual interest to remain co-operative friends and share the food, although Sherman would sometimes be tempted to lick Austin's sweets before giving them to him. In the wild chimps do not, as a rule, share food. Mothers share food with their infants and when the males occasionally hunt a monkey the meat is shared among the hunters, with the alpha male getting the lion's share. Any little bits of flesh that fall to the ground are fought over by females and juveniles. A group of chimps will sit by a fruiting tree and wait their turn; when the higher-status animals have had their fill the lower-status animals can move forwards to eat what is left. Chimps usually sit with their backs to each other and eat alone.

The sharing of food between Austin and Sherman grew out of their captive condition. They needed each other day in day out, so they co-operated. In the wild they would have had a different relationship; at times they would have needed each other but at other times they would have competed with each other over females, social status, sleeping sites and food. They adapted well to captivity; not all their behaviour was genetically programmed. Like humans, they could learn new things, amuse themselves and live a contented life in a way other species of captive animals cannot.

Co-operative behaviour is seen among wild chimpanzees in the formation of alliances. These friendships are unstable; chimps remain friends until the individuals fall out during struggles for power and later either reconcile or remain enemies. If you are strangers to each other and are not going to spend time together, there is little point in co-operating, but if you know each other already or plan to get to know one another and expect to need each other, co-operation is your best option. Co-operation between kin is more commonplace because you share genes with kin, and helping them is almost the same as helping yourself. This is known as kin selection.

Austin and Sherman were playing a game of trust and

dictatorship – trust in that the chimp who handed over the key trusted the other to be fair, and dictatorship because for a moment one chimp had it all; he was the dictator holding the key to the sweets and also having the final say on how or if the sweets were shared. The key on its own has no value, but when given by one chimp to another who needs the key, it suddenly takes on value. Because this and similar games are played over and over with the apes swapping roles, the chimps agree the value of the key is half the total amount of sweets in the box. The second chimp could simply take the key and eat all the sweets, but if he did so the game would stop there and so would their mutually beneficial relationship. This game is a game of life. Game theorists first analysed Second World War battles in this way, and today relationships – right down to those between competing cells in our body – are studied by evolutionary game theorists and sociobiologists.

As part of their daily routine Savage-Rumbaugh's apes are engaged in various intelligence tests to determine a theory of mind, such as the false belief test. Sometimes called the Sally and Ann test, it is also used to test for autism in children. The point of the test is to see whether the child realises that in the dolls scenario 'Ann' has information that 'Sally' does not have. The psychologist introduces the dolls to the child – there is Sally, who has some sweets, and there is Ann. The psychologist manipulates the story. Sally decides to hide her sweets under a chair, and she does so in front of the child and Ann. The psychologist then takes Sally away. In her absence Ann decides to play a trick on Sally by taking the sweets and hiding them in her own pocket. Then the psychologist brings back Sally and asks the child the crucial question, 'Where does Sally think the sweets are?' Young children of about three years of age and people who lack a theory of mind – that is, do not understand that their own reality is not necessarily someone else's – will say that Sally thinks the sweets are in Ann's pocket, because they know they are there, rather than under a chair (which Sally would still believe).

Using this principle, Savage-Rumbaugh tested her chimps. Austin and Sherman were back in their divided rooms but this time Austin's keyboard was turned off. The chimps had to communicate in another way. Savage-Rumbaugh had left some food labels in Sherman's room; Sherman's keyboard was left on and Austin's was off. The chimps always helped to prepare their own meals, handling jam jars and packets of biscuits. They had not been taught to recognise the food labels, but had absorbed the knowledge of which wrapper indicated which type of food into their memory through observation and experience. Savage-Rumbaugh entered Austin's room and put Peter Pan peanut butter in a food container, then picked it up and led Austin into Sherman's room and departed again, leaving the chimps with the sealed box and all the various food labels.

Austin at first tried to use his keyboard to tell Sherman what was in the box. When he found it was switched off he turned to the labels and picked out the correct Peter Pan label, showing it to Sherman. Sherman then used his keyboard, which was switched on, to tell Sue he wanted some of the peanut butter. The symbols used on the keyboard are abstract in shape and not figurative. The peanut butter symbol did not look like the peanut butter label. Savage-Rumbaugh reversed their roles back and forth and tested them thirty times that day. Each time the experiment was just as successful. Both chimps had a knowledge of food wrappers that Savage-Rumbaugh was unaware of.

She watched her apes play imaginatively many times. The majority of autistic children cannot play imaginatively with other children. Austin loved to play with pretend food. From a dish and spoon made of air he would savour the imaginary food on his tongue. Watching himself in a mirror he would lick his lips, probably remembering the taste of his favourite flavour.

Sherman loved to pretend his King Kong doll was biting his fingers and toes and that his dolls were fighting each other. One day both chimps were watching a video of *King Kong*. Kong, now in New York, was breaking his chains and freeing himself from his cage with the screaming human hordes running for their lives. Austin and Sherman became so excited that they started to pretend that Kong was actually in the small empty cage in their room. They started to threaten the empty cage and throw things at it. Sherman even got out the hose and sprayed water at the cage. It was time for Sue Savage-Rumbaugh to stop the play, even though it was fascinating to watch. Once, when Sherman saw a chimpanzee being taken into a van by people unknown to him, he came to Sue and told her there was a 'scare' outside.

I visited Sue Savage-Rumbaugh in 1996. It was February and Atlanta was in the grip of a freakish freeze. I drove slowly down the long drive to Georgia State University's Language Research Centre. Pine and silver birch trees line the route and make up most of the woods in the open compounds where the chimps and bonobos play. Intimidating signs – 'No Unauthorised Personnel' – confronted me. When I arrived Sue and the other researchers were feeling very blue. Austin, aged twenty-two, had just died from cavities in the cerebellum. His illness had started with symptoms like those of a cold. Apes cannot fight human diseases and a simple cold is often fatal. In the winter Savage-Rumbaugh often puts jackets on her apes if they go outside, not because she likes dressing them up but because she has to keep them warm. They come from Africa after all.

Austin was monitored around the clock but died suddenly. He belonged to the university, and as a research subject his body should have been transported to the department of anatomy where students studying animal biology could have dissected him. But Savage-Rumbaugh loved Austin and could not think of him in those terms. She told me, 'I'd helped raise him since he was two. It was a terrible shock to me and all the staff. You see them lying there and you think they are going to get up and walk around and display and tell you to get out the way, but it doesn't happen. You remember all the fun times. People have brought flowers, have written poems; there is no culturally set way for how to deal with this. It's not like a pet dying; it's like a human being dying. In some cases it's like a retarded child you tried to help all your life dying. It's a very, very deep personal loss for myself and for Duane. It's far deeper than we would have anticipated before Austin died. In the past when apes have died at the Yerkes Centre they have been biomedical subjects and parts of their body have been shipped about and it's been considered inappropriate to do anything. But we felt we needed to do something with regard to Austin, so the decision was made in part by myself and Duane and also in consultation with the other staff that we would have him cremated and scatter his ashes in the wood where we liked to walk.'

Savage-Rumbaugh's approach to her research is very different from the way most scientists treat their laboratory animals. Scientists and lay people alike have found her approach to the animals and the implications of her research threatening to their beliefs of human supremacy. It has been her study of the bonobos Kanzi and his younger half-sister Panbanisha that has caused the most controversy.

Sue Savage-Rumbaugh was thrilled to start working with three bonobos as well as Austin and Sherman. One of the bonobos was a young female named Matada. She would become a close friend. The bonobos had been caught wild in Zaire and flown to the United States. This was a traumatic experience; a bonobo mother would be killed by a bullet or a poison dart, then the dead mother's body would be sold to the bush-meat trade. Bonobo meat is a fashionable delicacy for rich city folk, in much the same way that in Britain we like to eat game hare or venison. It was Savage-Rumbaugh's first challenge, to win the trust of youngsters that were terrified of humans.

Bonobos are much more lithe and slender than chimps. They have longer legs and are quite happy walking upright. Their skulls are smaller than chimps', but their foreheads are higher. Bonobos have an eight-and-a-half-month pregnancy, whereas chimpanzees are pregnant for eight months. Bonobos have black faces, pink lips and their long thin hair falls into a neat middle parting on their heads, giving a just-been-brushed appearance. If you showed a layperson the skeleton of a bonobo and the skeleton of Lucy, *Australopithecus afarensis*, they would find it impossible to tell them apart, except perhaps for Lucy's large cheek teeth. It is quite likely that approximately 5 million years ago our ancestors looked and moved similarly to the bonobo.

After some months, Matada understood that Savage-Rumbaugh meant her no harm and together they embarked upon a friendship that would take woman and ape to new heights. Matada became the first bonobo to learn LANA, and was kept with the small colony of bonobos at the Yerkes Centre. In October 1980 a captive-born hand-reared naive female called Lorel, on loan from San Diego Zoo for breeding purposes, gave birth. Lorel and newborn Kanzi fell asleep together. Matada took this opportunity to sidle up and inspect the new baby.

In just the same way as women coo over each other's new babies, so do female chimps and bonobos. Matada wanted to hold Kanzi, even though she already had a youngster of her own, Akili. Matada pulled the sleeping Kanzi off his mother's belly and held him against herself with her own one-year-old baby, Akili, climbing on to her back.

Lorel woke up and begged for her newborn baby to be returned. Being young and new to the group, she held a low rank and Matada refused to hand him back. Matada ended up nursing both Akili and Kanzi. Though natural twins are unusual, they do occur in apes. Because of Kanzi's enforced adoption by Matada, he became one of Sue's charges. If Lorel had kept her baby the two of them would have gone back to the zoo and his talents would have gone unnoticed.

Matada was a slow language student. She watched Sherman and Austin using their LANA keyboards and knew that was how they communicated their desires, but she found it harder to comprehend that each symbol had a separate meaning. Often she would press any old key and stare expectantly at Savage-Rumbaugh, presuming her to be telepathic. Yet Matada was obviously intelligent with a forceful character. When she wanted food she would thrust her plate into Savage-Rumbaugh's hand and push her towards the fridge. Matada and Kanzi moved to the Language Research Centre and Kanzi delighted in the contact with Savage-Rumbaugh, who continued with Matada's lessons while six-month-old Kanzi clung to his adoptive mother's tummy. Savage-Rumbaugh directed all her attention towards Matada, asking her to solve problems and requesting her to answer questions. All the while the little Kanzi was watching and listening, like a human babe in arms.

Sometimes Kanzi and Savage-Rumbaugh would go for a walk around the language centre. When they reached Duane's department, where disabled children were using the same LANA symbols board, Kanzi would stop and stare. Kanzi still loves to visit the children's department. When he was one year old he started to press symbols and wait by the vending machine for food, though the symbols, at this stage, did not correspond to the desired food. By the time he was two he would, during his mother's lesson, press the 'chase' symbol and run off laughing, looking over his shoulder in the hope that Savage-Rumbaugh would chase him.

Most chimpanzees and bonobo mothers are tolerant of their baby's whims and tantrum. Matada would not allow Savage-Rumbaugh or the other researchers to reprimand Kanzi if he misbehaved; if he bit Sue, Matada would not allow Savage-Rumbaugh to chastise the infant. If a researcher ever tried, Matada would defend him by becoming hostile to the human. When Kanzi was eighteen months old the Yerkes Centre decided it was appropriate for Matada to breed once again, although in the wild a female bonobo would not have another baby so quickly. Matada was taken away and Kanzi remained with Sue and Sue's sister Liz, who was helping to care for him at this time.

That first night Kanzi was scared to be alone and climbed into bed with Austin, but he couldn't get comfortable. Savage-Rumbaugh, expecting trouble, had made up a camp bed for herself so she could offer Kanzi twenty-four-hour help if he required it. For a few nights, while Kanzi readjusted, they slept together. Sue started to teach him the symbols board but was astounded to discover he already knew most of it! While Sue had painstakingly gone over the symbols with Matada, Kanzi had been the one spontaneously absorbing the information, like a human child.

Because of this revelation Sue Savage-Rumbaugh and her team realised that language lessons with apes needed to start at the earliest age possible in order to achieve the best results. Because Kanzi was young and playful, the intelligence tests Savage-Rumbaugh set for him were in the form of games. Many of them were played out of doors, in the woods – hide and seek was a favourite. It was arranged for the now pregnant Matada to return, but Savage-Rumbaugh worried that Kanzi might regress on his mother's arrival; after all he had not used the board before she left, even though he had understood how to. Matada and Kanzi were delighted to be back in each other's arms, laughing and hugging. Then Kanzi, using the LANA board, spoke to Savage-Rumbaugh and asked to visit the children's department to look around. He gave his mother some bananas and juice before he left, but was happy to leave her. Savage-Rumbaugh realised Kanzi was content to oscillate in between the ape and human worlds.

Not only has Kanzi spontaneously acquired comprehension of words, but he actively eavesdrops on conversations. Kanzi listens to Savage-Rumbaugh as she speaks and interacts accordingly. For instance, if she is talking about the light to someone, Kanzi jumps up and puts the light on. Like with the parents of a young child with an increasing intelligence, things had to be spelled out in letters if people didn't want Kanzi to know what was happening. Kanzi, now four years old, had spontaneously acquired the comprehension of at least 150 spoken English words; today, aged nineteen, he comprehends well over 700 words.

Like most of us after a hard day's work, Kanzi likes to watch a bit of telly. *Quest for Fire* is Kanzi's favourite film. He also enjoys watching human wrestling and the National Geographic Jane Goodall specials, but he especially loves watching movies that depict half-ape half-man figures. A favourite is *Harry and the Hendersons*, in which Peter Elliot plays a bigfoot. It must be flattering for ape-actors Peter Elliot and Ailsa Berk to know that their number one fan is a bonobo; if their performance convinces an ape, it must be good. I love to think of Kanzi and other captive apes relaxing in front of the TV set as they watch human actors portraying apes and hominoids.

When the apes grow bored of watching the same movie, Sue and her stream of female researchers put on ape costumes and masks and, camcorder in hand, play in the grounds of the university making home movies of hide and seek and tag for the bonobos and chimps to watch. Who said science was boring?

Kanzi is the Einstein of all apes. Some people are smarter than others and some bonobos are smarter than other bonobos – Kanzi is a dignified, brainy bonobo. When I met him in February 1996, I spent a day filming him. He didn't like my all-male film crew. He saw them as invading his territory, and kept displaying and running at them, throwing himself with all his might at the reinforced glass between him and us. It was a noisy and threatening display.

Kanzi didn't charge at me. I was a blonde white woman and he is used to them. I leaned against the glass and he did the same. Face to face, a transparent inch between us, we stayed like that for some time. There was a small hatch in the glass through which food or small objects could be passed and one could reach through to touch Kanzi, if one dared. I spoke quietly to him through the hatch and he listened to me, smelling me, feeling my breath on the top of his head. But he wasn't looking at me any more – he was nonchalantly averting his eyes. He knew if I had looked into them I would have seen his intent. I've no concrete proof, but I feel sure that if I had put my hand through to make perfect contact, as he wanted me to do, he would have bitten me, just to assert his male status. Kanzi looked innocent enough, but I sensed his anger and I kept my ten fingers intact (unlike Sue Savage-Rumbaugh and Jane Goodall).

Later that day, looking for Savage-Rumbaugh, I walked past Panbanisha's enclosure and the bonobo started to hoot at me. I went over and asked her if she knew where Sue had gone? Panbanisha emitted some more high-pitched hoots and pulled her symbols board to the glass barrier and started 'talking' to me by pointing at the lexigrams. I panicked; this nine-year-old female bonobo shared her brother Kanzi's linguistic skills and was trying to talk to me but I didn't speak LANA. I urged her to stay where she was and that I would be right back. I ran into an office and luckily found a symbols board, and took it back to Panbanisha's enclosure. I watched carefully which symbols she pointed to and read what they meant on the symbols board I held. Panbanisha was saying 'Get Sue.' When I asked, 'Do you want me to get Sue?' Panbanisha replied 'Yes.' 'Shall I get her now?' Panbanisha hooted and jumped about; her answer was obvious, she didn't need to point to 'yes' again. I took her request for granted and at face value. Why should-n't a bonobo talk to me? It was all perfectly natural. Only later did I realise that I'd just had a conversation with a dumb animal that I'd never met in my life before, and I certainly hadn't cued.

In 1984, five years after Herb Terrace's declaration in *Science* that no one had discovered language acquisition in apes, Sue Savage-Rumbaugh wrote her first scientific paper on Kanzi, entitled

'Spontaneous Language Acquisition by a Pygmy Chimpanzee'. *Science* returned the paper, informing Savage-Rumbaugh that she had nothing new to add; as far as they were concerned there was no debate. Savage-Rumbaugh continues to suffer resistance to her research. I asked her about this problem.

'Magazines like *Science* have returned my papers saying, "This is a trivial finding, there's nothing new here." What they were really saying was, "Why don't you look for another journal? We don't want it." *Science* published the Gardners' work, then Terrace said it's all a fantasy and *Science* felt burned. They felt language research couldn't be a true science if the scientists couldn't agree on it. If you look through the magazine today you will find physics and molecular biology, but you will not find much behavioural science. If you have new facts, you have to look at these facts in a new light as other things are discovered.

'With regard to man–animal relationships, people have been banging their heads together over it probably before written history, but certainly since the time of Aristotle. And many of those bangings have had no facts whatsoever – they are just opinions based upon what I, this human being, think about myself, and this animal that I'm eating, or this animal that I have little respect for, or that I've killed, or hunted, or that maybe I share my home with, but I certainly don't consider them capable of communicating. In this case we have got new facts. You have to have the intellectual argument, but you also have to be responsible to discoveries in science, and that hasn't happened.'

The planum temporale, an asymmetric part of the brain, larger on the left side of the head, is integral to the control of language in humans. Recently it has been discovered that the planum temporale's positioning in chimps is the same as that in humans. This neurological discovery gives even more grist to the Great Ape Project's mill. It indicates that apes are in possession of language. Human language may have evolved to become more sophisticated, but the basic wiring is the same. I once heard a novelist comment on the ape languages studies: 'I like the fact it's only humans that can understand language, it makes me feel special.' Could his resistance to this research stem from human male speciesism, or are authors more possessive of language than other people? Certainly, Savage-Rumbaugh's most vociferous critics have been men.

I asked Sue Savage-Rumbaugh to explain what she thought we could learn from ape studies. 'We share 99 per cent of our DNA. For years man has drawn a big boundary between ourselves and other animals and hesitated to let studies of animals throw light on his own behaviour. Except for the establishment of city-states, the raising of armies and the establishment of written laws in books we can hold up a mirror and look through it to the ape to learn almost everything else about ourselves. We can learn about love, about aggression, about fear, about in-group out-group dynamics, about the development of mind, how social relationships are formed, how cognitive concepts about space and time are formed, about the complex process of communication and how it modulates the day-to-day life of all individuals in a social group and binds them together.'

I asked her to describe to me how she started on this quest for perfect communication with apes. 'I was very interested in the social development of mind and the development of personality. I suppose I had a rather naive idealistic view that if we could understand ourselves and how we developed as we do we could create better societies and do away with war. We could do away with many problems. Although technology helped us have a better day-to-day life, it didn't help us to get along with each other.

'I felt that by the time a child was four or five years of age they were pretty set in many of their aspects of personality, and that you had to study them between birth and three to four years of age if you wanted to understand how we shaped human beings that in turn shaped our society. I was focusing on the areas of child development and was very interested in the entire field. We control behaviour, we modify it, we shape it any way we want, and then we build the kind of society we want.

'I suppose these ideals are a function of the way my parents raised me. I thought that everyone should grow up and try to make the world a better place and leave it a little better than how they found it. My only question was how could I best do that, given my skills. I must admit it has been a great sadness to me to learn that other scientists don't always have that goal and many times their only question is: How can I further my career? How can I get another publication? How can I get a better position, or how can I beat the argument of this other person! Maybe I'm unusual, but at the time it was all I knew.'

Like Jane Goodall, Sue Savage-Rumbaugh had one child, a son, and a lifetime of first-contact narratives with great apes. I questioned her about the life of a woman primatologist: 'I think women have a great disadvantage because people are less likely to believe what they say. I think women have done well in this field because it is a brand new field and there haven't been many people in it and women have been prepared to be patient. They've been willing to watch, they've been willing to wait. Women are always willing to do that in all disciplines and this happens to be one where that gets rewarded. And men are often in a hurry to move on to the next thing, and miss something that had they been patient they would have seen. I have been accused of over-empathising with my apes like a proud parent might, but if people had taken time to read my work they would see I have learned from what the apes cannot do as well as what they can.

'Some of the best work in language acquisition and the development of cognition has been done by parents. Take the psychologist Jean Piaget, for example. No one accused Piaget of finding these things in his children because he was their father. Now I wonder if Piaget had been female if he would have been accused of finding these things in his children because he was the mother? I think there is a bias against females. It's okay for males to have empathy and be objective, but females are presumed to have some kind of difficulty in being empathetic and being objective at the same time. I think that is a false dichotomy

'The basic evidence that I have that is the strongest is the fact that these apes are learning spoken human language without anybody training them. We don't say "Ball", "Ball", "Ball" while holding up a ball or "Dog", "Dog", "Dog" while showing them a picture of a dog. We say, "Do you want to play with the dogs? Well, put your ball in the back pack and let's go!" The ability of Kanzi to understand new sentences – sentences that he has never heard before in his life – is the strongest evidence that they process language the way we do. They don't have a vocal tract, they can't speak rapidly like we can, but if they did have a vocal tract I've no doubt they would be talking. Kanzi can understand relative clauses, he can understand questions, he can understand inversions. So can Panbanisha. They can understand complex sentences that not only use inversion but also involve inference about the relationships between people and what those people know or don't know, have or have not got.'

Sue Savage-Rumbaugh then went on to tell me about a Sally and Ann theory of mind test she and her sister Liz put Panbanisha through. Without being in Panbanisha's presence, they had carefully rehearsed a scene together in advance of testing the ape. Then, so Panbanisha would not anticipate an intellectual test, Sue and Liz took her for a casual stroll in the woods. As they walked along with an unsuspecting Panbanisha between them, Sue ostentatiously gave Liz some M&Ms. Delighted, Liz put them in a box she held. Liz then stated she had to take a walk but promised to return for her sweets, and gave the box to Sue and left. In Liz's absence, Sue grinned at Panbanisha and said, 'I'm going to play a trick on Liz!' Making sure Panbanisha was watching, she took the sweets out of the box and put them in her pocket, then put a creepy-crawly in the box and shut the lid.

Liz then returned and asked Sue for the box so she could eat her sweets. Before opening the box Liz made sure she gave Sue enough time to ask Panbanisha the crucial question, 'What does Liz think is in the box?' Wide-eyed and obviously worried, Panbanisha looked at Sue and said, 'Bad.' Sue asked her if she thought she was being bad to Liz. Panbanisha emitted a quiet hoot which meant yes – she did think Sue was bad. Sue asked the question again, 'But what does Liz think is in the box?' Panbanisha gently pointed to M&Ms on her keyboard.

Later that day Sue and Liz tried the same test on Liz's four-year-old daughter, Heather, who had an emotionally similar response. When Sue asked the little girl what Liz thought was in the box Heather said, 'You're being bad to my mummy.' Sue asked again, 'But what does your mummy think is in the box?' Heather quietly answered 'M&Ms.' Ten out of ten for both bonobo and child. Autistic children, who it seems lack full theory of mind, would say that Liz thought there was a bug in the box, because they knew that to be the truth and would not be able to backtrack to mind-read Liz's thoughts.

What do wild bonobos have to say to each other? They are very vocal animals. Sitting up in the trees together, they are constantly talking to each other in high-pitched hoots. But they quieten when they climb down to the ground; bonobos are eaten by leopards and by men. Savage-Rumbaugh has been on several trips to Wamba, the Japanese bonobo research site in the Congo. She observed the wild bonobos and watched them quietly climb down out of the trees when it was time to eat. Small parties would wander off to forage, and sometimes individuals were left behind or a party would become split in two. But then Savage-Rumbaugh witnessed something extraordinary; at times like this the bonobo in the lead would take a small branch or a leaf and lay it on the ground pointing in the direction that they were intending to go. The branch or leaf was a signal that gave stragglers the chance to catch up. Sometimes the leaf would be torn or the branch broken to make an arrow that was left to signal the correct direction.

Savage-Rumbaugh observed this behaviour on ninety occasions. This is an indication of both tool and language use. Not only was the foliage being modified by the bonobos, but it was being used as an arrow to symbolise where they were going. So far no other researchers have witnessed this extraordinary phenomenon.

Bonobos do not fish for termites with grass like Goodall's chimps or crack nuts like Hedwige Boesch's chimps, as food is plentiful rather than because they cannot make tools, as has sometimes been suggested. Captive bonobos are just as handy with tools as chimps are. Wild bonobos are lucky to have the enormous, succulent and highly prized *Anonidium mannii* and the *Treculia africana* fruits growing in their neighbourhood. There is more than enough to eat. Bonobos spend a quarter less of their day than chimps in foraging for food and therefore have much more time to sit chatting to friends. Unlike chimps, who mostly sit alone to possessively eat their meals, the bonobos have big dinner parties where everyone shares – after the high-status females have eaten, naturally.

Louis Leakey was right in predicting the great leaps of understanding that have unfolded from the synthesis of ape behavioural research and paleoanthropology. Studying the capabilities of apes has taught us so much about the apes, and the apes, as human models, have taught us so much about ourselves. Knowing how excited he would have been to hear many of the more recent discoveries, there is one experiment with Kanzi that would have pleased him no end. Kanzi has been making stone tools, much like the ones Leakey dug up at Olduvai. Kanzi was shown that by knocking two stones together he produced sharp flakes that could cut through rope used to tie up a box that contained M&Ms. Kanzi was never required to do this task. He was encouraged to watch and Savage-Rumbaugh hoped he would learn by example – and he did.

Manual percussion of stones to make tools takes skill. Kanzi was at times frustrated by the length of time it took him to get to his M&Ms. One day Kanzi stood up on his legs and threw one of the rocks on to the ground. The force caused the stone to shatter, producing many razor-sharp flakes. Delighted with his own success he quickly cut through the rope and opened the box. Kanzi had not been shown this technique but from watching how he negotiated the problem palaeontologists have reconsidered the way 3-million-year-old artefacts like these were made. Perhaps archaic humans threw rocks to make sharp tools instantaneously in just the same way as Kanzi chose to, rather than painstakingly knocking two stones together as has been presumed.

Sue Savage-Rumbaugh says working with her apes has taught her what it means to be human. By this she means she understands which evolutionary adaptations divide us from other apes. But what does a life of captivity teach a human about being an ape or even an ape about being an ape? It is not a natural life, even if you are loved and treated like a replicant human. The sex life of the animals, for instance, becomes perverted. As white women many of them blondes care for the infants almost exclusively, the maturing male chimps and bonobos have this particular female image imprinted psychologically, and it becomes hardwired. This means the male apes, as adults, find blonde white women sexually attractive.

Carelessly, in the past there was too much inbreeding among captive populations of apes. Consequently most of the males Savage-Rumbaugh works with, including Kanzi, will not be allowed to breed. She has no influence over this matter; it is an issue of whose genes are over-represented in the stud book and Kanzi's father had been allowed to breed too many times. Animals that are not allowed to breed – the majority – are given magazines to encourage masturbation in a bid to alleviate their sexual frustration. Magazines depicting what? Ordinary women's magazines are a big turn-on for a sexy male chimp or bonobo if they contain pictures of blonde, white women, especially if they are dressed in fur coats. It is not an image of a naked woman, but simply a white woman covered with fur that helps to give Kanzi satisfaction. This fact tells us a great deal of how these animals perceive us. They feel close to us because they know they are, even if some of us reject their kinship.

We can see that apes look like us and Jane Goodall proved they behave like us. Sue Savage-Rumbaugh is determined to show us they think like us too. She will not give up the fight, and demonstrates the long-term commitment to her study that Leakey encouraged in the trimates. If you want to know about apes intimately, you must devote more than a few years to the challenge. The longing to understand them can take over your life.

The debate that surrounds the intelligence of apes will probably never be resolved, and apes are not going to be able to vocalise their rights. If a successful scientific breeding experiment between human and ape produces a living hybrid, then and maybe only then we will have troubling, tangible proof of our genetic and intellectual compatibility that could not be argued with. Until that time women will continue to care for captive primates and try to bond with them on a human level. Perhaps we need to study their own language more, and try to speak to them in their forests on their terms, instead of colonising apes, 'civilising' them and trying to turn them into surrogate, honorary humans.

9
The Battle of the Sexes (and the uses of the female orgasm)

Certain aspects of human behaviour are the outcome of the evolutionary process. Some sociobiologists would say that all areas of behaviour are predetermined and primates just choose the strategy that best suits them at that particular moment in time from their own innate repertoire of behaviours. Within the competitive area of reproduction we find a wide range of behaviours that contribute to the battle of the sexes. Primatology offers us scientific models from other species of primate from which we can make comparisons about our own sex lives. Although there are many parallels between us and them there is also much variation between ourselves and our nearest relatives (and there is much variation in behaviour between species of non-human primates). When similarities between us and them occur it is startling and we are presented with another clue to the meaning of life. But as evolutionary psychologists admit, there are still many gaps to the big picture. There is much more to learn and generalisations, however tempting, must be avoided.

Feminism and humanity have taken a progressive step forwards because of the collaborative research by women scientists in the area of female behaviour in primates. Charles Darwin, father of the theory of natural selection, described females as innately sexually passive: 'The female, with the rarest exception, is less eager than the male, she generally requires to be courted, she is coy, and may often be seen endeavouring for a long time to escape from the male.' Subsequently, any behaviour by females that contradicted this coyness was described as 'promiscuous'. It was only very recently that female non-human primates were thought of as sexual creatures with the ability to have orgasms.

Langurs and Sarah Hrdy

Anthropologists and animal behaviourists used to interpret acts of infanticide among the animals they studied as perverted aberrations. Infanticide used to be explained as a result of males killing babies when animal groups were over-populated. It was thought that a dominant force in nature was behaviour where an animal's actions were for the good of all. We now know that the individual animal, according to its age and social status, has its own triggers for making it behave the way it does. It does not usually act for the good of all, unless the group is under attack; most of the time it acts on behalf of itself and its immediate kin.

Primatologist Sarah Blaffer Hrdy revolutionised our view of the natural order when she discovered that the behavioural phenomena of promiscuity and infanticide were often linked. We now accept that infanticide is in certain circumstances a male reproductive strategy and that female primates often proffer their sexuality to circumvent male violence. Hrdy's interpretation has become the new orthodoxy.

Sarah Blaffer Hrdy grew up in Houston, Texas; she is a tall, blonde oil heiress. Her childhood and adolescence were privileged, though she always felt at odds with the patriarchal Texas mentality. In 1971 while a graduate anthropology student at

Harvard University, Sarah Hrdy heard the perplexing stories of infanticide committed by male Indian langur monkeys. No one could explain it; Hrdy decided to try. During her first summer at Harvard, with her mother's financial support Hrdy visited India to seek out the nature of the sacred monkey. Traditionally, primate studies had been undertaken by men who had focused their attention on the flamboyant males, but, as with other significant women primatologists, when Hrdy went into the field she made a point of watching both the females and the males before coming to her conclusions.

In 1971 Hrdy headed for northern India with the question 'How does high population density affect behaviour in primates?' In the desert land of Rajasthan and the medieval towns of Jodhpur, Jaipur and Mount Abu, black-faced, long-limbed hanuman langur monkeys sit confidently in rows outside temples, hoping to be fed. The langurs loop their arms around the monkey in front, their long tails carelessly entwined on the floor like piles of rope and dextrous black fingers lazily parting the light grey hair of their friend as they search for ticks. Hrdy spent ten years flying back and forth between America and India, accumulating data on the langurs, and over that decade formulated her radical theories on infanticide and the evolution of sex, especially the female orgasm. When Sarah Hrdy entered the scientific arena, the reigning paradigm of contemporary anthropological theory stated that every individual had a role to benefit the group as a whole. With that idea firmly in her mind, Hrdy settled into her study. Every day she rose at dawn to follow her langurs as the troop climbed down out of the trees and set off foraging. The monkeys would rest during the heat of the day and Hrdy would find some shade in which to sit and watch them. As the afternoon cooled, the animals would wake up and be off again, Hrdy hot on their tail.

It was not long before she noticed that whenever a new male langur took over the troop he systematically killed all unweaned infants belonging to the earlier and now ousted male. The new male would chase the nursing females and steal their suckling babies from them, biting their heads in two. But how could infanticide like this result from overcrowding, when no overcrowding had occurred? In addition, Hrdy was at a loss to see how this behaviour benefited the group as a whole.

Hrdy noticed that the changeover of the dominant male occurred roughly every year or so. These interloping males lead an insecure existence. They never know what tomorrow will bring and must remain alert and opportunistic. When a male langur reaches his prime he takes action, for this is the moment he was born for. He approaches a group; if the resident male tolerates him at a distance he hangs around, bides his time and weighs up the best plan of attack before pouncing.

Sarah Hrdy would watch the subsequent behaviour of the female langurs with interest. The females, whether sexually receptive or not, would notice the new male. If he seemed like a genuine threat, they would not run but approach him and, somewhat surprisingly, solicit him for sex. Hrdy saw pregnant females simulate oestrus behaviour by bobbing their heads, shuddering and sticking their rears in the male's face, behaving as though highly sexed even though they could not be, as they were already pregnant. They were in fact faking oestrus. Sometimes this strategy worked and the male was aroused into mounting them. At other times, no matter how persistent the female, he knew the performance was fake.

Hrdy watched both the male and female behaviour and made connections that no one else had between the threat from the newly arrived male and the sexual behaviour of the females towards him. Hrdy remembers that early on in her career Louis Leakey had said, 'You can send a man and a woman to church, but it is the woman who will be able to tell you what everyone had on.' Hrdy was a patient, intellectual observer of detail and even after being taught incorrect theories at university she was still able to recognise how the complex behaviour of the male and female langurs actually fitted together.

For many years the 'promiscuous' behaviour of the female primate had confused and disgusted male scientists. It didn't occur to them that this promiscuous behaviour was the pregnant or nursing female primates' way of trying to confuse the new male into thinking her existing baby was his, and this was Hrdy's radical interpretation. If the females could encourage the new aggressive male into having sex she could probably fool him into thinking her infant was his, and male primates very rarely kill their own flesh and blood. The females weren't promiscuous, but trying to save the life of their babies by outwitting the new male. As a female primate herself, Hrdy related to the female monkey's predicament.

Thanks to Hrdy, it is now recognised that the male reproductive strategy of infanticide is common among mammals. From the timid house mouse to the fiercest African lion, male animals will, when approaching a female who is nursing unweaned infants, kill those young if the babies are not theirs. Males have a built-in trigger that switches on at the moment of ejaculation; it stops them from killing their own but is not fail-safe and females are able to fool males into paternal commitments for babies that are not theirs.

Hrdy sometimes saw groups of female langurs gang up together in a mutually beneficial attempt to chase a potential infanticidal interloper away. Occasionally this would work, but once an ambitious new male takes his chance and chases off his predecessor he is unstoppable. Over a period of a few days he proceeds to kill all the unweaned babies he can get his hands on.

With the infanticidal male at the heart of the troop, individual female monkeys display various defensive strategies. Some mothers take their almost weaned infant to the previous dominant male and his male friends. These males are now on the periphery, and one of them is probably the baby's father. She knows they will not kill the baby and they may be able to look after it. But more often than not the baby misses its mother and follows her back to the bosom of the group – straight into the clutches of the infanticidal male.

Some females attempt to leave and join a group where the alpha male is already known to her, and where she feels safer. In the past she has probably had a baby from him and he is less likely to kill one of his progeny's half-siblings. But, in leaving her group, the female langur leaves her female friends behind and has to start at the bottom of the new group's female hierarchy. Her new lower status helps neither her nor her infant. Some pregnant female langurs will cut their losses and spontaneously abort their foetuses as there is no point in investing further resources in creating a baby that is doomed. Hrdy was sickened by the sight of these partially developed tiny bodies discarded on the ground. At a primate conference, Hrdy told Biruté Galdikas that catching the gaze of a male hanuman langur was akin to looking into 'the soul of a shark'.

All of this carnage selfishly benefits the newly self-appointed leader. He has waited for this moment; he may have tried before, but now he is the victor and he will revel in his time. The male must quickly reproduce himself as many times as possible, for his status is precarious. To achieve this he must, to the detriment of the females, subvert and control their reproductive lives. A third of baby langurs already born die at the hands of the new alpha male.

The harem of females now belongs to him. An existing suckling infant prevents the mother from coming into oestrus, but with her infant murdered, lactation abruptly stops and eventually the female comes back into season. It is then that the female mates with the infanticidal male – it is not in her interests to do otherwise, or her sons would never be able to compete with other males who have inherited the infanticidal trait from their fathers. A third of fertile females find themselves pregnant again before they have had time to replenish their fat reserves and recover naturally from the previous pregnancy, birth and lactation.

This form of infanticide should not be seen as out-of-control aggression, but as a specific reproductive strategy that has been selected for. Killing another male's brood not only enables the male to mate with the bereaved mother but also eliminates his progeny's competitors. Non-infanticidal males will not produce as many offspring as those that are infanticidal, so over time infanticidal behaviour becomes a dominant trait and the behaviour, however cruel, continues through countless generations.

Hrdy's research has helped to explain a male primate's feelings of sexual possessiveness. Our taste in and selection of a mate is a process that is directed by similar primal desires to those of our primate cousins. Male primates do not want to waste either resources or time on an infant that does not carry their genes. In humans we can see a similar strategy played out. Some men usually prefer to date childless women smaller than themselves, whom they think they can control, while women prefer to date men bigger than themselves, whom they think will protect them. Like the other apes, humans are sexually dimorphous; men, on average, are 5–20 per cent bigger than women. This difference in gender size can be greater in humans than it is between male and female chimps, though it is not as great a disparity as that found in gorillas. If we look to the hanuman langur as a model we can see that if the alpha male is big and strong, he can defeat other males and protect his females and young. At the same time his strength will also prevent his wives from leaving him.

Sexually dimorphous species are also usually polygynous to a lesser or greater degree, which can enable big, strong males, such as gorillas, to bully their way into achieving more than one wife. In humans the culture of money and the contents of a man's wallet can make him metaphorically bigger and stronger than other men around. Some rich men have mistresses; in Western societies wealth equals power and high status, making him sexy to women because ultimately he can offer beneficial resources to the woman's children if she bears any with him during their liaison.

Not all species of primate commit infanticide. For example, it has never been observed in bonobos or orang-utans, even though infanticide occurs regularly among their cousins, the chimps and the gorillas. One quarter of infant gorillas die as a result of infanticide when a new silverback takes control of the breeding females. In so rare a species this is an additional tragedy.

So why don't we see infanticide in orang-utans? Orang-utans are polygynous and sexually dimorphous, but as they are absent fathers they are unsure, when they happen upon females they have mated with, whether or not the babies the mothers nurse are theirs. To be safe they do not kill them, but they do not help to look after them either; if they are not theirs they have not wasted any personal resources on them. The main reason we do not see male-perpetrated infanticide in patrilineal bonobos is that they are the only matriarchal ape. Matriarchy leads to a number of other differences between them and other apes. The bonobo requires careful examination.

Female Bonobos

Some scientists believe bonobos are more like us than they are like chimpanzees. They are certainly quite different from chimps in appearance and especially in their behaviour. A bonobo female, so long as she is in good health, leads a fine life. She is not bullied or beaten or sexually coerced by the male of her species; she's her own boss. Bonobo females are what is known as female-bonded. Sisterhood is the most powerful force in their lives.

The bonobo was officially discovered in 1929 from a selection of ape skulls sent from the Congo to the Tervuren Museum in Belgium. The difference between the chimp and bonobo was evident just from the examination of its skull. Bonobos were originally known as pygmy chimpanzees, but this is no longer a favoured term as they are not small, they are not chimps and it forces us to refer to chimps as common chimpanzees – which are not common, but endangered. Bonobos were first studied in the wild in 1973 by Takayoshi Kano, a Japanese primatologist who established the Wamba field site.

One reason bonobos are considered to be more like us than any other living animal is that they engage in erotic sexual acts for pleasure and not just for procreation.

The unique bonobo is only found in a small and inaccessible region of the Congo. In the wild they do not live alongside chimps or gorillas; the only other apes they meet are humans. About 5 million years ago tectonic activity created a depression that we now call the Congo Basin. The subsequent earthquakes caused Lake Tanganyika to overflow, sending the water into the Congo. The resulting rivers and marsh land, over a million years or so, caused the separation of the contemporary chimpanzee and bonobo species from their joint ancestor. Between 5 million and 2.5 million years ago the Zaire/Congo river in the north and the Sankuru river in the south diverted the bonobo's evolutionary path from that of other apes (coincidentally just about the same time that *Homo*, our direct ancestor, evolved 2.5 million years ago).

Jo Thompson's bonobos in fact are unique amongst all other bonobos, being the most geographically southern. Approximately 8,000 years ago they were separated from the more northern communities of bonobos by the Lukenie river, and have since adapted to their mosaic habitat of forest and savannah in ways different from the bonobos further north, such as those at the Wamba field site. Thompson was amazed when she saw Scarlett, an extraordinary sexually mature adolescent female bonobo, and Ruby for the first time, as no other primatologist had come across cinnamon-coloured red-headed bonobos before; we think of them as having only dark brown or black hair. Scarlett and Ruby's red hair may be an adaptation to living in a partial savannah habitat where there is more exposure to sunlight

Scarlett has become Jo's favourite bonobo. Apparently the young bonobo, aged between twelve and fifteen years, behaves like Scarlett O'Hara from *Gone With the Wind* (1936). A very clear and animated communicator, she is a 'beguiling coquette'.

Thompson is still working on habituating the bonobos; until they are all completely comfortable in her presence they will not allow her to follow them freely as they interact with one another. So far she and her field assistant Mvula have observed Scarlett's overt and subtle demonstrations towards them. Scarlett's most obvious feelings of annoyance that Thompson is watching her yet again are shown when she stamps her foot, throws sticks, urinates, defecates and vocalises in their direction. Scarlett bobs her head, yawns and scratches when she spots Thompson, and sometimes she also leans and stares back at Jo, as if to say, 'What the hell are you staring at?' Ruby is an older female, possibly Scarlett's mother or aunt, mother to a small black infant that she carries everywhere. Thompson has tried to take some family photos of these magnificent specimens but she doesn't want to lose the animals' trust, fearing that they may think the camera lens is the barrel of a gun or that later they may mistake a hunter's shotgun for a lens. There is a market for bonobo babies as pets, and when Scarlett has her first baby – which will be soon – she could easily become a target for a poacher's bullet. Thompson told me that adolescent female bonobos are slim and well-toned but with each subsequent pregnancy their figure gets worse. With each baby, another 4 inches go on around the waist, their muscle tone and trim figures disappear for good and their figures start to resemble those of gorillas.

Bonobos have often been compared to early hominids. Anatomist Adrienne Zihlman, who has famously superimposed both a bonobo skeleton and an *Australopithecus afarensis* (Lucy) skeleton over the form of modern man, believes the bonobo is the closest living representative of our earliest known ape/hominid ancestor. Thompson's bonobos are especially useful for such comparisons as, unlike other bonobo communities further north living in forests, they have adapted to moving through partial savannah; in open areas they walk bipedally between fruiting bushes. (They also wade bipedally into waist-deep water to pick at and eat aquatic vegetation.) The evolving savannah persuaded our ancestors to come out of the trees and walk in much the same way.

The geoxylic suffrutice bushes that the bonobos walk to and feed from are an ancient species of tree that evolved during the development of the Great Rift Valley. Millions of years old, they only grow in savannah regions. Three million years ago Scarlett's bonobo ancestors and Lucy's kind would have fed from these fruits. When Thompson is out in the forest with the bonobos she eats all the same food as them. The ngungu fruit found on the geoxylic suffrutice bush is a cross between an apricot and a plum in appearance and taste. This is the original tree of knowledge. Female hominoids have bitten into its tempting fruit since the first female was born and Thompson assures me it tastes 'delicious!' When Thompson observes her bonobos walking on two legs she must think she is seeing something from the dawn of man.

The fossilised skeleton of Lucy is more than 40 per cent intact and from it we can tell much about her. Lucy stood only 4 feet tall, but was very strong, much stronger than modern woman, with large feet and muscular, long toes, able to walk over rough terrain. Lucy's legs were more like modern woman's; the knee joint was similar even if, proportionally, her legs were a little shorter than modern woman's. She could walk as well and for just as long as modern woman, though she probably could not run as fast. The shape of Lucy's pelvis was similar to modern woman's but narrower, as her baby had a smaller cranium at birth. Her face looked much more like a bonobo's than like modern woman's. Her sloping forehead was smaller and her protruding jaw bigger, and although her canine teeth were reducing in size, they were bigger than ours. The size of our canine teeth is significant. They have large roots and take a great deal of stress; these are the teeth we use to puncture and tear with. Our teeth were once our main tool, and we still have the instinct to use them to open or crack things. We connect humans with large canines with something inhuman – Bram Stoker named them vampires – and anthropologists would say extra large canines are a throwback. Lucy represents the link between woman and ape. When parts of the body evolve at different rates as they did with Lucy, it is known as mosaic evolution. Above the neck Lucy looked and thought like Scarlett. Like Scarlett, Lucy wasn't talking but she was probably capable of having orgasms and probably lived with the threat of male-perpetuated infanticide. From the neck down Lucy looked and moved like a small, hairy woman.

Lucy could climb with ease and probably made night nests in trees, just as Scarlett does. She no doubt took refuge in trees from prehistoric lions, but she was terrestrial for much of the day. Unlike Scarlett, Lucy may have scavenged the bones of dead prey once the pack of lions had moved on. Using a rock as a tool, she could split open the femur and share the marrow with her baby. Her people would have xenophobically killed other Australopithecines but they would have co-operated with each other and shared fruit and meat.

Lucy was in her early twenties when her life ended. She probably died of parasitic disease, collapsing at the side of a lake and her body sinking into the mud. When she died, Lucy possibly left behind two or three motherless young children. These early people probably had a forty- to fifty-year lifespan if they were lucky, but most probably died in their twenties, giving them just enough time to reproduce. Within our dormant junk DNA the genetic sequences for Lucy's species can be found.[1]

The bonobo is known as the, 'make love not war' ape because sex is used by bonobos as a substitute for aggression. When things get tense between males, they stop themselves before things get really nasty and they rub their penises together. A similar tactic applies to females. For instance, if a female grows tired of another female's infant, she may hit it, and in turn the youngster's mother may strike the other female back in revenge, but then, to smooth out the resulting tension, the females have lesbian sex, known as genito-genital rubbing, or GG rubbing. Bonobo groups range in size from 40 to 120 individuals, and even though inter-group aggression is very rare smaller groups do avoid larger groups.

Bonobos openly share food, whereas chimps do not. Bonobos sit in large groups, continuing to vocalise while they share food among their members. Adolescent bonobo females will also exchange sex for food, whereas adult females have the social stature to take whatever food they want. If a male bonobo has found some prized fruit, a young female will encourage him to give it to her in exchange for sex, an offer he cannot refuse, irrespective of her oestrus cycle. Fortunately within the bonobo's limited geographical range food sources are plenty. The pith of much of the fruit eaten by bonobos is rich in protein. But female bonobos *do* eat meat.[2] Wild bonobos have been observed cuddling and playing tag with colobus monkeys, whereas a chimp would make mincemeat out of a monkey given half the chance.

Because bonobos do not hunt co-operatively, like chimps, they do not have the ritualised greetings of chimps or those used to enforce a hierarchy by other social animals such as hyenas and hunting dogs. Bonobos are more laid back; although they can squabble competitively and the males can kick each other from time to time. Chimps in a fight are more aggressive and likely to pull their opponent towards them in order to bite the skin.

Food sharing is a rare trait in wild primates. Usually there is never enough food to share. Humans have an instinct to share, and babies automatically try to feed mothers with their mushy infant food. This co-operative habit of food sharing would have contributed to our species' successful evolution. Rhesus macaques have a certain food call that an individual should emit when it finds a food source, but it has been observed that sometimes a foraging monkey who happens upon food chooses to keep quiet and very quickly tries to eat as much as possible surreptitiously before any other members of their community appear. If other members of their troop discover an individual sitting in a heavily fruited tree and eating alone, they beat that monkey up for being selfish and failing to co-operate in alerting the troop to the fruit.

But why is female solidarity among apes so unusual and how did it evolve in bonobos? Scientists have deduced that the 10-million-year-old shared ancestor of contemporary apes, including humans, lived in gorilla-like family groups, with males being much larger than females. Over the last 8 million years the world's climate has cooled and in Africa a drier, mosaic savannah with patches of smaller forest evolved, especially in eastern Africa. The Miocene apes that did not become extinct at this time would have evolved with the changing habitat, starting to live in larger polygynous fission-fusion communities consisting of multi-male groups and unrelated females. Those ancient ape societies, home of our ancestors, would have been very similar to chimpanzee and bonobo society today.

Cambridge-based primatologists and palaeobiologists Phyllis Lee and husband Robert Foley, have, using contemporary apes as models, outlined traits they think we share with our ancestors.

One of the oldest traits is sociability. All primates have a compulsive need to make and maintain relationships and the hormones governing aggression and reconciliation are the same in primates and humans.

According to Lee and Foley, the evolution of menstruating female primates has created the need for a constant male presence in their lives. This has meant that fertile females evolved to be a resource that males had to monopolise. As group size increased, our male ancestors evolved to support their blood relatives in the control of females and to defend territory against other groups of males. As social groups enlarged because of predator pressure and as male–male competition escalated, patrilineal male kin-bonding evolved in the ancestors of man and chimp, though it did not evolve in females. There are no ape social structures that involve a core of related females, as the incest taboo forces single, female apes to leave their natal group.

Groups of male hominids would have been brothers and would have defended their territory from other males belonging to other communities. Males would have attempted to steal territory and fertile females from alien males. The origins of the tribal mentality, war and patriarchy go back millions of years.[3]

Humans and bonobos have both developed an ability to form extra strong bonds between individuals within a smaller sub-unit, that is, a family. We cannot be sure whether this trait belongs to the joint *Homo* and *Pan* heritage or whether, as a coincidence, bonobos and man developed this preference simultaneously but independently. All the apes have also evolved a correlation between kin and social status. Some family networks have high status and corresponding advantages, and others do not. It has evolved that all primates will always try to mate with, and so mix their genes with, someone of a higher status. Primates are highly competitive with each other and an intrinsic element to their sociality is that, usually, one individual's gain equals another's loss.

During the last few million years of human evolution adult females have been isolated from their female kin while their infants have evolved to become all the more dependent on their mothers, isolating her still further. It is a somewhat remarkable feat that women have coped. Living in a world where sociability was essential to making friends and climbing to the top of the pecking order, large brains would have been selected for. Men and especially women would have chosen sociable partners. Robin Dunbar has noticed from analysis of lonely hearts ads that men are more concerned with physical attributes, whereas women are more interested in the type of personality of their prospective mate, though both men and women usually request a partner with a sense of humour. The more amusing and empathetic you are, the higher your social intelligence, and people with a sharp, social intelligence are usually successful. Those who don't think they need friends usually don't go far.

Scarlett may resemble Lucy in appearance, but her life is quite different. Although the female bonobo is not kin-bonded as the male bonobo is, the females are nonetheless bonded to each other and this means the males cannot control them. (In addition, the sexual dimorphism is not that pronounced, so the males cannot easily push the females around.) Vervet monkeys are female kin-bonded and if the male vervet tries to control a female he suddenly finds himself surrounded by her angry female relatives. The high-ranking female vervet uses her female friends and sisters to help her rule and any irritating male who attempts to mount her without prior invitation gets a slap around the face. Female lemurs lead a similar lifestyle because they are also female kin-bonded; they are able to boss the males around not vice versa.

Adolescent female bonobos are the sexiest of all. They have been observed to engage in more heterosexual and more homosexual activity than the males or any other age of female. They have to use the power of their sexual attractiveness to make their way in the world.

When an adolescent female bonobo tries to ingratiate herself into a new group of bonobos, she looks for a senior female and tries to become her friend. She sits on the periphery of things for a while and sizes up who is who in the hierarchy. The young female bonobo then tries to cement a bond with a high-status older female by engaging in homoerotic acts with her. The younger and older female bonobos hug each other face to face, kiss, groom, share food and GG rub. If the older female finds this new young female sexy, friendly and co-operative, she will welcome her into the fold. This is a crucial period of initiation for the young females. They need to get along with the high-ranking female so that they get enough high-calorie food to eat to enable them to start to cycle, gestate and lactate. Their lives and their babies' lives depend on having older, influential female friends. It is the powerful force of female bonding in bonobos that prevents male-perpetrated infanticide from occurring in this species.

An adolescent female bonobo may work her way through several groups and many high-status females before she settles in one community. She needs to feel she has the best deal available. Adult females have a priority for access to food; they take their progeny and their female friends with them as they climb into a tree laden with fruit. Males wait their turn. These female bonds between bonobos are strong and males cannot divide and rule the females as male chimps or men occasionally do. Females have not been seen to form coalitions and support each other in fights, because they do not need to. During the odd competitive skirmish between individuals, female bonobos have never been observed to gang up and join either side; fights are usually worked out between the two individuals, although mothers are observed supporting their sons in conflicts.

But aggressive coalitions of female bonobos have been observed to gang up and attack lone males. This is very unusual primate behaviour. Vervet females help each other out and gang up against males because they are literally sisters; lacking a blood tie, female bonobos have evolved a cultural solution to help them through their isolation from their kin. To keep their authority female bonobos can get physically tough with males from time to time.

Amy Parish is a primatologist who has studied the behaviour of captive bonobos for years. Parish started studying chimpanzees with Barbara Smuts and then took her Ph.D. in 'Female Relationships In Bonobos' with Sarah Hrdy as her supervisor. Parish told me the extraordinary bonds between female bonobos are not seen in any other primate. The nearest comparison would be the female kin-bonded spotted hyena, who, like the female bonobo, also has an extra-large clitoris.

Throughout the 1990s Parish has spent time observing bonobos at Frankfurt, Stuttgart and San Diego Zoos. Recently, at Stuttgart Zoo the resident single, adult wild-caught male, called Masikini, suffered a vicious attack and had his penis bitten off by a high-ranking female. Luckily for Masikini, the zoo vet is an expert in micro-surgery who sewed the penis back on using a series of special stitches that would allow erection to take place. Since the attack Masikini has made a full recovery and has gone on to breed successfully with a number of females, including the female who bit off his penis. The story at the other zoos is the same. At Frankfurt Zoo a male has had eight fingers and toes bitten off by high-ranking females; his testicles are also badly scarred where they have been bitten many times. At San Diego Zoo, on occasions the three resident males are so terrified of the females that they will refuse to come in at night for food, preferring to go hungry and stay out in the cold until the females are put in a separate enclosure.

When Parish first started to observe this behaviour the male zoo keepers would say the males were being beaten up because they were wimps and had been cared for as babies by women who had spoilt them. But this is not the case. The zoo males occasionally suffer these beatings because in a confined space they have nowhere to flee. In the wild groups of angry females have been seen to threaten and chase a male in just the same way, but in the wild he can scream in fright and run away, avoiding having his extremities bitten off. The coward male will then remain on the periphery of the group until he feels safe enough to come back. Sons of high-status females are never beaten this way, as their mothers would never allow it.

A wild-caught male was sent to Stuttgart, to the delight of two highly sexed adolescent females, who copulated with him almost non-stop as soon as he arrived. According to Parish, he was covered in ejaculate and she feared he might have a heart attack, but then the behaviour of the two young female bonobos dramatically changed – their desire for him turned to aggression. First one female became violent and hit him, and then she encouraged the second female to join her. Initially unsure what to do, she was cajoled by the first female and joined in the attack. Paris interpreted this behaviour as a way of the adolescent females letting the new male know he could copulate with them but not undermine their friendship.

Relations between females and males are usually friendly, but occasionally the peace-loving female bonobo shows her demonic side in order to retain the balance of power. The beatings the males receive are far less frequent than the beatings female chimps endure from males and female bonobos do not murder, as male chimps do, but the aggression is significant and essential to maintaining the matriarchal hierarchy.

Parish has one son, Kalind, named after a bonobo she once knew. On one occasion Parish took newborn baby Kalind with her to Santiago Zoo, visiting her favourite bonobo, Lana, who by coincidence had also just had a baby. Like a mother in a crèche, Lana spent some time inspecting Kalind, and then Lana reciprocally made a point of allowing Amy Parish to inspect her little baby too. Since studying female bonobos, Parish has discovered her menstrual cycle has synchronised with the bonobos. Parish believes menstrual synchronicity is an ancient primate trait that gives menstruating females equality and prevents males from undermining female co-operation by showing favours to individual fertile females over others who are not at their fertile time.

A bonobo male remains close to his mother throughout her life. They never venture far from her side and remain devoted because she holds the social status and is the only one who can help him. As their mothers give birth to younger siblings, the older brother takes on a parental role and handicaps himself in play fights so as not to hurt the boisterous babies. As part of the agreed rites of passage in allowing immigrant adolescent females to join the group, the older female passes her new, young, sexy female friend on to her sons. Sons of high-status females mate exclusively with their mother's protégées. The higher the status the mother holds, the more sex her sons get and probably the more grandchildren they give to granny.

When an adult male's mother dies, his status plummets. They are almost worthless without their mother to back them up. They lose their breeding opportunities and, unless they are very lucky, their reproductive life is pretty much over though they are not ostracised from the group. At this phase of the male's life he adopts a new set of behavioural priorities and concentrates on minding his younger siblings and his progeny.

Adolescent male chimps have a different reproductive life. They also initially inherit their mother's status, but they have to join the male coalitions, leaving their mothers and travelling through the forests with the big adult males, working their way up through the male hierarchy. They are often beaten when a larger chimp flexes his muscles and redirects his aggression on to a smaller male, at which point they leave the male group and return to their mother. She grooms and comforts them and, after a few days, once they have recovered confidence, they are back out with the males again. If their mother dies they will miss her, but if they are healthy her death will make no difference to their reproductive lives.

Infanticide

Sarah Hrdy has identified five separate areas of infanticide in primates: infanticide for nutritional benefits, where the cannibalistic killer could be of either sex and of any age; infanticide that results in increased resources for the killer of either sex and the killer's kin; infanticide that results in increased breeding opportunities, with the killer usually being male (for example, langurs); infanticide perpetrated by a parent, usually a mother, who at this stage in her life cannot cope; and infanticide perpetrated by an adult, typically an unrelated male, who kills the baby through frustration.

The description of infanticide resulting from frustration describes killings by foster fathers, stepfathers or boyfriends of unrelated children living in the home in human step families.

From the early 1970s and through the 1990s, Margo Wilson and Martin Daly of McMaster University have studied mortality rates for children living in Canada, Britain and the USA. They discovered that children living with one step-parent had an elevated risk of dying compared to children living with both biological parents. A child is more likely to die at the hands of a step-father, especially if the child is of pre-school age; as a child gets older it becomes less vulnerable to fatal attacks.

Stepchildren are sometimes neglected by step-parents in favour of that parent's biological children. This is not to say that all stepchildren suffer neglect, but neglect does occur and it can be interpreted as a strategy to channel resources to the child that carries your own genes. From questionnaires filled in by Boston's middle classes it has been noted that parents would much rather invest financially in their own children's particular school than in all schools in the system on an equal basis. Parents don't want all children to do well, just their own. Parents even go so far as admitting that they would prefer their children to score 70 per cent in a test where all the other kids scored 50 per cent, instead of all children doing equally well and receiving 95 per cent.

The infanticidal human stepfather has no genetic investment in the child, but neither does the child's existence prevent him from having sexual relations with the mother as it does with langurs; so this sort of human infanticide is not a male reproductive strategy. If like most children it is noisy and demanding, the man may respond to the child's demands with lethal battery. If the mother does not come forward as a witness to her child's death often the killer escapes prosecution. Some women protect the murderer of their baby, as she may by now be pregnant with his child and he may be the family bread-winner. Women who have been found to have lied to the police are treated just as harshly by judges and juries as the killer. When we hear that another child has been murdered by a man we may not feel the same sense of uncomprehending horror that we feel when a woman is directly involved with the murder of children.

In some cases very young mothers seem willing to neglect their child in favour of their new boyfriend. A young woman wants the chance to start over again after the failure of her first family set-up. If she has many years left before her menopause, she may nurture her new boyfriend's needs before her children's because she doesn't want her children to stop her from having a full life. This maternal dilemma is known in primatology as a 'fitness trade-off', fitness being defined by the number of babies you produce. An older woman, nearer to her menopause, is less likely to have time left to start again, and is thus more likely to put her children before her new male partner.

If we look at the maternal strategies of monkeys and apes we find in various species different forms of female-perpetrated infanticide.

Female monkeys and apes can find themselves in a fitness trade-off dilemma at different stages in their lives. Some species of female monkeys, such as South American marmosets, will suppress ovulation until they have reached a higher social status. When they reach their prime they will breed because their infants will then have a better chance of survival because of better access to important resources such as food and sleeping sites. The subordinate female must stay alive, socialise and push her way to the top of the female hierarchy before taking on motherhood.

Marmoset and tamarin monkey groups usually only have one dominant breeding female during the breeding season. The other members of the group help to care for the babies, often twins, that the dominant female produces, the males carrying her babies and subordinate females, possibly aunts, helping to suckle the twins. Other group members, often relatives, will assist in the process of weaning by finding tasty morsels of insects for the youngsters to cut their teeth on. If a subordinate female marmoset does breed, it is likely that the dominant female will kill the subordinate female's babies, which, although possibly cousins, represent competition with her own babies for the group's resources. For most primate mothers the personal cost of gestation, subsequent lactation and general care of an infant is so expensive that only the most favoured female can pay the price. If a subordinate female cannot push herself to the top of the female hierarchy she may never breed and will be individually a reproductive zero, even though she was a supportive aunt.

Recently Jane Goodall, together with primatologists Anne Pusey and Jennifer Williams, from the University of Minnesota, have been studying the parallels between dominance rank and reproductive success in the Gombe female chimps. They found that babies of high-ranking females were more likely to live and the daughters of high-ranking females were likely to start cycling some four years before low-ranking females. High-ranking families dominated certain core areas of territory with the highest quality of food. But access to food was not the only factor. The scientists found that a number of low-ranking females were losing their babies through infanticide perpetuated by high-ranking females. Jane Goodall once thought that chimpanzee Passion's infanticidal behaviour was psychotic, but now suspects that Passion's murderous behaviour towards lower-ranking Gilka's babies might not have been pathological, as she first thought, but adaptive.

In committing infanticide, high-ranking females reinforce their social rank, killing their infants' competitors and having a meal of fresh meat in one fell swoop. Goodall has seen a number of female chimps behave this way, including Flo's daughter Fifi, observed trying to snatch a baby chimp away from its middle-ranking mother.

Sarah Hrdy has also studied infanticide perpetrated by a child's mother. Globally this is probably the most frequent form of human infanticide and in Third World countries it is a common form of birth control. There is a phenomenon known in the West as 'concealed pregnancy' where women who do not want to be pregnant and do not want to become a mother deny to themselves they are pregnant. Their psychological state, known as 'splitting', helps them to suppress the signs of pregnancy, which would normally entail gaining 30–40 pounds around the girth and a rapidly expanding bust size. More often than not, no one knows these women are pregnant. They give birth alone, successfully delivering their own baby, but once the infant cries out the denial is broken and the woman silences the baby's cry by smothering it. These variations of maternally inflicted infanticide are a regrettable form of birth control used by women all over the world and by other non-human primate mothers at times of stress or starvation.

It has been documented that some hunter–gatherer women and certain South-East Asian female peasants may kill their first sons if they are young mothers with more reproductive years to live through. These mothers want a daughter first to help them to care for the son that ideally comes next. This favoured second-born son effectively has two older females to care for him, meaning he will be safer and better fed with a much greater long-term success rate. Often in families where a girl is born first, the mother goes on to raise more children than other mothers in her community whose first-born children are sons because the oldest daughter traditionally helps to care for her younger siblings.

By 1870 the British Raj had outlawed the practice of socially constructed infanticide in India, but they could not raise the status of women so the practice simply went underground or changed its face to that of lethal child neglect. Sarah Hrdy researched census data collated by colonial authorities in the 1800s. In the very highest castes, like the Rajputs in northern India, no daughter ever survived and throughout the caste system boys were five times more likely to survive than girls. There are a number of ecological reasons for this. Marrying your daughter honourably was, and still is, essential for an Asian family, and a daughter born in the Indian subcontinent carried a bride price, a dowry, which the British tried unsuccessfully to outlaw. If the wealthy Rajputs had girls, on the day of their wedding daughters would take family wealth with them to the new family. If the Rajputs only had sons dowries would always be paid to them, keeping them rich and thus socially attractive. Only daughters bringing with them the fattest dowry would be lucky enough to marry a desired Rajput man.

Today, when we hear that a woman has abandoned her newborn baby on the steps of a hospital (in the past it would have been the foundling home), the authorities try to trace her and reunite mother and child. Social services gather around, offering support. That mother knew she could not cope, and we must not under-estimate her reasons. When we hear a mother has killed her own baby we feel dismay that so young a life has been snuffed out, but many of us also feel deep sympathy for the woman, instinctively knowing that she must have felt very alone and very depressed to behave that way. In this context, abandoning and smothering your child are very similar acts from the mother's perspective.

The Infanticide Act was passed in England and Wales in 1923 to stop judges from sending women to the gallows for the murder of their baby. Only the biological mother who murders her own baby of under twelve months of age can be convicted of the specific crime of infanticide. If anyone else murders a baby, depending on the circumstances, it is a crime of murder or manslaughter. Today a woman convicted of infanticide would be given the average and more moderate penalty of three years' probation with psychiatric care.

The United States does not recognise the crime of infanticide as anything other than cold-blooded murder. American jails house young women given life sentences for smothering their unwanted newborn babies. The United States has rough justice in store for women who cannot be perfect mothers in an imperfect world. Even foetuses have rights over the mother. It is estimated by the National Association of Parental Addiction and Education that every year some 375,000 babies are born addicted to the same drug that their mother is addicted to. An increasing number of these women, who obviously are not coping and have low status within their society, are being arrested by the 'uterus police' and if their baby dies or is prematurely still-born, the women are charged with the crime of first-degree murder.

Data reveal that parents bias investment in their children to enhance long-term reproduction. It has been suggested that a human mother might, depending on her own status, evolutionarily favour boys over girls, or vice versa. If a woman is high-status it seems likely her children will be high-status too. Theoretically high-status men usually have more opportunities to reproduce than high-status females. Ottoman emperor Moulay Ismail the Bloodthirsty, who fathered hundreds of children, is a prime example, but the majority of high-status males will have the possibility of a polygynous reproductive life. For each high-status male who monopolises the fertile females, some low-status males have to go without (although, on average, males and females reproduce the same number of times). In evolutionary terms it is an imperative to have as many descendants as possible. If we accept that humans are patrilineal, dominant women benefit from sons because they have the potential to grow up to be high-status men and have more chances to reproduce.

Lower-status women will have lower-status children and because women usually have a chance to reproduce, whereas low-status males do not, it is better for a low-status mother to have daughters as a means of having descendants. Sociological studies have shown a trend for low-income single mothers to have daughters. It has been discovered that on average, boys born into wealthy families are breast-fed for twice as long as girls and girls in poor families are breast-fed for twice as long as boys.

Sex ratio studies go back to the 1920s. The *Who's Who* publication and the children of America's presidents have been studied and a predominance of sons has been found within these high-ranking families. In 1953 Marianne Bernstein found that women working in 'masculine' occupations were more likely to have sons and women working in 'feminine' occupations were more likely to have daughters. In 1966 Valerie Grant found that New Zealand women who had independent and dominant personalities were more likely to have sons. There is a vast amount of data to support the theory that high-status women show a slightly higher tendency to have sons, but why this is the case no one knows. So far no scientist has figured out the working of the biological mechanism that allows a woman to subconsciously favour the sex of her unborn child. Since some sperm carry an X chromosome, designating the female gender and some sperm carry a Y chromosome, designating males, one theory suggests that the ovum has the ability to favour one type of sperm over another.

The biological triggers that select one gender over another and influence infanticidal behaviour in adults are often far more flexible than the social and cultural pressures found in, for instance, the Indian caste system or China's one-child policy. If those Chinese or Indian families belonged in another society, where girls were not second-class citizens and more than one child was not frowned upon, so many little girls would not be abandoned to die.

Sexism and Science
In 1974, after extensive research, Sarah Hrdy wrote her first paper on sexually selected infanticide in langurs and subsequently upset many people. Hrdy considered herself to be a sociobiologist.

Edward O. Wilson lectured Hrdy at Harvard and his theories influenced her and countless other students.

Many 1970s feminists identified a male bias in sociobiology that they believed to be biologically deterministic. They wanted to replace this inherent bias with a politically corrected one, a sort of conscious partiality. Philosophically, they did not appreciate Wilson's theories and began to demonstrate at his lectures. On one occasion protesters who accused sociobiology of justifying racism and sexism yelled out to Wilson, 'Wilson, you're all wet!' as they poured a bucket of cold water all over him during a lecture.

Hrdy also considered herself to be a feminist, but when her allegiances were tested she chose science. 'By and large feminists objected to underlying determinism, i.e. males by nature will be one way, females by nature another way. But as I see it this is a real misconstruction. What sociobiology has to offer to evolutionary perspectives on behaviour is the main paradigm shift to the study of the individual and individual strategies. No longer did we just have one type of male or one type of female; we had males and females that had a whole range of strategies they could adopt. Females could do very different things with their lives and pursue very different strategies at different phases of their life. Sociobiology is anything but essentialist, although some people did stick to old essentialist stereotypes. If feminists had looked deeper into this field that distressed them so much they would have been intrigued by what they would have learned.'

Plenty of cranks and modern-day eugenicists have selectively taken elements of sociobiology and incorporated them into their fascist beliefs. The identification of a genetic underpinning to human behaviour is provocative and can be dangerous. The extent to which our overall behaviour is driven by our genes is debatable. Some believe there is a fifty–fifty split between nature and nurture. Certainly, we do have a choice, especially if we are enlightened to the reasons behind our choices. On this point Hrdy says, 'I feel for a woman to live an informed, rational, ethical life she's going to have to draw on resources and make decisions other than those that evolved. She's not just going to fall into place in line with what was evolutionarily stable. She's going to have to make conscious decisions.'

A favourite quote of Sarah Hrdy comes from John Huston's classic 1951 movie *The African Queen*, where steadfast Rosie, played by Katharine Hepburn, turns to a long-suffering Charlie Allnutt, played by Humphrey Bogart and pronounces, 'Nature, Mr Allnut, is what we were put in this world to rise above.' Hrdy believes we do not have to be slaves to our genes if we are self-aware of their influence over us.

In the mid-1970s Sarah Hrdy provoked heated intellectual debate with a number of different factions, which intensified with the publication of her book *The Woman That Never Evolved*. Anthropologists were forced to accept that the 'for the good of all' paradigm was wrong and that infanticide was not unnatural behaviour. Sociobiologists also had to accept that they hadn't been seeing the big picture; they had only really observed the males and so had employed sexist stereotyping to favour males in their analyses. Feminists felt let down that a woman was suggesting a biological underpinning made male and female behaviour different, instead of social conditioning being responsible for boys liking slugs and snails and girls liking sugar and spice.

People tend to think that natural equals 'good', but this is a misleading distortion. Hrdy says, 'No one who knows what the lives of monkeys and apes are actually like would hold up these animals as a paradigm for how we should choose to live. These are not admirable or desirable lives that these creatures are leading in terms of our moral sensibilities.'

Orgasms, Sexy Females and Violent Males
Not one to shrink away from challenges, Hrdy moved on from the emotive issue of infanticide to a now hotly debated subject – the

female orgasm. Hrdy has hypothesised that the orgasm is a reward system that encourages female primates to solicit sex with many male partners, thus confusing paternity and hopefully saving the skin of unborn babies. Not only women have orgasms – all female primates do and our non-human primate sisters are invariably having better sex than modern women are.

Many of these animals experience multiple or 'cumulative' orgasms frequently. It was obvious to women primatologists that the female primates they studied were at times enjoying sexual gratification, but men found this harder to accept. Hrdy says: 'Men might say, "I want you to prove to me she has a libido – I question the assumption that it exists. Prove it first, and then I'll question the stereotype." Women said, "I don't need anyone to prove to me the existence of the female libido, therefore I shall question the stereotype now." Women proceeded with the necessary data collection and argumentation to make the point.'

No one doubted male primates had orgasms – the male orgasm is necessary for ejaculation to take place – but until the sexual revolution of the 1960s, women's sexual nature was a taboo subject. So even though non-human female primates have clitorises, because women's orgasms seemed to be a rare phenomenon and rarely spoken of, the possibility of female monkeys and apes experiencing orgasms was overlooked and the reason for the existence of the female orgasm never questioned. But since the debate over the orgasm in female non-human primates has been put on the agenda female primates have been tested in invasive laboratory experiments to ascertain the physiological processes that occur during simulated intercourse.

In the early 1970s, captive macaque monkeys in oestrus were strapped by the ankles to keep them still. Their clitoris was stimulated and a simulated macaque penis was inserted into their vaginas. These macaque monkeys climaxed, as wild monkeys, apes and women do. Non-human female primates experience the same stages of orgasm as women experience: excitement, plateau, orgasm and resolution. Like aroused women, these animals can oscillate between plateau and orgasm many times, unlike male primates. The female non-human primate has an increased heart rate, her vagina secretes mucus, she reaches for her partner, she pants, she utters what are known as 'ejaculation calls', there are uterine contractions, vaginal contractions, body spasms and, to coincide with this, the female makes an open-mouthed face.

For Hrdy to understand the function of the female orgasm she had to examine the overall sexual behaviour of female primates. Female non-human primates like to have face-to-face contact with their male partner during intercourse. Macaque females, for instance, will turn their head around and make eye contact with the male partner that is mounting them. Female bonobos' large, elongated, prominently shaped clitoris is located at the front of their genital area and their vaginas are found between their legs. This makes the location of female bonobo genitalia more like women's sexual anatomy than that of female chimps. A chimp's vagina is located a little further back, encouraging mounting from behind. Even so, male bonobos prefer to mount from behind and often try to copulate in that position, but female bonobos, encouraged by the size and positioning of their clitoris, favour the missionary sexual position, known by primatologists as the ventro-ventral position, and the male bonobo usually complies with the female's wishes.

During copulation, the female bonobo holds the face of her male partner and forces him to make eye contact. If he doesn't want to and appears more concerned with his own feelings and pleasure, the matriarchal female bonobo invariably pushes him off to seek a more agreeable partner instead, which more often than not will be another female bonobo. Male bonobos never rape; being of a lower status they must make even more of an effort to please the females if they want to have sexual relations.

Group sex often occurs. In chimps missionary sex, group sex and homosexual sex are not as commonly observed. Orang-utans engage in face-to-face copulations, as do gorillas very occasionally. Like the bonobo, orang-utans also have a great array of sexual techniques, but none of these other apes have orgies between groups of different aged individuals the way bonobos do.

Oestrus in female primates is a time of intense libidinal activity because of their increased amounts of the oestrogen hormones. The females of some twenty-two primate species have an ostentatious reddening and swelling of the genital area during oestrus. For example, both baboons and chimp females have a large, engorged posteria to help advertise their receptivity. As in humans, the apes' and monkeys' eyesight is a more dominant sense than smell, and this red, bulbous beacon can be seen over great distances by eager males. To help them judge the timing of ovulation, males also respond to pheromones emitted by the females and sniff the females' urine and inspect their vaginas, but the visual cue of a red posterior seems to be the most significant trigger.

Female baboons and chimpanzees frantically copulate during oestrus, which lasts for two weeks in the middle of an average thirty-five-day cycle. Women also have an increase in libido during the time of ovulation, but they do not have to endure the rapid swelling of the perineum and vulva while ovulation takes place. (On top of PMT that would be too much!)

Before we started to look to bonobos for clues to human evolution, chimps and baboons were our human models. The question arose of what became of engorged, swollen genitals in women during the oestrus period. It was presumed that ovulation became concealed in woman to stop men fighting over sexual access to women with large red backsides. It has been suggested that concealed ovulation in women has encouraged the evolution of a monogamous pair-bond between man and woman. In this male-fantasy pair-bond she rewards him with constant and exclusive sexual access and he, in return, helps her to raise their baby.

If a male chimp takes a fancy to an oestrous female, he may stand bipedally, roughly shake branches to gain her attention, hold eye contact and wave his penis at her. If she is attracted to a male she will confidently stroll towards him and present him with her highly attractive red rump. It was a generally held belief that woman's ancestors behaved like wild female chimpanzees during oestrus, mated with eight different males within a matter of minutes and over a day copulated fifty times. This is the sort of sex life that Jane Goodall's favourite female chimp, Flo, used to look forward to.

Female gorillas, gibbons and orang-utans do not have large, swollen genitalia during oestrus. A female gorilla has only a one- to three-day oestrus in which she is receptive and a very modest swelling that only a devotee could spot. Galagos are African nocturnal prosimians, more commonly known as bushbabies. There are forty species of galagos and they are small with fluffy fur, large eyes and live concealed in trees. Female galagos have a period of oestrus limited only to a few hours; outside of that window of opportunity the curtains are firmly closed. The female galago has a membrane that seals her vulva, making copulation impossible, even though the male galago tries to mount outside the oestrus period. By contrast female bonobos are genitally swollen for 75 per cent of their menstrual cycle and have sex whenever and however they fancy it. Bonobos make love purely for fun and as a part of social discourse; for them reproduction seems almost incidental. Female bonobos have been observed to solicit sex sixty-nine times within an hour.

Concealed evolution, breasts that are permanently large instead of only swelling during lactation and the menopause are the most striking differences between women and our closest ape sisters. Even though an ape reaching old age has a fading fertility, female apes can, theoretically, keep reproducing until they die. Birth is a far more demanding and dangerous experience for a woman than it is for an ape. Before Western modern medicine, women died in child-birth with boring regularity. Death in childbirth is more unusual in apes. Their labours are shorter and their babies smaller, mainly because of their smaller craniums.

There may be other reasons for oestrus swellings in primates. It has been noticed that swollen female chimps and bonobos can transfer to other groups without hostilities. The incest taboo forces them to leave their natal group and mate with an unrelated male. But if a chimp tries to transfer when flat (unswollen), males show her no favours; they may attack her and will not protect her from any resident hostile females, who do not like the appearance of a new adult female, a competitor for sperm and food. If the female is sporting a red rump and therefore obviously in oestrus, she is hun-grily welcomed by the male chimps, who also protect her from any aggression from resident females. After a female has become preg-nant she sometimes transfers back into her natal group. Missing females often reappear pregnant.

When a female chimp is in oestrus she is the centre of attention and her status peaks; when she is flat she spends much time alone, foraging with her youngest child. Female chimps spend more time alone than female bonobos, partly because food sources are more scattered in chimpanzee habitats than in the bonobo range. Foraging expeditions can be dangerous. In search of food, a female chimp can find herself face to face with hostile chimps who are trying to increase their territory, though if she is in oestrus she is usually safe from potentially hostile unknown males.

Sarah Hrdy hypothesises that oestrus swellings as seen in female chimps and bonobos may have evolved in these primates as our common ancestors separated some 6 million years ago. Ostentatious ovulation has evolved in several different primate species. Hrdy suspects the idea that oestrus swellings have been lost to women is a false belief, like various other interpretations of woman's nature. One is the false belief that modern woman is continuously sexually receptive; Hrdy instead believes that women experience a 'situation-dependent' sexual receptivity. A favourite evolutionary theory among men anthropologists is that the primitive woman who was continuously receptive to sexual advances from her mate had a mate who found her to be continuously rewarding. This encouraged the contented Stone Age man to stay and look after his woman and the subsequent children, which he thought must be his, thus sealing their pair-bond.

Some anthropologists believe we can find this pattern of behaviour in modern people. But it is hard to believe that a man in a heterosexual relationship could think that his female partner is continuously sexually receptive to him. If he accepts this, he is deluding himself (and no doubt making a thorough nuisance of himself). Yet generations of mostly male anthropologists and biologists have even suggested this pre-hominid quadrupedal fantasy Jezebel was so alluring that her mate stood up and walked bipedally, freeing his hands, enabling him to carry food back as a reward for all the good sex he was getting and thus bipedalism evolved in men first.

Sarah Hrdy was not prepared to accept the man-the-hunter propaganda that evolution would select bipedalism in one gender first. She saw this as another false belief. 'You have a much lower reproductive potential in women than you do in men. Moulay Ismail the Bloodthirsty supposedly produced 888 offspring after marrying several wives and having many concubines. But there can be a great variance in male reproductive success, with some males producing no offspring and other males finding plenty of opportunities to breed. This led people to assume that natural selection was somehow working more strongly on males than it was on females, and this was the source of a lot of misunderstanding.

'We see this argument coming up all the time in terms of bipedalism evolving in humans because it was selected for in males so they could provision their females waiting in camp, and [in terms of] intelligence [being] selected first in males because males were hunters. (Today in hunter–gatherer societies it is the women who find most of the shared food eaten.) In fact selection weighs heavily on both sexes and there's also important variance in female reproductive success. A lot of theoretical contributions about female reproductive strategies flow naturally from these insights. For example, in non-human primates – and it is very easy to see – we know there's variance in female reproductive success. Some females produce very few or no offspring, and others produce many more. It is actually conceivable that females have the same variability in reproductive success as males do. If you think about rank-related variance in infant survivorship, when the female first begins to breed, if you magnify these effects on the individual females through her lifetime and over subsequent generations, you will see females inherit their rank from their mother. Selection will weigh very heavily on those traits that allow females to breed rather than on those traits that make them less likely to breed.'

In university Sarah Hrdy had been taught the male-centred dogma that the female orgasm had evolved in women to help enhance the pair-bond between innately monogamous *Homo sapiens*. We know now that non-human female primates are not monogamous and experience orgasms and multiple orgasms. Not even the famously 'monogamous' gibbon is monogamous; both the male and female will solicit sex with other partners if they have the chance. Females do not mate with just one male. Even if the male tries to dominate her sex life, he cannot do so 100 per cent of the time. The female orgasm actually encourages females to solicit sex with various males.

At the same time a male primate does not want to be a cuckold. Jane Goodall has told of the sadomasochistic love story of chimpanzees Evered and Passion. Evered was the beta male at Gombe and Passion was a high-ranking female, who bullied feeble Gilka, Evered's little sister, and murdered Gilka's babies by eating them.

During the time between Passion's oestrus swelling, Goodall observed Evered try to coerce Passion to accompany him exclusively for a few days so they could mate without interruption from other males. But she was very reluctant to go anywhere with him and resisted. This provoked Evered into attacking her. Over a ninety-minute period Evered beat up Passion five times, rendering her lame on the ground. Her screams attracted the attention of some other males, and Evered left with them without looking back at her. As the second-highest ranking male, Evered could not be reprimanded by the other males for beating Passion; they could only offer themselves as a distraction.

Two weeks later Passion, recovered from her beating, was now in oestrus; Evered found her and gave her a look, then turned and walked away into the forest. Obediently Passion kept her head down and followed him. Goodall believes the beatings outside the oestrus period are used by the males to help exercise proprietorial rights over the female when she is in oestrus. Making her lame when she was in oestrus would not be a smart move. As a reproductive strategy it was a success. Through fear he was able to monopolise her body during ovulation and later was sure that her baby was his.

Goodall watched Evered use the same tactic of forced sexual access on a young female called Winkle. Over a five-hour period Evered attacked Winkle six times, and again his treatment persuaded her to follow him without making a fuss. Perhaps the brutalising of the females by the males encourages the females to bully. Passion's beatings could have encouraged her to beat Evered's sister and kill her babies later on. She held on to her high status through fear, and the other females were scared of her. Passion's dominance made her sexually attractive. Male primates usually prefer to mate with dominant females as their infants have a better survival rate.

In 1975, during the four weeks that Barbara Smuts was at Gombe before she was kidnapped, she spent her days rushing around the park following females. She was shocked the first time she saw a male severely attack a female in the early stages of oestrus. The unprovoked violence left the female cringing and whimpering in pain. Later, when Smuts transferred her study to baboons, she witnessed the same unprovoked bullying and beating of fertile females by males, which seemed to increase the perpetrator's chance of mating with that female and also seemed to lessen the chances she would mate with anyone else. This reproductive strategy is very different from the male baboon strategy of forming friendships with a female in return for sexual access. This agenda is like saying you might as well rape them if they won't be your girlfriend.

This scenario reminds me of a self-help book on 'speed seduction' for men by Ross Jeffries entitled *How to Get the Woman You Desire into Bed – A Down and Dirty Guide to Dating and Seduction for the Man Who's Fed Up With Being Mr Nice Guy*. Another of his books is *How to Nail That Girl Who Just Wants to Be Friends*. Jeffries recommends that men use hypnosis and humiliation to coerce an unwilling woman into bed. If all else fails – including seductive one-liners such as 'I just wanted you to know that 99 per cent of the women who walk in that door would kill their own mothers to look half as good as you' – you 'cry like a baby, sob uncontrollably and try to guilt-trip her into bed'. Sexual coercion can be seen practised by males of almost every species alive from insects to man and sometimes their strategies are remarkably similar.

But few people have made this particular link between humans and non-human primates. Barbara Smuts has said of men's brutality towards women 'Although scientists investigated this kind of behaviour from many perspectives they mostly ignored the existence of similar behaviour in other animals. My observations over the years have convinced me that a deeper understanding of male aggression against females in other species can help us understand its counterpart in our own.'

During the late 1970s and early 1980s, while Barbara Smuts was studying sexual coercion and friendships in primates, feminists were studying male violence towards women. Male sexual aggression towards women is a behaviour innate to all cultures and classes. Sociobiologists have suggested this misogyny can reveal its ugly face in three areas of male behaviour: firstly the man wants sex but doesn't want a relationship or wants children but is unwilling to give much, if any, support; secondly the man may want a secure relationship with the woman but he is hostile and violent during conflicts with her; and thirdly he has regular conflicts with the woman to assert his interests over hers and he expects her to nurture his will and desire. These three patterns of male behaviour often combine in one individual, facilitating sexual aggression.

Human beings practise various forms of female subjugation. In medieval times women were forced to wear chastity belts if their husband was away warring. An extreme contemporary example would be clitoridectomy and circumcision, which involves a young girl's genitalia being mutilated and her vagina being almost completely sewn up. A young woman who has suffered this mutilation is less likely to seek extra-marital sex so her husband can be confident that her babies – if she is able to give birth successfully after such injury – are his.

Chimpanzees do not practise genital mutilation, but the males do beat the females into submission. One reproductive strategy of male chimpanzees is to encourage an oestrus female to leave the group with him and the two of them disappear into the forest together; the arrangement is known as a consortship. These sexual expeditions can last a couple of weeks. Sometimes the female is happy to go along and the consortship can be seen as a honeymoon. At other times she feels bullied and gives loud cries known as pant hoots, to call for help. If she's lucky other males may come to her defence.

There are three main mating patterns to chimpanzee life. The consortship is one. During their trip they will mate frequently and by the time they return the female will no longer be in oestrus but most likely pregnant with the male's infant. If, secondly, the oestrous female is not being particularly monopolised she makes herself available to anyone she likes and the males line up and take turns in mounting her. This sexual activity will be heightened if the group of chimps are visiting a fruiting tree at the time – sex and food go well together in chimpanzee, bonobo and human life. The third mating pattern in chimps is usually seen when a popular female comes into oestrus. At this point a high-ranking male will step forward and want her for himself and try to dominate all the other males into leaving her alone. He will not take her away from the other males on a consortship because he feels confident that his high status alone will keep other males at bay.

Female chimps, bonobos and women have sex long after conception. Chimps and bonobos continue to swell and women experience an increase in libido in their first six months of pregnancy. But why have sex after fertilisation. What is the function of non-reproductive sex? Perhaps this encourages a male to stay close and provision the mother and child, or perhaps it's a lucky by-product of the female primate's increased hormonal output. Or it could be a behaviour intended to continue to confuse other males over paternity in case her particular male deserts her or dies?

Jane Goodall remembers a time when Figan, Flo's son – who was Gombe's alpha male for some years – tried to monopolise a high-ranking female's period of oestrus. Figan had his favourites, and when they were in oestrus he would be very attentive over the ten-day period, especially over the last four days in which ovulation takes place. But on one occasion, during the last four days Figan was distracted by some colobus monkeys in a tree. Figan touched the arm of his second-in-command while keeping his eyes on the colobus troop. The beta male followed Figan's gaze and off they went hunting. When males leave the group to hunt they reaffirm their macho bonds, which is important to help them individually retain their place in the hierarchy. Copulation often encourages hunting expeditions, and the male will share the meat with his favoured female. While Figan was away his favoured female furtively fornicated with six other subordinate males. Had Figan taken this female on a consortship, he would have had her all to himself, but in his absence another ambitious male may have plotted his downfall. Figan chose the reproductive strategy that best suited him at that time.

It is thought that sperm competition rather than the monopolisation of one female by one male, is responsible for most pregnancies. The theory is that if the sperm has to compete with that of another male, the female will have the very best sperm for her precious egg. This is an example where high status does not ensure high reproductive potential. Two British scientists, Mark Bellis and Robin Baker, author of *Sperm Wars*, have hypothesised that sperm from competing males will fight and kill each other inside the female's reproductive tract as they race to reach the precious ovum first. Robin Baker told me that Sarah Hrdy's interpretation of the female orgasm evolving purely as a reward system does not explain the full picture. He believes that female primates can, through a subconscious evolutionary trigger, time their orgasms. In this way the female orgasms with the male of her choice and his seed is assisted up into her fallopian tubes by her climactic uterine and vaginal contractions. Baker suggests a female primate can bias whether or not she conceives by whether or not she orgasms. Not having an orgasm is just as important as having one.

But Sarah Hrdy is not so sure. 'Although it is possible, we need more information first. But I don't see this as a likely reason for the evolution of orgasms originally, because what point would there be to select for sperm retention unless females were mating sequentially with a number of different males in the first place?'

Baker and Bellis have studied the secret sex lives of women. Their research has proved that married women having extra-marital affairs who arrange to see their lovers during ovulation, are less likely to use contraception and are more likely to orgasm with their lover than with their husbands. There is a good reason for the saying 'It's a wise child that knows its own father.'

Among chimpanzees, the older, high-ranking females are usually considered the sexiest females by the males because they are well fed and their offspring usually survive. The older Flo became and the more out of shape from the babies she produced, the more her sex appeal increased. A male chimp hopes its baby will have the best mother, and the most maternally experienced of chimpanzees are the most sexy. This is where chimps and humans deviate. A big-hipped forty-year-old woman with five chubby children is not usually considered a pin-up. Men favour slim-waisted, slim-hipped women with symmetrical faces and figures. Young, symmetrical bodies are usually healthier and therefore more fertile. Asymmetrical faces and bodies, before the recent development of modern medicine, indicated the individual was diseased and probably not fertile. In chimps, as a female gets older and produces more babies and her waist expands, her social rank increases. Older, high-status female chimps keep younger, virginal inexperienced female chimps away from the sperm of the alpha males. This behaviour is not seen in women. Older women cannot suppress the next generation of younger women in the same way. Men seek younger women as mothers for their children because children are dependent on their mothers for many years.

It has been discovered that women of a certain age and shape become pin-ups. Pin-ups are typically attractive, shapely women in their fertile years, but what is it that makes some women seem more sexy and beautiful than others? Evolutionary psychologist Professor Devendra Singh has studied sexy beautiful women through history, from the Venus de Milo to photographs in contemporary magazines such as *Playboy*. Singh has found that these celebrated women have a waist to hip measurement in common. If a woman is ovulating regularly, the oestrogen in her body will distribute her body fat away from her waist and instead deposit it on her hips, thighs and buttocks. Singh says that bust size does not indicate whether a woman is ovulating, as adolescent girls develop breasts before they menstruate and post-menopausal women are no longer fertile even though they have a bosom. Post-menopause, the amount of oestrogen in a woman's body declines rapidly. The storage of fat changes to linger around the waist. Singh has divided the waist measurements by the hip measurements of famous fertility symbols from ancient sculptures to Marilyn Monroe and he has calculated the average waist-to-hip ratio of a range of sexy women to be 0.7

Singh believes a woman's waist-to-hip ratio, WHR, is an evolutionary visual cue that tells men whether a woman is fertile or not. Women with waists as big as their hips are far less likely to be fertile. Supermodel Kate Moss is often thought of as too skinny, but Singh has found that her WHR is 0.66, equal to the WHR of icons Marilyn Monroe, Brigitte Bardot and Elizabeth Taylor when they were in their prime. Individually the sizes of the waist and hips are not as important as their relationship to each other. According to Singh, Marilyn Monroe's vital statistics were 36-24-36; Kate Moss's are 33-23-35, but if we divide 24 by 36 or 23 by 35 we get WHRs of within one decimal place of 0.66. Singh believes these women's celebrity and high status is due at a primal level to the magic 0.66 WHR, which in the subconscious eyes of men defines the women as fertile and therefore sexy. No amount of dieting can give women a slim waist when compared to her hips; only her hormones and possibly exercise or alternatively corsets can do that. Singh argues that Kate Moss's shape is not an anorexic one, as a malnourished woman does not ovulate and therefore would not have a 0.66 WHR. Singh goes on to say that a woman's WHR is only a starting point in a man's attraction to her; personality plays a great part in the mating game.

As a sexy woman grows older and produces children, her status peaks and then starts to decline. Once a woman goes through her menopause, her social status falls even further and her waist expands. Hence the popularity of HRT, an oestrogen supplement that helps to maintain a fertile figure. A career woman aged fifty has high social status as a result of her salary and peer-group flattery, but she is unlikely now to produce another baby and the production of babies biologically defines a woman's prime. In cultural terms her career may peak long after her children have grown up but evolutionarily the production and survival of babies gives a woman importance. As Jeanne Altmann said, the mother and infant relationship is the palpable interface of evolutionary impact.

Primatologists have observed that casting in Hollywood is run along these lines. In just about any Hollywood movie, an actor like Michael Douglas – who is not getting any younger – will find himself in the lead role, though his leading ladies will not have such a long and illustrious career. In each successive movie, Douglas finds himself cast opposite increasingly younger women, sometimes young enough to be his granddaughters. (These thespian alpha males are also paid much more than the women they act opposite.) Because the movies represent generalities and use basic triggers to appeal to a mass audience, we rarely see actresses in their fifties or sixties act in a sex scene. They are past their reproductive prime and their loss of reproductive value mirrors their loss of box-office value. Great and beautiful actresses, if they are lucky to still get scripts, go on to play character parts as mothers and older wise women. But slowly the status quo is changing. Increasing longevity and HRT are helping to keep women younger for longer, and many older women have sex appeal for men of all ages. Trips to the gym and courses of HRT are not so much exercises in narcissism, as essential tactics in exploiting one's sex appeal and sustaining a high social status. Interestingly, although Western men want to make love with beautiful, symmetrical women with an average 0.7 WHR, they apparently fear catching STDs from those same women. Men surveyed presumed that a beauty is more promiscuous and diseased than her dowdy but cleaner sister!

Men have it the other way around. By the time a man is fifty, if he has been successful in pushing his way up the male hierarchy, he will have accumulated some wealth, which makes him sexy. (But this fifty-year-old man should avoid complacency. According to Singh women also have a favoured WHR for men; a beer gut tends to be a turn-off.) In Singh's surveys women show a preference for tall men, but they will dump a tall man who does not have a healthy WHR in favour of a shorter, slimmer man.

Looking at the sexual strategies of non-human primates, Barbara Smuts noticed that when a female baboon was sexually harassed by a male, one of her special male friends would come to her rescue. But it then became evident that these male friends believed their gallant protection warranted unlimited and exclusive sexual access to that female. If the female did not want to mate with her friend he would attack her and he would also attack her if she tried to distance herself by forming a new alliance with another male. Ultimately, having male friends actually left these females all the more isolated within their community, because males who were not her friends would not offer her protection from aggression coming from her male friends. Men often behave as though a woman owes them a sexual favour if they offer her their kindness in some way. Conversely women can feel obliged to repay male kindness with sexual favours.

In the early days of their research a number of women primatologists had wanted the friendships they observed between male and female primates to be of more value to the female than they probably were. Because primatologists like to think that the natural order will teach us how to behave, they will see what they want to see. In the same way that male primatologists want high male rank to be equated with high numbers of progeny, women primatologists want to find pure, platonic friendships between male and female primates, because they know from personal experience that modern women would be better off if men were their friends rather than their enemies. But there are no easy answers in the battle between the sexes. Relationships between male and female baboons or chimps are as complicated as relationships between men and women. Individuals in non-kin relationships are constantly re-evaluating the relationship. A male may be a friend to a female one day, but the next day he may subconsciously view the friendship as a strategic investment that is no longer worth his while and other male strategies will supersede.

Male sub-adult orang-utans invariably use forced copulation or rape as a reproductive strategy. An adult female orang-utan travels alone with her child for much of her adult life. Fully adult male orang-utans – males who have developed large cheek pads, throat pouches and muscled bodies, are impressive beasts. They do not rape females and, similarly to male chimpanzees, prefer to mate with experienced mothers. They are attracted to the older female, who has proven maternal skills.

Adult male orang-utans have overlapping ranges with other adult males and within those ranges reside their females and the sub-adult males. When two adult males meet they invariably fight, often injuring themselves. Sub-adult males never challenge a fully fledged adult male over a female, even though the two males may actually be of the same age. The presence of a fully adult male suppresses the release of hormones in the other developing males. Only when out of the range of an adult male can a sub-adult mature, metamorphosing into a simian version of the Incredible Hulk in a matter of days.

Sub-adult males are not attractive to females, so they rape. Biruté Galdikas observed Nick on a consort with a young oestrous female she'd called Noisy, who was very attentive to Nick and very keen to copulate with him. As the two of them moved through the trees they were followed not just by Biruté Galdikas but by three sub-adult males. When one of the trio came too close to Noisy, Nick chased after him through the trees, leaving Noisy vulnerable in Nick's absence. One by one the sub-adult males seized their opportunity, raping Noisy, who repeatedly tried her hardest to fight them off. Sub-adult males will spend fifteen minutes or more fighting with a female in their attempt to rape her. Adult males sometimes try to force copulation, but if the female tries to wriggle free they usually let her go. Adult males are desirable to the females, so they do not need to waste their energy fighting.

It seems that most first-born infants may well be the result of rapes by sub-adult males. This initiates the virgin to motherhood and allows the sub-adult the chance to reproduce with an inexperienced female. There is a possibility the progeny may not survive (like motherhood in women, it's a skill that improves with time), but the sub-adult does gain sexual experience. The older the male orang-utan, the more sophisticated is his copulatory repertoire and the more desirable he is to the females.

Because female primates have orgasms and multiple orgasms, it does seem possible that females with a choice of males – such as the common ancestors to modern woman, chimpanzees and bonobos – favour males who could give them satisfactory stimulation. The human penis is longer and thicker than all of the other great ape penises. On average the length of a man's erect penis is 5 inches. The next in size are chimpanzee and bonobo penises, which when erect are 3 inches long and pink. Out of the great apes the massive gorilla has the smallest penis, only 1.25 inches long, but then the female gorilla does not have as much choice as females in multi-male breeding set-ups and therefore has less chance of influencing the size of the next generation of male gorillas' penises. Did our female hominid forebears favour males that could give them good sex and at the same time did that sex confuse those well-endowed males into offering paternal investment to babies born in their community? Has it taken 4 million years of female choice for modern man's penis to evolve to its present size?

The size of testes also reflects the reproductive lifestyle of a primate. Chimpanzees and bonobos have the largest testes; men's testes are less than half the size of chimps; and by comparison those of gorillas are the smallest. As a male living in a multi-male society you need to produce copious amounts of sperm to compete with the other males.

The Debate

The penis and the testes are easy organs to measure, but the clitoris is not. Evolutionary theorists continue to argue over the female clitoris. Stephen Jay Gould, the famous Harvard biologist, thinks that the female orgasm and clitoris are 'incidental' and only appear because erectile tissue has been selected for in males. He has argued against Hrdy's evolutionary theory that the female orgasm is an adaptive reward to encourage females to solicit sex from males to confuse paternity and avoid the infanticide of their children. Gould sees many physical traits as by-products evolving from mutations and accidents of nature. He does not believe that everything has been selected for. Gould is interested in exceptions. After the Cretaceous/Tertiary extinction 65 million years ago, the mother of all primates, *Altiatlasius*, survived when many other species of placental mammal died out. We could so easily not be here. A mass extinction is unconnected to natural selection and Gould claims that extinctions and genetic mutations have helped create life on earth as much as natural selection has.

Richard Dawkins is a British gene-selectionist Darwinist, who believes that all traits are adaptive. Nothing in life is left to chance. Right down to our genes the selective programming is firmly in place. Gould, who believes himself to be closer to Darwinian thought than Dawkins, does not agree. Gould's main point is that we do not know which of our interesting traits have been achieved through accidents or selective pressures that in turn have given rise to other traits, so how can scientists like Sarah Hrdy and Richard Dawkins 'reverse-engineer' human evolution? Many human traits may be the result of mutations and ecological pressures rather than sexually selected ones and the human female orgasm may just be a by-product.

Orgasms aside, Gould's reticence over claims that all traits and behaviour can be explained through our genes may stem from his concern that neo-Darwinist 'fundamentalist' theories could become a lethal weapon in the hands of eugenicist ideologies. He thinks Sarah Hrdy is as much of a product of her time as she thinks Darwin was. He sees her work as stemming from a modern social construction – feminism. Gould believes it is in Sarah Hrdy's feminist interests to endow women with 'female choice' over paternity. He sees both the male nipple and the female clitoris as accidents and by-products that cannot be explained through adaptation.

But according to 1998 research by Melbourne urologist Helen O'Connell, the clitoris is no accident. Most of the anatomy diagrams used in medical textbooks today were originally based on bodies dissected in the Victorian era. Today, after invasive surgery to their uteruses, vaginas, urethras, or bladders many women complain that they suffer sexual dysfunction and no longer feel any pleasure from sexual intercourse. It seems that those Victorian diagrams of the clitoris and the descriptions of it as a 'poor homologue' of the male penis are wrong. After dissecting the bodies of ten adult women of different ages, O'Connell has documented that the external tip or 'glans' of the clitoris that can be seen just above the urethra is just the tip of the iceberg. Most of the clitoris is internally located; it is 100 times bigger than most people realise. Inside the woman's body, the erectile tissue of the clitoris forms a pyramid with arms 9 centimetres long. The clitoris surrounds the urethra on three sides and runs along the anterior wall of the vagina, and two prominent clitoral bulbs are also located on either side of the vagina. From now on men may suffer clitoris envy. When a woman is sexually excited, this mass of erectile tissue fills with blood, helping protectively to close off the urethra, preventing bacteria from the man's penis from entering the woman's urinary tract and holding the vagina wall rigid, aiding pleasurable penetration for the woman. From dissections of older women O'Connell discovered that, post-menopause, the erectile tissue of the clitoris shrinks. In the past gynaecologists have not known where the nerves of the clitoris are and have severed them permanently during invasive surgery.

In the West over the last thirty years we have tentatively begun to speak about women's sexuality. Victorian attitudes towards women and female animals have been very persistent. A nineteenth-century British medical doctor wrote about women's sexuality: 'The majority of women (happily for them) are not much troubled with sexual feelings of any kind.'

Darwin was stumped by the way some female apes and monkeys advertise their fertility with swollen and engorged genitalia. 'No other case interested me and perplexed me so much as the brightly coloured hinder ends and adjoining parts of certain monkeys.' Over the last thirty years the influx of women into the scientific arena, who have related to the female of the species they were studying, has opened matters up.

I mentioned to Sarah Hrdy that I'd heard it said that she was far more theoretical than most other women primatologists. Many other women primatologists, when asked, will say they love observational field work much more than desk-bound musing. She is not a primatologist who later turned to the study of human behaviour; right from the beginning she was studying behavioural phenomena in both human and non-human primates. 'I was interested in my own nature. I found out as much about these topics as

I could both in humans and in animals and in people like myself . . .

'I didn't come into science and primatology because of a love for non-human primates. I was first attracted to the problems and to things I wanted to understand, so I was drawn into primatology for much more intellectual reasons than some of my colleagues, who would have been drawn in at a more emotional or passionate level. I do understand myself better now. I was raised in the South and women really were thought to be quite chaste creatures; this was a combined legacy of Freud and Victorian sensibilities. I really do feel some of the issues I set out to understand about my own nature, such as parental investment strategies, sibling rivalry and female sexuality. I now understand. I'm much more open-minded and have a general level of understanding that I never would have achieved otherwise. So I would certainly say I now feel closer to understanding myself.'

Hrdy continued: 'I see science proceeding in phases. Not only does it proceed in phases for almost every project, but different scientists are going to emphasise different phases of the project. What I enjoy is this speculative construction of possibilities. I like to immerse myself in everything that is known about a phenomenon and then try to understand for myself what's going on here. This is at one level the creation of imaginary worlds. That's just one phase of science. At the observational, data-collection phase, some people combine both into one; at different phases of my life I have combined both. People have said I speculate too much because what I am really interested in is the imaginary phase, when I try to hypothesise what is going on. There is really no one way of doing science. I think that we certainly need people to stress theory as much as observation.'

There was not much difficulty in getting women to believe in the female orgasm, but Sarah Hrdy had problems in communicating her research to the scientific establishment. Senior male figures were resistant to change. Hrdy says, 'I have felt like a savage on the outskirts of society trying to rediscover the wheel all by myself. I was often resentful at the steps that were necessary to make this discovery. But it was beginning to understand what feminists were trying to say that helped. There was nothing in my traditional education in physical anthropology to give me any tools to help me understand male bias, so I had to discover it on my own. It hurt me. It caused great anxiety and stress and undue misery. I have written letters to the President of Harvard complaining about sexism and then burned them and flushed them down the toilet to be sure no one would see them. And I have seen what's happened to the women who have complained and paid big prices.'

But at the end of the 1980s affirmative action was infiltrating academic establishments. While only in her thirties Sarah Hrdy was made a full professor and some six years later, in her early forties, she was elected to the National Academy of Sciences. Of her promotion Hrdy commented, 'I knew at the time there were men who deserved it more. They wanted women.'

It has been said objectivity is an impossibility. Scientists are people first, bringing with them their conscious and subconscious baggage. The politically correct amongst us have said, as biases are inevitable, that it is our moral duty in science to replace stereotypes with positive biases. But Hrdy, who believes science is self-correcting, disagrees. 'At the assumption/hypothesis stage of science, you need a vast array of scientists. You don't just need privileged people; you need people who come from under-privileged backgrounds, because they will ask different questions differently. I think it is true to say that learning about biases in science humbles scientists. There may be personality types who are not sufficiently humble about their work, but in my opinion a good scientist should be extremely humble about what she can learn and about the need to check and replicate. If what's been done here has engendered more humility all the better. But has it challenged the scientific enterprise? My hope is that it's strengthened it.'

Hrdy describes her struggle for scientific integrity: 'I am afraid of my material being hijacked and used in the battle of the sexes. One of the reasons I no longer lecture in frameworks that are self-identified as "feminist" is my fear that I will distort my own independence by trying to be, quote, "politically correct". I think it's incredibly important to be one's own person, to not feel I'm part of a movement. This is very hard. You certainly want approval from your audience, and in my case a segment of the feminist audience really would prefer that I didn't exist, because it's much easier to claim that an area of evolutionary perspective is deterministic and sexist if a feminist does not exist using an evolutionary perspective to show how flexible behaviour in males and females can be. There is also a desire to paint all feminists as rabidly anti-Darwinian. These other people have a stake in me not existing too, because I clearly don't reject an evolutionary perspective. I'm simply critiquing specific biases that have plagued an evolutionary perspective from the nineteenth century onwards.'

Hrdy told me more of her determination to keep her research culturally uncontaminated. 'In fields like biology I don't think women are under special pressure to be politically correct. By and large they have to be especially independent. If you are talking about field primatologists, we are by nature very independent people, loners, non-joiners – imagine the kind of person who is willing to go off for years at a time forgoing a lot of social contact. I think that these women are more likely to be caught by surprise by political correctness than to be pressurised by it.'

Hrdy never managed to prove genetically that male-perpetrated infanticide in langurs was a reproductive strategy. She intended to take blood samples from infants and infanticidal males to confirm that the male langurs did not share DNA with the infants they killed but did share DNA with the infants they tolerated. Unfortunately the Indian authorities accused Hrdy of spying and refused to allow blood samples to be taken from their sacred monkeys. Abruptly, after ten years Hrdy returned to the United States. It has always rankled with her that she has not been able to prove her theory that male langurs are genetically programmed to kill non-kin infants, even though a mass of research now backs up her claims. But because she never took any DNA samples to confirm paternity, her theory that the female orgasm is an intermittent reward system to encourage females to mate sequentially with different males, so confusing paternity and avoiding infanticide, is also just her theory. This has meant that the debate has never gone away and people continue to argue with her about her interpretations, even though some years back she moved on to other research. Hrdy told me that she has been aware many times that men have chosen to discuss the origins of the female orgasm with her as a way of openly talking 'dirty' with a woman.

Darwin's coy female was his feminine incarnation that never actually evolved and a creature that no real woman could ever become. Naturally sexy women are often considered to be bad and unnatural, as the femme fatale in the oestrus-red dress of fiction reveals, whereas aggressive and sexy men are doing what comes naturally. We would never describe a heterosexual man as a nymphomaniac or promiscuous, so why should a woman be thought of this way?

It is Sarah Hrdy's reasoning that the human female orgasm and a woman's ability to experience a multiple orgasm are legacies from her pre-hominid past. Orgasms evolved before man did. To avoid infanticide, females would mate with several different males whether the society was multi-male or polygynous, allowing the female to have a group of possibly kin-bonded males around her who would make some paternal investment in her baby as there was a chance they were related to it.

Hrdy adds, 'I don't know any human society where female promiscuity is an adaptive female reproductive trait. I don't know any human societies in which females are rewarded for mating sequentially with a series of males. In fact in our own society and in most patriarchal societies, women would be gravely punished. I'm leaving aside female prostitution, which is a special economic gain and no woman's first choice. So I would say women were not using that strategy today, but women's physiological legacy may well have been derived from these kinds of selection pressures from our evolutionary past.'

The female bonobo has used her orgasm to move her farther forward than her other ape sisters. Through lesbian sex, her ability to give and receive orgasms unites her to other females. This female-bonding prevents male aggression and infanticide. She doesn't need the orgasm to encourage her to mate widely in order to confuse paternity any more, because the female bonobo does not live in fear of males' aggression as her other ape sisters do.

Today women live in stable societies, usually paired monogamously. They do not en masse need to confuse the men in their lives about the paternity of their babies. If Sarah Hrdy is right and the female orgasm is a legacy from our distant female ancestors, this does suggest it is an adaptation that is on its way out. Could this explain why women's magazines are full of self-help guides to achieving the elusive orgasm, while our primate sisters still have orgasms so easily? Or is it, as Robin Baker hypothesises, that we are not orgasming (and sucking sperm up into our fallopian tubes) because something deep inside is telling us that we do not want that particular man's baby?

10
Woman's Work

While writing this book I found the easier way to introduce it to non-scientists was to mention the film *Gorillas in the Mist*. Even though Jane Goodall was Dian Fossey's role model (eventually the two of them became Biruté Galdikas's role models) and women primatologists know that the 'Jane Goodall Phenomenon' is found at the heart of primatology, the general public seem to know more about Fossey than about Goodall and they have gathered this knowledge from a Hollywood movie.

Since the release in 1988 of *Gorillas in the Mist* the film has gone on to have a greater influence on the public perception of primatology than perhaps any other single piece of popular culture devoted to primates, women and nature. Approximately forty years old, primatology is a massive and diverse discipline that includes the study of the evolutionary anatomical, behavioural and psychological make-up of two hundred species. And this wide-ranging subject is generally thought of as a woman's subject – specifically white middle-class women – and something women do well. It has become the only scientific subject to bleed into popular culture and leave a permanent mark of gender that in turn has gone on to colour the science. Many of the newest generation of women primatologists have been spurred on after watching *Gorillas in the Mist*.

As with those of Jane Goodall and Biruté Galdikas, Fossey's story consists of adventure, discovery and idealism. But it was Fossey's gruesome murder that became the final ingredient to a ripping yarn. Hollywood producers knew the self-sacrificial element to Fossey's story would have popular appeal. Many people have speculated on what drove Fossey on: why did she intentionally make so many dangerous enemies? Why, after a series of death threats, didn't she leave? Bizarre rumours have spread to the effect that she was in some way sexually satisfied by the gorillas. I never spoke to Fossey – she died when I was a teenager – but if I could ask her why she chose to stay on after her paradise had turned to purgatory, I suspect she would tell me what other women have said: that she felt solely responsible for the gorillas, that she loved them and feared more for their future than for her own.

Fossil hunter Louis Leakey was convinced field primatology was women's work. He believed in the human female's eye for detail and he was especially turned on by Goodall's (then a secretary) 'uncluttered mind, unbiased by theory'. He wanted untrained women to take advantage of their own femaleness to become patient, persistent observers of nature. Richard Leakey commented, 'Jane Goodall made science popular and interesting and science should be popular.' Because Goodall *did* make science popular and accessible we now have many women primatologists, and again this wouldn't have happened without *National Geographic*'s coverage of the beautiful young Goodall. Primatology is the only area of scientific endeavour that has grown to fit its cultural image; life imitated art.

In the early years of Goodall's career she was depicted by *National Geographic* as an intrepid 'girl guide' figure, camping in the woods, cooking over a camp fire, sleeping under a starry

African sky and all the time absorbing nature. Goodall was seen to be as innocent and as natural as the creatures she studied. After she married and her son was born, she was viewed in a more sacred light, as a young woman with child living in peace alongside chimpanzees, discovering her primal roots in a primitive landscape. Today Goodall is very much the 'wise woman'; she is a sage, and someone we turn to for answers. Goodall has relinquished her early roles and over the years she has distanced herself not only from those earlier incarnations but also from Leakey, Fossey and Galdikas.

Goodall knows her work with chimpanzees is associated with that of Leakey and her fellow trimates. And those three people have all had bad publicity, justifiably or not. Leakey was a known womaniser and had adored Jane, but Goodall does not like to refer to Leakey's love for her. She was not just his passing fancy. In retrospect Fossey is thought of by many as an irrational, sentimental and defensive woman who put animals before people. Goodall is reluctant to share that reputation. As a spokeswoman for chimpanzees, it is important that her judgement is trusted.

Maybe Goodall would have lost herself in the work like Fossey and Galdikas if she had not had so many students so early on to care for and if the 1974 kidnapping attempt hadn't forced her to leave Gombe for two years. After that Goodall made frequent but random, low-key visits to Gombe and the day-to-day monitoring of the chimps was taken over by her African staff. Being physically detached from Gombe prevented Goodall from losing herself totally in the life. Today Goodall is a chimpanzee purist; she is against human physical contact with chimpanzees. Goodall's ideal world would be a place where all chimps live free in Africa, 'uncontaminated' by humans. She believes that humans ruin chimpanzees through cultural brain-washing. She feels that if there is an influential human presence in an ape's life a chimp will metamorphose into a creature that continues to look like a chimp but will never behave like one again. Goodall does not think the Gombe chimps are contaminated because today researchers do not make physical contact with them.

But the young Goodall felt differently. In her book *In the Shadow of Man* she wrote,

> One day, as I sat near him [David Greybeard] at the bank of a tiny trickle of crystal-clear water, I saw a ripe red palm nut lying on the ground. I picked it up and held it out to him on my open palm. He turned his head away. But when I moved my hand a little closer he looked at it, and then at me, and then he took the fruit and, at the same time, he held my hand firmly and gently in his own. As I sat, motionless, he released my hand, looked down at the nut, and dropped it to the ground. At that moment there was no need of any scientific knowledge to understand his communication of reassurance. The soft pressure of his fingers spoke to me not through my intellect but through a primitive emotional channel: the barrier of untold centuries which has grown up during the separate evolution of man and chimpanzee was, for those seconds, broken down. It was a reward far beyond my greatest hopes

The passionate young woman who wanted to tame the wild beast that visited her camp to steal her banana and who was 'thrilled' when he allowed her to sit next to him and groom him lives on only in memory. Young women students may go to Gombe hoping to emulate Goodall in every sense, but they will not be able to make contact and communicate telepathically with the chimps the way Jane used to. It's no longer allowed. Researchers must resist the urge to reach out and emotionally break down the evolutionary separation between us and them; they must do so intellectually instead. (When the war allows researchers back to Karisoke, the situation there today is the same; as Liz Williamson says, gone are the days of sitting in a gorilla's lap.) At Gombe Goodall's chimpanzee adventure is Goodall's alone. The unique, sensual bond that Jane developed with David Greybeard is no longer considered appropriate in the politically correct world of contemporary primate field research.

In Borneo, after her first husband and son returned to America, Biruté Galdikas not only chose to stay on with the orang-utans but she also immersed herself in the spiritually ancient world of the Dyaks. As a result, Westerners saw her as rejecting their ways for another, far more mystical way of living that felt alien. Some seem to think that in marrying a Dyak Indian Galdikas has absorbed too much Dyak spirituality. Goodall's writing occasionally also has a certain Christian morality about it – especially evident in her autobiography, *Reason for Hope: A Spiritual Journey* (1999) – Christian sensibilities remain acceptable to Western tastes. But Galdikas, who is seen by some as having already gone native, is now further distancing herself from Western standards by forgoing scientific endeavour to fight instead a far too emotional war with logging companies. Galdikas has also been criticised for encouraging eco-tourism and allowing the mostly women tourists to run amok. Her scientific reputation has suffered over claims that she is rehabilitating too many orang-utans into a forest that can hardly sustain the resident community.

Because Goodall was the first woman to habituate wild apes and because she discovered tool use and hunting in chimpanzees, we are encouraged to think of her as a cut above the rest. That, along with Goodall's stoical British reserve, today helps to give her a sane and trustworthy image in the eyes of a fickle public. The trimates were always different types of women with different approaches to life and science. At present, even with Fossey dead, they continue to have thousands of equally devoted yet very different types of followers. Goodall's' supporters are primarily women with an air of sensible superiority about them, mainly because Goodall is so revered and chimpanzees are genetically closer to us than orang-utans or gorillas. Galdikas's followers are drawn to the sanctity of a rainforest and the majesty and mystery of the only Asian great ape. Fossey's fans feel the vulnerability of the natural world in the hands of careless men. In their minds the tragedy of Fossey's ultimate sacrifice and the threatened extinction of a gorilla species blur into one almighty symbol of good against evil.

Not every woman could have done what those three women achieved, but Leakey's trimates were exceptional women. Womaniser or not, Leakey was a great judge of character. He may have generalised about women's ability to read non-verbal cues and their non-threatening presence when compared to a handful of men who had previously tried in vain to habituate wild apes, but he recognised these abilities in Goodall, Fossey and Galdikas. As three uniquely different role models they are hard acts to follow, but followed they are. Today primatology is the only area of science where women dominate men in numbers. Ninety per cent of primate sanctuaries are run by women, 62 per cent of members listed in the World Directory of Primatologists are women, and numbers of women researchers are growing.

On the surface women field primatologists seem to get more satisfaction from their research with the animals than they do from other human relationships. Boyfriends, husbands and children of women mentioned in this book have lost the embrace of their woman or mother to an ape. When asked to choose, women have chosen apes. Knowing that your animals will probably die without you there to protect them tugs hard at the women, who cannot turn their backs and free themselves of the responsibility they have assumed. The trimates could not and neither can the new genera-tion of women field primatologists.

Liza Gadsby belongs to this new generation. Because of her

passionate commitment to her chosen primate species, the drill baboon, Gadsby is seen by many as the new Dian Fossey. She has centred her conservation work with drills in the wilderness of Nigeria's mist-covered Afi Mountain. Drills are even more endangered than the mountain gorilla.

Gadsby originally comes from Oregon. In her late twenties, after completing a BA in biology, she decided to visit Africa with her 'high school dropout' boyfriend, Peter Jenkins, and tour around. While travelling through Nigeria in 1988 Gadsby discovered the bush-meat trade in drill baboons. She decided to stay and do something to stop it. The bush-meat trade in Africa is devastating for endangered species. Practically anything that walks, creeps, crawls or flies will be hunted, trapped and butchered in the forest in spite of its protected status. Later, the free-range meat is sold on as a highly desired delicacy. The trade in bush-meat is today the greatest threat to wild primates, greater even than the threat of habitat loss.

Drills, similar to mandrills, are forest baboons. They are found only in Nigeria, south-western Cameroon and on the Equatorial Guinea island of Bioko off Cameroon. They have smooth black faces, unlike their cousins the mandrills, which have bright blue and red muzzles. Poachers track the drills with dogs and wipe out groups of forty or fifty at a time. Gadsby tries to rehabilitate orphans who end up as loggers' pets after the mother has been butchered for bush-meat. Faced with the knowledge that a species of primate was about to be wiped off the face of the planet, Gadsby felt compelled to stay and fight on the animals' behalf. Nigeria is now her home.

For the first few years of her work she was entirely self-funded, but as her own money ran out the conservation charity Fauna and Flora International, FFI, stepped forward to support her Pandrillus Conservation Project. Gadsby lives with Jenkins in a tiny house, surviving on a salary of £75 a month and her rehabilitation centre for rescued infant drills and chimpanzees runs on donations alone. The two of them have organised an education project so that nearby schoolchildren can come and learn about the drill. In 1996 Gadsby was the recipient of the prestigious Whitley Award for conservation, which she used to employ a team of ex-poachers as gamekeepers.

Entirely self-taught and unbiased by theory, Gadsby is now the world expert on the drill. She believes drills are not territorial as originally thought, but nomadic, and that they cover great distances in a day, looking for fruit, roots, insects and invertebrates to eat. Although they are comfortable in the trees and make night nests up in the canopy, in the daytime these short-tailed monkeys are primarily terrestrial.

Like Dian Fossey, Gadsby feels angry with researchers who are not also conservationists. She spoke to me about Fossey. 'Researchers still smirk about Dian. They say, "I wouldn't want to be like Dian." They just want to be seen as cool, calm professionals. Too many primatologists don't give a damn about conservation, too many Ph.D. students just want to get study done before the species is wiped out; there is no concern for the species, only their own career. In southern Cameroon Japanese researchers developed a feeding site for chimps in order to study them. Once they'd satisfied themselves they returned to Japan, they made no provision to de-habituate the chimps. After they left hunters arrived, put down food, waited for the chimps to come and then shot them all dead.' Gadsby told me more about her feelings. 'I now have this massive responsibility; it's terrifying. I can't leave, although I expect my boyfriend will in the end. There's no one to take my place; I can't get out of it. People think I do this work because I'm sterile. I'm not; I've stopped myself from having babies. I don't want children; I've more than enough to cope with as it is.'

Looking over the lives of women primatologists mentioned in earlier chapters we can see how their feelings of responsibility and emotional commitment to their primates also rooted them to the spot. Women become attuned to the emotions of primates, and it is this sensitivity that encourages them to commit their own lives to their animals' cause. I asked Stella Brewer to tell me what keeps her devoted to the chimpanzees she rehabilitated on to an ecologically self-sustaining chain of islands in the River Gambia National Park. 'I've been accused of loving my chimps too much, but it's more practical than that. I do love them, the way I love my family and friends that I've spent a lot of time with. A job never ends if it involves living things because life keeps evolving. Babies are born to our chimps and it is the survival of the new generation that shows me the twenty-five-year project is a success. I see the continuation of a rare species and I feel very proud. I keep Bruno and Rene salaried and a camp on the island not because the chimps need daily care but because tourists and poachers would take over if I didn't. I worry what will happen to the chimps after I die. I just hope chimpanzees will by then hold their rightful place in the minds of people and be respected and protected.'

Janis Carter, who collaborates with Brewer, had completely lost contact with her boyfriend, family and friends for most of the years she had been with chimpanzees in Africa, particularly during the seven years on the island when she was living more or less as a chimp while rehabilitating the chimpanzee Lucy. When faced with the needs of Lucy and the other chimps she had adopted, Carter felt she had no choice but to stay in the forest with them until they no longer needed her constant support. But living alone with wild chimpanzees affected her own social and emotional development. During the brief times she left the chimps and returned to the capital of Banjul she found it very hard to talk to people and behave the way a woman is expected to. She had become a chimp and felt more at home with her group of apes than with her own primate species. It wasn't until 1996, nineteen years after first leaving for Africa, that Carter finally saw her sister in America again and in the summer of 1999 she saw her parents again after a gap of fifteen years.

Carter also has a rehabilitation centre in Guinea with forty infant chimpanzees all rescued from the ape and bush-meat trades after their mothers were killed. But she has no official financial support to help pay for chimp food and vets' fees. At present she is entirely dependent on donations to look after the growing numbers of babies. 'It's desperate, it's causing me much anxiety and stress. How am I going to keep them alive? I'm running out of money!'

Working with apes, monkeys and prosimians on location is physically and mentally isolating. After spending months habituating the animals the primatologist begins to feel morally responsible for the lives of the individual primates studied and for the species as a whole. But these are animals that possess a dangerous bite and they do not like humans they do not know. Primatology is work that is not easily shared or understood by outsiders. Yet it gives a geat deal back. The long term commitment seen in women primatologists (but in very few male primatologists) must have been encouraged by the behaviour of the animals. Women find this human-like behaviour especially fascinating to watch.

When a primatologist first embarks on a field study she usually has a question she wants to answer. But as time passes the newly acquired observational data will throw up questions of its own. A patient woman primatologist, who has been accepted by the animals and allowed to get close, often closer than men manage to get, may see something no one has seen before, as Goodall did when chimpanzee David Greybeard made and used a tool. Or she may interpret a behaviour in a way no one has before, as Sarah Hrdy did after watching male langurs murdering infants. Evolutionarily the animals' behaviour presents primatologists with personal and general insights. But it is the physically arduous nature of the work that also tests them as women and reveals something to them of their own nature. Hrdy has said of women primatologists, 'I think we are an unusually independent lot.' Not only can a woman primatologist be her own boss but she can change the way we think of these animals and ourselves.

Discoveries by women primatologists have helped to shape the way we define the human species and the terms in which we see our own evolution. The women mentioned in this book have all been extremely persistent over their revolutionary and often unwelcome beliefs. Their dogged tenacity since the beginning of their careers has changed the science completely. Perhaps in a male-dominated hierarchy such as the academic world, some junior female scientists have felt there was nothing to lose by standing their ground. This quiet self-confidence in the long term has gained these women widespread respect. Jane Goodall insisted in the early days that her chimpanzees' gender mattered and she persisted in seeing them as individuals and in naming them. If Goodall had not fought to have the significance of individual behaviour taken into account when she did, it would be much harder today for women psychologists, such as Sue Savage-Rumbaugh, to have their research acknowledged. If Sarah Hrdy had not interpreted a particular male-perpetrated infanticide as a male reproductive strategy we might still think of the female langur as a mindless nymphomaniac instead of an animal acting out her own counter-strategy to deflect male oppression – and the female orgasm would have remained irrelevant.

Women scientists automatically appreciated the fact that female primates were sexual, libidinous creatures capable of orgasms and that their sexuality gave them power and choice. Men did not want to accept this. Women such as Barbara Smuts and Jeanne Altmann taught us that primate mothers had to budget their time carefully between all of their daily activities, and male friendships were a necessary support, but only if there were no strings attached. Women primatologists have watched bargains made between males and females and have observed that once an investment of time has been made males will become aggressive if they are not rewarded with sex. But proving to the rest of the scientific community that the nursing mother was the palpable face of evolutionary impact, rather than those flamboyant males, took time.

Western women primatologists and Japanese primatologists observed in species of matriarchal monkeys and prosimians that rank is inherited directly from mother to daughter. But during the 1970s high-ranking Western male scientists were busy teaching that primates were patriarchal. Many women have had to fight with the establishment for what they believed in because the male establishment felt comfortable with their false beliefs. Some men found it satisfying teaching students that evolution was working on the male gender but not the female and that it was the male who was the instigator of change. Women scientists have proved that evolution works equally on both genders. It is accepted today that the two genders engage in a perpetual fight for choice and dominance. Within this sex war each gender has a competitive counter strategy to its opponent's every move, the sexes' ingenuity being as much a part of the force of evolution as predator–prey relationships and climate changes.

The expansion of primatology has coincided with, and contributed to, the dismemberment of general Freudian beliefs. Our own genetic programming has been underestimated for years. The behaviour of individual primates has been used to back up Darwinian theories of genetic selfishness. Accounts of human twins separated at birth and reunited after forty years showed many behavioural similarities and always seemed to have something of the supernatural about them. Western culture has absorbed Freudian analysis without realising it, so when we see two people who happen to be twins but who have lived through starkly different environmental experiences we are amazed that they behave alike. Today a primatologist would tell you that identical twins share the same genes and that they would be amazed if the twins, even though separated at birth, did *not* behave alike. From the long-term study of the behaviour of groups of primates genetic programming has been understood. This is not to say that environmental factors have no bearing. We know, for example, that if an infant primate is separated from its mother it will not receive the social guidance it needs to survive. But we can see from observations of primates (including humans) that we inherit the programming for a series of reactions to any situation we find ourselves in. It is an individual primate's strategic yet often subconscious choice as to which impulse it acts upon. Our genes motivate us as much as or possibly more than environmental experiences; to a significant degree we are born who we are.

Women have observed and theorised on primate behaviour, especially female behaviour, in original ways. Without primatology we wouldn't have appreciated the full significance of social rank. We now realise that social status affects the gender of infant primates depending on whether they are born to a patrilineal or matrilineal species. Scientists have observed that all apes (including humans) appear to be patrilineal and high-ranking mothers are more inclined to bear sons, while low-ranking mothers are more inclined to bear daughters.

Women scientists have also pushed scientific boundaries on the internal world of primate intellect. Sue Savage-Rumbaugh and Sally Boysen's research into the minds and language abilities of chimpanzees and bonobos has challenged contemporary beliefs. These women testify that the apes they work with are self-aware, reflective, calculating individuals, capable of counting, communicating their thoughts with humans, of feeling and showing love and hate and much, much more. Breaking down the boundaries between us and them is a theme common to the work of women primatologists – not because women think the lives of apes and monkeys are desirable or more utopian than human behaviour, but because women scientists cannot deny to themselves or to the rest of us what they have discovered.

Women's work in primatology has pushed the philosophical issue of animal rights and the intellectual ability of apes to the forefront of science. It is today one of the most important and most controversial of areas. In theory-of-mind tests the Machiavellian nature of primates, especially the great apes, has become evident. Apes are smart and sneaky; they can manipulate others. Tactical deception has been measured in apes and there exists a level of intuitive intelligence not found in autistic humans. Autistic people and those with Asperger's syndrome can have a higher than average academic intelligence but nonetheless will often behave incompetently, as they cannot instinctively feel their way through what life throws at them. Autistics cannot tactically deceive because they find concealing their true feelings hard and this contributes to their vulnerability. In a group these sorts of people cannot think on their feet and spontaneously say and do what's best; they are unilaterally reliant on favours and good advice, whereas apes have an emotional intelligence that allows them successful, regular cheating against one another. Apes are aware of when they have knowledge that another ape does not have and they often use that advantage selfishly.

Since these revelations there has been growing support for the Great Ape Project. This movement wants to see great apes being given rights instead of being treated like mere animals. One argument says that just as we would find it unacceptable to experiment on autistic humans, so we should find it unacceptable to experiment on apes. From recent studies on the structure and workings of the ape brain it has become apparent that there are more similarities between the human brain and the ape brain than there are differences. In humans the planum temporale is an asymmetric area of the brain essential for language comprehension. It has recently been discovered that the positioning of the planum temporale in apes is the same. All of these scientific developments suggest that apes have the intellectual ability for basic language at the level of perhaps a four-year-old child, but an ape's underdeveloped larynx prevents it from telling us whether this is so.

Sue Savage-Rumbaugh has said that working with apes for twenty-five years has been the most intensely intellectual experience of her life. Her research in ape intelligence and language acquisition has shown her what differences exist between us and our sister apes, the chimps and the bonobos. In evolutionary terms Savage-Rumbaugh has, by studying our closest relatives, learned what particular physiology it is that makes a human a human. Each new chimpanzee study seems to find even more evidence that these animals are only a few evolutionary steps behind us. The impulse to understand the mind of a primate, which may think similar thoughts to you but will never be able to tell you so, seems to be a female urge stemming from a female skill.

For some women, a career involving work with children can satisfy the maternal instinct enough for them to not have children of their own. Researching the lives of apes means women primatologists are spending hours every day in the company of animals with a mental capacity similar to that of a young child. Although many of the animals are adult and monstrously strong there is something of the overgrown chaotic child about them. They are creatures that need to be understood, forgiven and protected and certain women have the perception to understand, the patience to forgive and the instinct to protect. Jane Goodall has asked how we should relate to 'beings who look into mirrors and see themselves as individuals, who mourn companions and may die of grief, who have a consciousness of 'self'. Don't they deserve to be treated with the same sort of consideration we accord to other highly sensitive beings – ourselves?'

The average mother feels the needs of her infant, she defends the child's interests, knows what the child wants and what the child's utterances mean, and she informs other members of the child's world of her child's intent. As it naturally occurs between mothers and babies or very best friends this mother-infant *complicité* is also satisfyingly achieved between women and primates. Women pimatologists are the voice of their primates. Women's skills in non-verbal communication were understood by Louis Leakey. He also realised that once immersed in the animals' world the women would become as emotionally embroiled as they become with children (Dian Fossey had worked with disabled children before moving on to gorillas). Leakey suspected the harsh living conditions would not be enough to make the women give up and he predicted, as the animals were long-lived with long childhoods, it was research that would take a number of years. Long-term commitment was essential for long-term success, and *complicité* elicits commitment.

But ultimately Leakey underestimated what the trimates and the apes had to offer. He thought Goodall's study would last a few years and then, as an experienced field researcher, she could go on to study the bonobo for him. He also suggested these options to Fossey and Galdikas. But once the trimates rolled up their sleeves and got stuck in, and the animals started revealing their nature, none of the women wanted to move on. There was so much to see, things no one had ever seen before, that the bonobos would have to wait. Leakey suspected the bonobos would be uniquely fascinating but he died before their behaviour was fully appreciated.

An average woman has the necessary emotional and intellectual apparatus to 'read' the behaviour of others. This perception means women feel the feelings of the animals they study; if you cannot feel a fellow primate's feelings, you cannot understand its nature. Primatology is a sub-discipline of biology but it cannot be compared to the study of less sophisticated animals such as fruit-flies, where empathy would not be crucial to research. Goodall has said that when she first entered the world of academia an ape was considered to be a basic creature that reacted to circumstances with mechanical responses, as a fruit-fly might. Comparing the gestures of apes with humans, such as the action of one chimpanzee hugging another, was, back in the 1960s, sinfully anthropomorphic. The young Goodall was told that although it might seem that two chimps enjoyed hugging each other and consciously chose to do so, just as humans might, it only seemed that way and was not actually so. Endowing primates with feelings and thoughts was impossible before women studied apes. Prior to women's research, apes and monkeys were little more than hair-covered automatons in the eyes of male scientists.

Many thought Leakey was mad sending women to jungles to study apes; if qualified men couldn't do it, how could untrained women possibly succeed? Before Goodall proved it was possible no one thought you could habituate wild apes. But today most of the species belonging to the vast primate family have been studied. In the early years and in order to get the photos National Geographic were paying for, Goodall provisioned the chimps with bananas. She was criticised for doing this by Leakey and others and she now regrets it. For a time it affected the chimps' behaviour and caused more aggression between individual chimps and between baboons and chimps than was usual, but in the long term it did no harm. Today most women primatologists are proud to announce that they do not provision the primates they study as a way of speeding up the habituation process. Rather than giving fruit to the animals, women prefer to give their time. It's better for all involved, apes and women, if the animals do not become dependent on the primatologist for food.

Leakey was a natural historian and animal lover but professionally he was a palaeoanthropologist. He wanted to know about the behaviour of our hominid ancestors and that is why Goodall was originally sent to Gombe. From observations of prosimians, monkeys and apes primatologists have made many informed guesses on the lives of our ape-like ancestors. We can speculate much about the lives of early people by comparing the fossil record with the environment, ethology and morphology of contemporary primates, especially apes. For example, sexually dimorphic species, such as gorillas, live in patriarchal harems. A newly sexually mature female gorilla must leave her natal group to look for a large, strong silverback in his prime to protect her and her infants from other males. It is thought that our 3–4 million-year-old ape-human ancestors were sexually dimorphous and probably lived in gorilla-like harems.

By comparison chimpanzees, whose sexual dimorphism is much less exaggerated, live in multi-male breeding systems and not male-led harems. Here an oestrous female chimpanzee will probably mate with most of the males, and male chimps' sperm must compete with other sperm. It is thought from the fossil record that other 2–3 million-year-old humans, whose sexual dimorphism was not so exaggerated, probably lived a social life more akin to that of the chimpanzee. Primate studies have shown us that in many species of primate some sexually active females, in spite of living in patriarchal societies, are often sexually attracted to a stranger, a new male unknown to her and an enemy of the resident males. Female primates have been observed to risk a beating from their own males by slipping off into the bushes to mate with a stranger. It is speculated that our female hominid ancestors were probably also attracted to sexy strangers.[1] But from primate studies we can only know for sure how contemporary primates behave, and as the matriarchal bonobo is such a striking exception to the patriarchal ape rule, we will never be certain how our ape-man forbears lived – unless perhaps Debbie Martyr finds her orangpendek.

The orangpendek is a bipedal hominoid (human and ape), a missing link, that apparently lives in the Sumatran jungle. Had Leakey heard of this creature he would, I'm sure, have gone himself or at least have sent one of the trimates there, because this creature is a living relic of archaic man. In the tradition of Goodall and Fossey before her, Martyr has no formal qualifications in biology;

she was influenced by the trimates' reputation and, just as they had, allowed her instincts to take her on a quest for knowledge.

To find a bipedal species of ape would link us even more closely to the natural world. If we share 98.5 per cent of our genes with chimpanzees and bonobos do we share 99.5 per cent with the orangpendek? The orangpendek can walk; can it talk? Debbie Martyr has been searching for evolution's missing link for ten years. Like many of the first generation of women primatologists before her, Martyr is a self-taught amateur who has crossed the boundary to become a professional primatologist through her determination and skill. From total obscurity working as a journalist on a local paper in Balham, South London, Martyr is now heading an FFl-funded scientific Sumatran expedition for the elusive (and some say mythological) orangpendek.

In 1989 Debbie Martyr was fed up with life as a journalist; she wanted adventure and decided to travel. She was single and leaving Britain for a life of back-packing in South East Asia was an easy decision. Martyr fell in love with the Indonesian island of Sumatra. Her guide book told of mysterious animals such as the cigau, or wild horse, which is supposed to live deep in the heart of the island, and of the orangpendek. Martyr wanted to see some wildlife and hired a Dyak guide. He knew where the very rare Sumatran rhino could be observed and the Sumatran tiger, also hunted to near extinction. He also knew the whereabouts of the rarely seen gu-gu, or orangpendek. Martyr was hooked.

She visited the National Park and was told that to see an orang-pendek footprint would be easy, but to see the actual animal was going to be very hard. She settled on a footprint. Sure enough they found one. She'd had the presence of mind to buy some plaster of Paris and took a mould. Martyr continued with her travels and eventually returned to Britain. Back in London she visited the Natural History Museum's extensive library to do some research, but to her surprise and frustration she found no information on the orangpendek. So Martyr contacted the Indonesian gibbon expert, David Chivers of Cambridge University. Chivers said the ape could easily exist and that the plaster cast of a footprint was not from a gibbon or an orang-utan. At that time Martyr knew she was the only Westerner interested enough in the orangpendek to try to find it before it became extinct. For the first few years Martyr used up her pension plan to fund her research.

But as the orangpendek has never been officially discovered, many people say it does not exist. They liken it to the legends of the yeti and the sasquatch. Probably this ape-man, a likely descendant of the ten-foot-tall *Gigantopithecus*, did once exist alongside *Homo erectus* and *ergaster* and possibly even *Homo sapiens*, but was eventually forced into obscurity in the mountains and ultimate extinction. But Martyr's orangpendek is only four-feet tall and could easily camouflage itself in ancient rain-forest. Doubters might say that Martyr is seeing fleeting glimpses of a gibbon or the Sumatran orang-utan. But having watched gibbons and orang-utans, Martyr is convinced she is not confusing these different species. 'Gibbons and siamangs have a different colour, hair length, morphology, size, body type, calls, behaviour pattern. Orangpendek are very similar to orang-utan in many details of morphology, coloration and behaviour, not least in that both animals are generally solitary and reclusive. But there are no large primates in southern or central Sumatra where I work. The orang-utan is now found only on the northern tip of Sumatra. It has not lived in central Sumatra for 15,000 years. The fact remains that everyone on the team is an expert at identifying what we see and the animals we have seen were animals of a species that should not be here – i.e., a large primate not recorded in this part of Sumatra and behaving differently to the almost entirely arboreal Sumatran orang-utan.'

Martyr believes that another, scientifically unidentified terrestrial primate has colonised the central forests of Sumatra. But being terrestrial, the animal needs to inhabit even more hectares than an arboreal species would to survive.

On 3 September 1994 Martyr had her first sighting of the orang-pendek. By that time she had teamed up with wildlife photographer Jeremy Holden, who had coincidentally arrived in Sumatra to photograph birds. But when the orangpendek appeared Holden was looking in the wrong direction and after five years Martyr was so stunned by the sight of the beast that she was unable to whisper to Holden to pick up his camera. Six years on, they still have no photo. The orangpendek was dark brown with shorter hair than an orang-utan's. Martyr says it stood four feet tall, was heavily built with a broad back and shoulders. It was completely bipedal and could run, 'like bloody blue blazes', swinging its arms like a human. 'It was like looking at a person through the wrong end of a telescope."

Martyr said that seeing the beast for the first time after feeling close to it for so long caused her to cry. She felt a strong pull of kinship towards the animal; it was a relative. Bonding feelings of kinship and solidarity towards their animals are mentioned time and time again by women studying primates. Since that first sighting Martyr has seen various individuals and small groups on many occasions, always an emotional and very special experience. She now knows the orangpendek's hair can range in colour from dark brown to a 'gorgeous, pale, dappled tawny gold'. The visceral shock of kinship has never waned, and Martyr now feels utterly committed to protecting the hominoid from extinction.

Debbie Martyr will not leave Sumatra until she has accomplished her mission. Nothing is more important to her now. Holden and Martyr take intensive three-week-long expeditions on foot through the jungle of Kerinci Seblat looking for tracks and organic matter and setting and checking camera traps. At night they sleep on the ground with a sheet of plastic tied to foliage over them and a sarong or sleeping bag around them. In South East Asia new species are still being discovered, such as the muntjac deer in Vietnam. It is possible that Debbie Martyr's ape is cathe-meral: like many mammals, it may take a number of naps within a twenty-four-hour period but remain active at other times. Between naps it may travel alone or in small groups and cover miles of forest. As a result trying to find samples of bone, hair or faeces is almost impossible. The natural recycling of organic material in a rainforest is swift; even deer will consume the bones of dead ani-mals to top up their calcium.

As an experiment, Martyr passed a stool and timed how long it took before dung beetles and other creatures had dispatched it: twenty minutes. The area she is hunting for the orangpendek in is the size of Wales. It is like hunting a man on the run: how is the orangpendek to know that Martyr is friend and not foe? Debbie Martyr needs organic evidence, and the possibility of her stumbling across some fresh orangpendek dung or, even better, an orangpen-dek corpse, is slim. But she refuses to give in. Recently Martyr heard a poacher had accidentally caught an orangpendek in his snare, but fearing prosecution the man had destroyed the ape's remains. When confronted the man denied placing traps in the park. Martyr did find some 'very interesting looking' orange hair with the follicle attached caught in a prickly vine. Carefully she bottled the sample in alcohol to preserve it and using the Sumatran postal service sent it to a laboratory in Cambridge for testing. It never arrived.

When Martyr first started studying the Sumatran orangpendek in 1994, she estimated that there were 1000 individuals ranging through the forests. Since then, mainly due to logging, she thinks the numbers have fallen to 100. 'The area where I was first shown footprints in 1989 suggesting orangpendek was a real animal and not a myth/misidentification is now farmland far from forest. During our 1995 study, which was 12 km into the forest, all project members saw orangpendek during the five months we were there. That study area now has a 12,000 hectare (30,000 acre) palm oil plantation growing on it. Our 1996 study and photo-trapping area where we saw one animal has been totally clear-felled. A survey area where faecal matter has been collected and animals seen and heard is in the process of being clear-felled for a 12,000 hectare palm oil plantation. The orangpendek will be the first great ape to go into extinction in this modern time before it has even been scientifically validated and described. Bloody wonderful, isn't it?'

This ape is the only truly bipedal primate apart from man. If Martyr can secure its place in evolutionary history, our primal origins will not seem so distant. We will have to acknowledge this long-lost hominoid member of our family and the 'link' will no longer be missing, because it will be living. If Martyr conclusively discovers the orangpendek, human evolutionary continuity will no longer be a debatable issue in primatological circles and Leakey's theory of the hominoid line being a bushy one with many different species and sub-species will have been proved.

Many of the women mentioned in this book are striving for evolutionary continuity. In the same way that Goodall emotionally broke down the evolutionary barrier between her and David Greybeard, today women intellectually continue the fight for ape equality. The primates they study are close human homologues and they want the animals to be commonly acknowledged as our sister species, because they genuinely feel that they are. The women see apes as kin, not as study subjects from a distant species. They want to create a community of equals where humans recognise they are apes or, if that is not possible, a community that credits apes as being other types of humans. The fossil record has proved that humans and apes belong jointly to the hominoid family group. If we humans cannot step off the pedestal and become used to thinking of ourselves as apes and just another primate species on this planet, some women primatologists intend to have apes climb up on to the pedestal with us. For them, the physical achievements of speech and bipedal locomotion are superficial differences compared to all that we have in common.

If evolutionary continuity can be proved, perhaps humans can be spared the mental anxiety that pervades our species. The meaning of life will cease to be a philosophical issue, and the human uncertainty that fuels religions and the insecurities behind superstitions can be laid to rest. If we are prepared to understand that we are walking, talking apes and very similar to other species of apes, we will realise that we are a product of evolution. We have been designed and modified for a terrestrial, bipedal existence and an intensely social life that demands a spoken language. If we visualise ourselves as part of a great chain of being and stop drawing a line between ourselves and other primates it will take us to a higher plane. Once we know why we are inclined to feel and do certain ape-like things, appreciation of our ancient primate genealogy will follow and evolutionary continuity begin to make complete sense. Yet even though contemporary behavioural studies present a powerful argument, we are still some time away from closing the fossil record's five-million-year gap between ourselves and other apes.

Over the last twenty years primate research has expanded rapidly. The second generation of male and female scientists now confidently absorb women's research from the 1960s–70s and 80s. Ironically, much of that research when first published was controversial and it was hard to have it accepted. Today we are in possession of a great deal of data on the behaviour of wild primates which has been indispensable to the schools of sociobiology and evolutionary biology and psychology. Stephen Jay Gould refers to Goodall's work as 'one of the great achievements of twentieth-century scholarship'. Neo-Darwinists have hungrily used primate data to back up their theories. Primate studies have been essential in developing the theory of kin selection (primates always defend their relatives). Goodall, Fossey and Galdikas all observed their wild apes showing preferential treatment to close blood relatives over strangers. Primatology has also shown that homosexuality occurs in the wild between individuals from both sexes; we now know it is common and natural primate behaviour. Before Goodall studied the Gombe chimps we had no idea how wild chimps courted one another, either homosexually or heterosexually, and the chimpanzee consortship (where a male takes an oestrous female, sometimes against her will, away from the group for a few days) was unheard of.

Today Neo-Darwinists can make comparisons between our nearest relatives and ourselves with ease. We are the naked ape and the differences between us and our cousins amuse as well as inform. For example, an alpha male chimpanzee prefers to mate with a mature female who is an experienced mother over an inexperienced and only recently matured female, while human alpha males often seem to hold a preference for younger, childless women. Scientists can even go on to speculate on the origins of aesthetics and beauty and why the waist-to-hip ratio in human females makes some women especially desirable. But if Jane Goodall had not succeeded in her research at Gombe, information about what type of female primate generally turns on a chimp or a man would not have been available.

Before Goodall went to Gombe no one knew what types of wild female chimpanzee males generally preferred to mate with, nor what type of male females desired. Darwin speculated on female choice in 1871: 'Man can modify his domesticated birds by selecting the individuals which appear to him the most beautiful, so the habitual or even occasional preference by the female of the more attractive males would almost certainly lead to their modification; and such modification might in the course of time be augmented to almost any extent, compatible with the existence of the species.' But his argument for female choice was overlooked in favour of the dominant male being the force of evolution. Not until Goodall had shown us it was possible to study our nearest relatives in their natural habitat could the theory of female choice, by then over one hundred years old, be studied in wild chimps. Now we know that a female's choice of mate makes a major contribution to the evolution of the species. Goodall was the first Westerner to see many of the chimpanzee behaviours we now take for granted. If Goodall had not proved that wild apes could be habituated to human observers and studied continuously, it's likely no one else would have tried it.

Before Goodall's work the few male natural historians who had attempted to observe wild chimpanzees, such as Sherwood Washburn, had felt lucky if they glimpsed a chimp's backside as the animal fled. Male scientists generally assumed that studying captive apes in unnaturally assembled groups could show us all there was to know. But with hindsight, we can see that the 1930s colony of hamadryas baboons at London Zoo, which served as the source of information for influential anthropologist/zoologist Solly Zuckerman and his followers taught us nothing accurate about primate societies only how cruel it is to assemble a group of mostly male hamadryas baboons in a confined space.

Goodall's almost forty-year-long study on the lives of the wild Gombe chimpanzees forms the basic foundation of every other chimpanzee study. Her research into the life histories of the Gombe chimpanzees belongs to all of us, amateur naturalists and Neo-Darwinists alike, and her data are used by all scientists asking questions of apes or of human evolution as no one else has such rich, intergenerational data available on natural chimpanzee behaviour. Any hypothesis or theory can be reinforced or contradicted by the Gombe research and therefore scientists cannot afford to ignore it.

Andy Whiten's research on culture in chimpanzees would not have been possible without Goodall's collaboration and the years of observation on the typical and atypical cultural and ritualised behaviours of the Gombe chimpanzees. Whiten and Goodall have compared the behaviour of the Gombe chimps with other geographically isolated groups of chimps. From a total of more than

150 years worth of observation notes from seven different chimpanzee field sites it became evident that communities of chimpanzees are not all the same. Seventy cultural differences between groups of chimps were discovered, thirty-nine of which could not be attributed to ecological parameters.

Culture is developed and modified in human societies by the ways in which the novel actions of individuals, who are responding to changes around them, inspire the rest of us to imitate their behaviour. It is the same for chimpanzees. That non-human primates have cultural abilities is not a new concept. Imo, the Japanese macaque (studied by the only known mother-and-daughter primatologists, Umeyo Mori and Satsu Mito), influenced the cultural traditions of her monkey community on the Koshima island when she washed sand off her sweet potato. This was an isolated example and alone couldn't be used to demonstrate non-human primate culture. But when the behaviour of geographically different groups of chimpanzees is compared, unique social behaviours are seen even if they belong to the same sub-species. For example, the process of relieving a friend of a tick varies between field sites. At Gombe once a chimp has removed a parasite it is squashed between a leaf. In the Taï Forest the grooming chimp squashes the tick on its own forearm. Ugandan chimps living in the Budongo Forest place the parasite on a leaf and inspect its merits before throwing it away or eating it.

The onset of the monsoon also brings out ritualised behaviour in different communities of chimps. At Gombe only males 'rain dance', standing bipedally in a trance, swaying, charging, slapping the ground and dragging branches back and forth. A male chimp will conduct a rain dance whether alone or with other dancers. But not all communities celebrate the rain. In the Taï Forest it is never celebrated; instead the chimps construct small platforms of leaves to try to keep themselves dry until the downpour stops.

Courting rituals also vary between communities of chimps just as they do between communities of people. In some communities a male will clip off pieces of a leaf with his teeth in a specifically noisy and mannered way to let a female know he is sexually interested in her. Elsewhere, a male might pull branches around and look intently at the female of his choice and in other communities the males knock on wood to ask for a date. How ancient these customs are we cannot know. But it is evident that communities of chimpanzees are as culturally different from each other as groups of humans can be, or as culturally different as Eastern and Western primatologists have been. From Goodall's continuous study we know the Gombe chimps have 'termited' with long grass and sticks for at least three generations. This analysis of chimpanzee culture has proved that some chimpanzee behaviour is passed on through a social learning and imitation process and not inherited. Individual chimpanzees adapt to problems and invent new technologies. Their unique personalities cause modifications to social customs and these trend-setters and inventors inadvertently lead their community forward. Only a continuous long-term study can supply this type of information and it has been women who have led the way in long-term field studies

Perhaps it is the male ego that prevents men from disappearing for years to live with apes. The women mentioned in this book do not seem to be looking for fame or recognition. Jo Thompson, who toils away anonymously in the Congo, says: 'My ethos is to take only knowledge and leave only footprints. I love to sit close by and see, hear and smell the bonobos. I could die like that. It's an out of body experience, it's like I'm not there. I feel so lucky to be there, just to eat, drink and survive day by day. I get so much contentment being a part of the wind and the rain. I'm an insignificant part of everything, and this feeling is worth every bad experience.'

The majority of women discussed in this book have had enough blood-curdling experiences to write a string of adventure books. Women primatologists make the most intrepid adventurer look tame, but few of them have written books about their amazing experiences and observations despite the fact that many are sitting on a treasure-trove of information. With daily observational research to be done, orphaned primates to feed and confrontations with poachers and loggers to overcome there just isn't the time to write it all up. Women publish fewer academic papers than men, but it is the women's papers that are a more reliable source of information and that are cited more often. Male scientists based at universities write more popular science books than women field researchers. But well-known male primatologists, such as ex-Goodall student Richard Wrangham, are heavily reliant on women field primatologists' data to fill the pages of their lucrative books. (Over the years Wrangham has been especially influenced by the research of Barbara Smuts.) Established male scientists undertake sporadic trips to research centres, but during these short visits it is rare for them individually to gather enough data to substantiate a scientific paper, let alone fill a book.

Male primatologists seem happier in the man-made safety of a Western university, where they can receive typed field reports or e-mails from women field researchers. Secure within the enclave of male-oriented academia and ensconced in a comfortable office they can pontificate and theorise on the women's observations. These men have egos and do not enjoy insignificance, but they do not feel the urge to put in fourteen-hour days of data-collecting themselves. Nor do they want to fight poachers or (understandably) catch malaria. Perhaps most significantly, they lack an impulse towards enjoying non-human primate companionship. But as men are the ones communicating the information, they are also the ones receiving the publicity, prestige and promotion. Yet women primatologists do not seem to feel cheated by this. They are happy to leave professional competition behind and instead enter another world where they can be anonymous and close to nature. Usually film companies in search of beauty-and-the-beast imagery raise the public profile of these women, rather than the women themselves ambitiously wanting to further their careers by courting publicity.

Contemporary studies on human primate behaviour theorise on the human need to seek out social information. Why do we care so much about the motivations of others? Most of us have an interest in what others are doing and we automatically speculate on why they do this or that. Fictional characters in novels, plays and TV soap operas hold our interest as much as real life can. Women primatologists have likened their experience of watching primates to watching TV soap operas. Perhaps evolution has seen to it that women have a more highly tuned interest in social activities than men. Gathering information on individuals around you can save your baby's life, as well as your own. It may be an especially female evolutionary adaptation to seek social information — one that encourages a mesmeric fascination for the behaviour of primates.

Evolutionary psychologists also speculate on the mechanism for the emergence of a spoken language in our ancestors. It has been suggested that women spoke before men. Archaic women with dependent offspring were vulnerable; they needed to bond with others, especially other women, for support and they needed to gather information on the safety of their environment. Language achieved this. Studies have shown in general girls' and women's linguistic skills exceed those of boys and men. This innate linguistic and communicative ability in females could fuel women's urge to understand the minds of apes and to try to talk to them. On average women have a superior emotional intelligence to men. Mothering children has meant that evolution has selected certain maternally advantageous adaptations over other behaviours that did not help to keep a baby alive. *Complicité*, non-verbal communication, and spoken language all tie into the maternal instinct. Women's talent for studying primates is a by-product of their femaleness. Women's sensibilities have been designed to help them understand other sensitive creatures, namely their own children and other humans. Why not use that specific empathy on our primate cousins? As Leakey suspected, all the elements that make up the maternal instinct can be exploited by women primatologists.

Primatologist Julia Casperd believes a maternal instinct helps motivate women to study primates. The nature of the female primate continues to intrigue the newest generation of women scientists. Casperd has been influenced by the famous women primatologists who have preceded her and like many of them she has chosen to study female chimpanzees. The emotional lives of the apes that have fascinated Casperd, and which Goodall has pushed to have acknowledged, are an area of study that has taken thirty years to become 'acceptable' science. Casperd has observed the processes that chimpanzees go through when they make and break friends. Apparently bonding and reconciliation only take place between individuals if there are mutual benefits to the relationship and therefore in resuming the relationship. But occasionally some apes refuse to reconcile, even if that seems to be to their detriment. Casperd has discovered that some female chimps will not forgive or forget once their friendship has broken down. Reconciliation only occurs in species that self-reflect, such as hyenas, dolphins and apes. Casperd has theorised the art of reconciliation evolved in primates in order to protect and perpetuate important social bonds.

Quite often we humans don't work at the restoration of a relationship but cut our losses, and non-human primates are just the same. Chimpanzees do not always try to restore friendly relations and will bear a grudge towards another individual for years. Casperd noticed that female chimps bore grudges against each other for much longer than male chimps would. Some female chimps would not make friends again even though it would help them to put their differences behind them and co-operate again. It seems that once a female friendship has broken down it is hard for the parties involved to reassemble it. They are unwilling to trust each other again because being female their relationship with others does not involve just themselves, they have to negotiate their infants' futures too. Females will not trust an old enemy because they fear a revenge attack upon their child. By comparison, male chimps who have fought each other will move with the political times of the group and will make and break friends according to the other's status. After a fight males can quickly make friends again if they need to for mutual advancement. The trait of male bonding is millions of years old – one fight cannot undermine that for long.

Casperd noticed that after a break down in a chimp's relationship, there was much anxiety and uncertainty, especially if the individual had never fought with the other party before. The primate with the lower status of the two was always left feeling especially tense. Monkeys with smaller brains reconciled after a fight within ten minutes, but apes, with a much larger brain, take much longer. Days and months can pass before individuals will let go of their grudges. Goodall witnessed the break-up of female friendships in her chimps. The calculated making and breaking of alliances is all detailed in the personal life histories of the Gombe chimps. It is an invaluable study and the Gombe chimps must be protected so that the 'soap opera' can continue.

Mature female chimps have to leave their mother and siblings and are therefore more isolated from their kin than male chimpanzees. They need their female friends, especially for mutual care of infants. But primates are competitive animals and since much of the competitiveness is sexually driven, it is directed towards members of the same gender. Once trust is gone and betrayal is felt female chimps consider the relationship over for good through fear that the other female may kill their infant in revenge. A physical fight between females can be more aggressive than one between males. Female chimps and women are the primary carers for their progeny; both have long memories and do not forgive or forget other females who have let them down. This behaviour is another by-product of the maternal instinct. Casperd noticed that several adult female chimps stayed completely out of each other's way. They never sat together, never groomed one another; theirs was a dysfunctional relationship. They had obviously had a breakdown in the past before Casperd had studied them and bitterness prevented them from reconciling their differences.

Casperd commented that this depth of feeling has also been observed amongst women. If two women who have been reciprocally close friends and felt a strong bond have a breakdown in their friendship for whatever reason, but especially over their children, reconciliation may become an impossibility. They may try, but once the trust is gone and the bond broken women may not reinvest in that friendship. A subconscious fear trigger can take over. As in the prisoner's dilemma, it is a mutual fear of the other female as an enemy that motivates behaviour. Suspicions that she may betray the friendship first and take your social place, mate with your partner and, most importantly, neglect your baby in favour of her own, can take hold. Mutually women can feel, and sometimes mistakenly, they must defect first before inevitably being betrayed. Ancient primal feelings of the stakes being too great remain with us, even if the situation for modern woman is no longer one of life and death.

Casperd was using evolutionary psychology to analyse chimps' behaviour. Rather than just noting down that she had observed reconciliation she asked why it had evolved. This is an example of the scientific leaps that have taken place in primatology, especially where the more frequently studied species are concerned. Today chimpanzees are being explored from all dimensions and so long as claims can be substantiated it seems the more anthropomorphic the better. A lesser-known primate may be observed in the wild to teach us about their group size and gender composition and also what seasonal food the animals eat. This information can help conservationists manage captive and wild populations better. But the species' methods of reconciliation would not be an obvious first choice. When Goodall first went to Gombe, one basic request from Leakey was that she take samples of all the vegetation she saw the chimps eating. In those days, no one knew exactly what they ate and no one knew that they ate meat. But after nearly forty years of data, studies on chimpanzees are today much more sophisticated.

The sex lives of primates have always captured the attention of primatologists, but it has been women who have shown the most interest in the female chimpanzees' sexual power. When a female chimpanzee comes into oestrus her social status shoots up, regardless of her social status at other times. When she feels sexy a female primate can manipulate others. The heightened sex appeal of a female chimpanzee means she can call upon males to help her in fights, to gain access to food and the best sleeping sites. During oestrus a female chimp can encourage males to attack her enemies. Females have been seen to start a fight with other females during their oestrus and that power will remain until the cycle has passed.

Female chimps will always initiate lesbian sex when they are in oestrus. The dominant female will not be upstaged; she will form a duet and quickly change partners when another female comes into oestrus. The dominant female is not the only female to show an interest in another female who is in oestrus; there is always increased, if not frenzied, lesbian activity when females cycle. Female–female genital rubbing is always present to some extent with chimpanzees but is more popular with certain individuals than others. The main difference between female chimps and bonobos is that female bonobos do not need to be in oestrus for homoerotic sex to fuse their bonds. Female bonobos feel sexy irrelevant of the time of the month. Female bonobos do not wait for the males to lead the way; they do not need a male's sexual approval in another female to see her as special. Female bonobos act autonomously and bond in spite of their cycling. Female bonobos avoid sexual coercion and male patriarchy and they are not dependent on a thirty-five-day cycle for fleeting moments of glory.

Indeed the bonobo is beginning to replace its sister ape in the hearts and minds of women primatologists. It may look similar to a chimpanzee but the bonobo is as genetically close to us as it is to chimps. Alone in 1992 Jo Thompson first travelled up the Congo's Sankuru river and into the uncharted Lukuru region after hearing stories that bonobos might be there. She walked into the forest and through the trees, and eventually heard the vocalisations of the bonobos. There is no mistaking their cry, a high-pitched 'ooooo' that carries on the wind. Bonobos are sensitive, vocal animals. If Thompson absent-mindedly hums a tune while she is sitting with them, or even if she coughs or sneezes, the bonobos will be triggered off by the sound and will all start mimicking her. If they catch the sound of a motorbike on a distant track they start to call out. When bonobos call out to each other they stretch out their arms and shake their hands. Chimpanzees do not do this. By 1993 Thompson had made conservation groups aware of the existence of the previously unknown bonobos and she established the Lukuru Wildlife Research Project.

Amy Parish and Jo Thompson adore this matriarchal ape. Parish seems to secretly relish the bullying she has observed the females inflict on the subservient males. 'The females always help to clean the male up after they have attacked him; they just wanted to reinforce the female hierarchy. It's heartening to see feminism occurring naturally.' Parish feels a unique kinship to these apes; her menstrual cycle has even synchronised with her bonobos' thirty-five-day cycle. She feels favoured that this ancient primate trait unites her physically to the female bonobos she studies. When her son was born she named him in honour of a bonobo she once knew and loved. Thompson has said that if she couldn't study the bonobo she would rather give up primatology than turn her attentions to another species of primate. 'Every moment with the bonobos is a high for me, a really awesome privilege.' The female bonobo's sassy style gives women primatologists strength and hope.

To be able at last to watch a female ape have the upper hand and not need to use her unilateral sexuality to achieve this, but instead achieve dominance through her female bonds, has left women primatologists with a good feeling.

As many primatologists choose to study apes specifically for answers to the question of who we really are, the discovery of true matriarchy encourages yet more speculation on our origins. Were we once a matriarchal species? Did archaic woman use physical force over males to retain her superiority? Was her ability to produce young held in awe by males until they realised it was their part in the sexual act that set reproduction in motion? Are we actually more female-bonded now than women realise? Does her emotional intelligence and the female trait for non-verbal communication subtly give modern woman the upper hand? Or have humans always been, as the fossil record to date suggests, patriarchal?

Primates are highly adaptable creatures. It is believed that they have a 'social plasticity' to their lives. This means that even if a primate species' evolutionary heritage is a male-bonded one, the species can adapt to the prevailing ecological conditions that will finally determine whether it is male-bonded, like chimps, female-bonded, like bonobos, or neither. Since the 1960s, feminism has made a philosophical impact on human beings but has it ecologically changed women's lives? Feminist teaching has told us that supporting your female relatives and friends is mutually beneficial. But it is women primatologists' research into the lives of female primates that explains to women why, in evolutionary terms, they should bond, and why at the same time, female-bonding can be hard to maintain. Primatology also reveals to us the probable results of women's failure to bond with each other. Women need to appreciate that if they are not evolutionarily female-bonded, as Foley and Lee suspect, they must work much harder to maintain their female alliances.

I asked Jo Thompson if the matriarchal bonobo was a good role model for modern woman. 'You cannot watch a great ape and not see yourself there, but modern woman must juggle everything in a man's world. Female bonobos do not live in a male world. Modern woman needs to be a partner to a man; bonobos are not like that. A woman who is a single parent has a tougher time than a female bonobo. The female bonobo has food, companionship, space, freedom, her own identity, security within the system, some power and control, autonomy. She has it all. She has a man if she wants one, or not. The female bonobo's social network is made up of female companionship and friendship. Women do not support each other in the same way.'

It probably took chimps and bonobos 4 million years to evolve oestrous swellings in order to save their babies' necks by confusing paternity. After being separated geographically from other chimps it probably took female bonobos another 2.5 million years to learn to bond to each other to avoid sexual coercion and infanticide by males. That female chimps never managed to evolve the strategy of female-bonding is to their detriment. It doesn't have to take modern woman 2.5 million years to change things for her benefit. To improve the future of the human female primate, women must take advantage of social plasticity to consciously re-invent themselves. Millions of years of patriarchy and male-bonding do not have to dominate women's lives. Women need to make themselves mutually trustworthy, reconcile differences and stop fearing betrayal. Helping another woman is about the best thing she can do. If there is just one useful lesson from primatology for women in general it might be: divided, like female chimps, women fall; but bonded and united, like female bonobos, women stand.

Clearly, women have brought a female style of science to primatology but in some cases not without criticism. Francine (Penny) Patterson, a psychologist, works with captive lowland gorillas in California. She is the primary carer of the famous twenty-eight-year-old female lowland gorilla known as Koko and has done a great deal to awaken American children to the gentle and intelligent nature of the gorilla and to the need to conserve this threatened species. Cuddly Koko dolls can be found in thousands of American nurseries. But since first meeting Koko Patterson has been shrouded in controversy.

Koko was born at the San Francisco Zoo in 1971. Working at the zoo, Patterson, then a student, pleaded with the zoo officials to be allowed to take Koko home at night; amazingly they agreed. This went on for a few years, during which time Patterson had started teaching Koko ASL, or American Sign Language. Patterson formed a deep and meaningful bond with the gorilla, and during these early years she made use of the publicity surrounding 'Project Koko' to make Koko and herself stars. After a few years the zoo decided that Koko should be back in the zoo full time, in order to breed. But Patterson refused to cut the psychological umbilical cord she felt between herself and Koko. As she had already won over the general public's imagination, she used the publicity to raise funds enabling her to buy Koko. A wealthy benefactor came to her assistance and Koko and Penny are now together for ever.

Whether Patterson should have kept Koko is debatable, but she is genuinely devoted to Koko and will continue to care and provide for the gorilla until the end of the animal's life, which could be at least another twenty years. In the 1970s too many psychologists became bored with their glorified pets and passed the apes on to other laboratories for more sinister experiments. This will never happen to Koko. But a few years after acquiring Koko Patterson found herself in trouble again. She wanted more gorillas. She contacted a well-known animal trader and asked him to acquire her two wild-born baby gorillas, including a male so that there would be a chance for Koko to breed. The animal trader did what she asked and two 'orphans' were captured from the wild. One died but Michael survived.

Conservation groups, such as the International Primate Protection League (IPPL), pushed for Patterson to be convicted for perpetuating the illegal ape trade. The case went to court but Patterson had the financial support of the public and she could afford a very good lawyer. Patterson's lawyer cleverly turned the case around, saying that we should convict poachers – not this nice, white American lady who loves gorillas.

But Koko and Michael's relationship is purely platonic. Even though they are not blood relatives they have grown up together like siblings and the incest taboo prevents the gorillas from mating. It has been noted that non-related children who have grown up alongside each other, such as children from a Kibbutz, do not feel sexually attracted to each other when they reach adulthood and do not want to marry, often to the irritation of their respective parents. Patterson has even had Koko go through IVF treatment, but so far the gorilla has not successfully conceived.

Patterson claims, not without some criticism from other ape language supporters, that during the years of Project Koko, Koko, who certainly does use ASL to talk, has advanced further with language than any other non-human. But language doubters, such as Steven Pinker, see Patterson more as a pushy mother and scientifically as a 'joke'. They do not take her seriously or believe that she is able to substantiate her claims, and suggest she is faking the whole thing.

Patterson, with the American public behind her, doesn't take much notice of her critics. To date her Californian-based Gorilla Foundation (founded in 1976 and currently receiving support from 30,000 members) has raised $1,700,000 from the general public to fund her project to translocate Koko, Michael and her new addition Ndume, a seventeen-year-old male gorilla from Cincinnati, to her new expansive gorilla preserve on the Hawaiian island of Maui. Patterson wants the new reserve to offer a captive breeding programme for wild lowland and mountain gorillas. It seems she has always felt ashamed that Koko never bred as the zoo originally intended.

Many captive apes have problems mating successfully. Finding a suitable partner is a matter of choice in the wild, and in captivity a human matchmaker may not fully appreciate the sexual requirements of the apes. Captive male apes often become sexually fixated on the women who keep them. Orang-utans have raped women and captive chimps and bonobos can become sexually aroused by photographs of women in fur coats. Some women primatologists at Gombe who, out of sight of any men, undressed in the heat of the midday sun felt leered at by the male chimps they observed and compelled to put their clothes back on. Chimpanzee Lucy was sexually aroused when given *Playgirl* magazine to read and when in oestrus she would throw herself at terrified door-to-door salesmen. Do apes view humans as ape homologues?

Like many others, Patterson is terribly worried about the dwindling population of mountain gorillas. She has a plan to translocate all the young mountain gorillas from Rwanda to Hawaii. But gorilla experts Kelly Stewart and Sandy Harcourt, who worked with Dian Fossey, and Liz Williamson, who runs Karisoke today, do not think this is a viable proposition. Patterson believes America is a safer location for gorillas than Africa. In some ways she is right, but a wild juvenile gorilla would become very distressed at being darted, taken from its natal group, caged, then flown to Hawaii. Gorillas are sensitive to disease and stress and do not breed well in captivity.

Conserving them in Africa, however fraught, must be the best option. For it is only there that gorillas' natural behaviour and culture can remain intact. In the right hands, Patterson's $1,700,000 would go a long way to help conserve the mountain gorillas at Karisoke.

Conserving wild primates is the aim that motivates most women primatologists today. Although they have grown spiritually and intellectually through their research, many are now devoting their careers to the conservation of the animals that taught them so much.

Species of Great Ape	Estimated population in year 2000
Chimpanzee	150,000
Orang-utan	17,500
Lowland Gorilla	100,000
Mountain Gorilla	350
Bonobo	12,500
Human	6,091,000,000

Figures from the Ape Alliance and the United Nations

The table above shows the population figures for the world's great apes. But all species of primate, except humans, are threatened with extinction. It is this threat of annihilation that encourages some women to undertake field work and set up primate sanctuaries in primary habitat. Upwards of 10,000 apes are hunted annually for bush-meat, killed by poison arrow, blowpipe, snaring, or increasingly by bullet. The animals are then butchered on the spot. Not all the bush-meat is consumed in Africa; some of it is smuggled out of Africa and sent to some of Europe's and America's best restaurants. The meat may not be listed on the menu but it's available if you know where to go, and a chimpanzee steak can sell for an extortionate price. The very rare sun-tailed guenon has been hunted and sold in local Gabonese meat markets for the equivalent of £12 and chimps and lowland gorillas for £20. Twenty pounds sterling equals an average monthly wage, so a little bit of bush-meat hunting can greatly supplement an African family's income. The trade in bush-meat for human consumption is rapidly destroying some primate populations for ever.

The bush-meat trade goes hand in hand with the timber trade and the timber trade earns millions of dollars in Africa. The loggers and their families are itinerant workers, always on the move and often without food. The impromptu bush-meat and fetish markets, held deep in the forest where the loggers live and work, provide the workforce with fresh meat. In return, the loggers provide the hunters with cable for their snares, guns and bullets, and transport by which to move their contraband around. Confiscated meat is often found hidden in timber lorries. In Gabon and Liberia the bush-meat trade is worth $24 million, which is more than the value of these countries' timber industries.

Jo Thompson's Lukuru Wildlife Research Project is in an obscure part of the Congo, but the bush-meat hunters are there too. When she first arrived, there were five bonobo groups in that region. But during Thompson's long habituation of the Bososandja group, the other communities of bonobos to the east of her project were entirely wiped out by bush-meat hunters.

The ancestors of the local tribes have practised a taboo on eating bonobo meat and older women still practise this taboo. This is probably why there are still bonobos living in the region. But now younger people are turning to bonobo meat and hunters from other areas, such as the Ba Tetela tribe, consider bonobo meat a delicacy. Duiker snares injure bonobos even if the hunter isn't trying to catch an ape. Many of Thompson's bonobos have scarred arms and legs and missing hands, fingers and toes after being caught in snares. Gangrene kills many animals that have been caught by a snare. Thompson is encouraging local people to eat domestic chicken meat rather than hunting for bush-meat. Her research project is sponsoring the inoculation and deworming of domestic chickens which suffer badly from disease.

Another of Thompson's primary concerns is the annihilation within the Lukuru region of the unique species of red colobus monkey known as Thollon's colobus. It is found only south of the Congo River and north of the Sankuru River. Because Thollon's is the second largest monkey in the Lukuru area, it is a favoured food target of hunters. Unfortunately, Thollon's colobus has a particularly detrimental defence behaviour: when alarmed by hunters group members will cuddle together and maintain a relatively immobile position, darting between a few trees within a close area and vocalising loudly. This behaviour means hunters can kill several individuals at a time. Occasionally whole groups have been exterminated by one hunter.

As with other species of primates, the trade in infants as pets is also a constant threat to the bonobos. On one occasion a European white man came up the Sankuru River by boat, trying to recruit hunters for the ape trade. He knocked on Jo's front door and asked her if she could get him a baby bonobo, for which he was willing to pay her a lot of money. As he spoke the man realised his mistake and his voice trailed off as he quickly got back on his boat and sailed away. Thompson never saw him again. Scarlett, Thompson's favourite red-haired bonobo, will be having her first infant soon. Animal traders would pay even more money for a red-haired bonobo infant.

Thompson fears for Scarlett's future because Malaysian loggers have arrived in her area of the Congo. Their hunt for tropical hardwood has taken them from their own decimated forests in South East Asia to the heart of Africa. Once loggers move in and cut roads into the heart of the Congo the bush-meat traders, in their hundreds, will surely follow and Thompson's bonobos, if they survive, will become as geographically isolated from others of their kind as Jane Goodall's Gombe chimps are now.

When Goodall first went to Tanzania it was just the Gombe chimps that lived near the stream and in the hills that Leakey advised her to habituate. It could have been any other community of chimps in that area of the world. In the early 1960s Goodall stood on the top of one of the hills and looked out at the view, feasting her eyes upon green forests that covered the land as far as she could see. These forests had stood there for millions of years and communities of chimpanzees could be found throughout equatorial Africa. Now the Gombe chimps are the only chimpanzees in that region of the world. The forests that once stretched for ever have been felled and human agriculture has moved in. Goodall's chimps live in an island of forest surrounded by farms and human settlements. Today the three separate groups of Gombe chimps, if we take into account births and deaths, total between 100 and 150 individuals at any one time. The sexually active females can only migrate between the existing groups. The female Gombe chimpanzees will never be able to migrate from that area to share their genes and their culture with other chimps; once upon a time that would have been a natural occurrence.

Some primatologists believe that small nocturnal prosimians, such as tarsiers and galagos, will outlive the great apes. New species of these small, inconspicuous primates are still being discovered. Good at hiding, with small appetites, they can survive in secondary forest, unlike the great apes. Being far larger, apes need primary forest in which to hide and find shelter from the elements and to sustain their far greater appetite.

Women primatologists have importantly introduced us to the fragile, complicated and often human-like lives of wild primates. Empathetic women have proven that long-term field work with wild animals is possible. Women have pursued animal intelligence and championed the importance of the female primate in her society. It has been women scientists who have revealed how important a figure a primate mother is to her infant and how tightly bonded they are. Women have also shown how our primate sisters juggle their roles of mother and political animal in a hierarchical community and the varied strategies the female primate employs to avoid male dominance. It is perhaps in their study of female behaviour that women have been at their most insightful. The insight into how a non-human female primate uses her sexuality to better her social position and to avoid an unwanted male taking advantage of her fertility to ensure a future for his own genes has been revolutionary. The ability to empathise with the hanuman female langur as her baby is snatched and murdered, or with a female chimpanzee, baboon or orang-utan as she is beaten and raped has strengthened the influence of women's theories and given women an advantage over men, who do not demonstrate this same empathetic response to non-human primates.

Without the crucible in which primatologists' research and Darwinian theories combine we would not understand ourselves so intimately. A number of women primatologists have said to me that before they studied primates they were puzzled by behaviour patterns, but now understand what makes humans behave the way we do. When we feel emotions such as insecurity, defensiveness, aggressiveness, security and happiness, we are reacting to primal instincts in the same way as non-human primates. Feelings of ambition, greed and commitment to one's biological family are all primal instincts necessary for our survival, and we share these and other traits with today's living primates. Through this personal transformation women primatologists have gained insights into their own psyches.

Confronting adversity causes many women field primatologists to fear failure, but surprisingly few have failed. Most women refuse to cut their losses; in a relationship with no threat of betrayal women continue to co-operate with their primates. Against the odds women have created charities, raised public consciousness, voluntarily set up sanctuaries and established wildlife reserves all over the world. Intrigued by evolutionary questions of human behaviour some women move in other directions. They study the lives of diverse populations of people around the world. After observing mother and infant bonds in non-human primates, and having three children of her own, it is the behaviour of modern women with children and the facts and myths that surround the idea of 'mother love' that Sarah Hrdy has recently put under the microscope. Since witnessing chimpanzee Ntologi's murderous nature Mariko Hiraiwa-Hasegawa now researches the behaviour of Japanese murderers. Those who stick with non-human primates are usually those who entered the profession because they loved animals. Those women see it as their duty to make people aware that all non-human primates are threatened with extinction.

Thankfully, women primatologists do find a great deal of support from the public. A great many of us do care what happens and if we are on holiday in the tropics we leap at the chance to see wild primates. Stella Brewer told me what happens if she ever takes inquisitive tourists in a boat over to the Baboon Island and the chimps come down to the shore: 'When I take uninitiated people over to see the chimps they become amazed at how human-like the chimps are. They often say to me it's like meeting an alien or coming face to face with E.T. It's uncanny; 'it's like meeting somebody from outer space.'

If the great apes and other primate species had become extinct in the wild before Louis Leakey had the idea of synthesising contemporary ape behaviour with the fossils of extinct species of hominids, we would probably still incorrectly think of ourselves as, 'man the tool maker', 'man the hunter', 'man the patriarch who walked first' and 'man the ape with language'. Without primatology the fossil record alone could never reveal to us the extent of non-human primate intelligence. The complex primate social life and culture and their bonds to their relatives would not be evident from fossils, and female-bonding, bonobo style, would be a matter of fiction not of fact. If the animals had never been studied by Jane Goodall, Dian Fossey and Biruté Galdikas we would have no idea of how ancient the origin of sentience is. There would be no concept of evolutionary continuity; in such a situation we could easily feel like aliens on this planet.

Women primatologists do not want their young science to be a short-lived one; they want it to go on indefinitely. But it cannot if the bush-meat trade, trade in primates for pets, trade in primates for biomedical experiments and deforestation do not stop. Primatologists have said that without public support and more professional input the great apes and many other monkey and prosimian species have no more than fifty years left in the wild.

Emotionally, intuitively and intellectually primates speak a language very like ours. Over a hundred years ago Darwin controversially wrote, 'The difference in mind between man and the higher animals, great as it is, certainly is one of degree and not of kind.' From Jane Goodall's inspired beginnings, the work of women primatologists continues to reveal that there really is only a degree between 'them' and 'us'. It is for this reason that women primatologists refuse to turn their backs on the dwindling numbers of non-human primates. With urgency Jane Goodall says, 'This is now pay-back time!'

Epilogue

War

Jo Thompson was forced to leave her bonobos in the Congo when the civil war between troops fighting for the late President Mobutu and the army of the late Laurent Kabila surrounded her field site. In March 1998 Mobutu's army of trained and ruthless killers passed through the outskirts of Yasa Thompson's Lukuru wildlife research project. They wiped out the main mission station on the Sankuru River and left a trail of pillage and destruction behind them. It was Mobutu's men who kidnapped and raped Dian Fossey back in 1967. These soldiers are rumoured to have eaten white people alive during the Congolese war, in which white mercenaries also committed atrocities to put Mobutu in power, before they turned against him. It is this same political unrest that today prevents Liz Williamson from returning full time to her mountain gorilla research at Karisoke. These soldiers are fighting Kabila's men; it was Kabila's soldiers who attempted to kidnap Jane Goodall and succeeded in kidnapping Barbara Smuts from Gombe in 1975. As Thompson is the only white Westerner in the region, her camp Yasa is not a well-known research centre like Gombe. Thompson has not had any students; she travels and works alone and feels safer that way. If Yasa had been full of wide-eyed, enthusiastic Western Ph.D. students, as Gombe was in 1975, Mobutu's men may well have given Thompson a visit.

In 1999 there was a partial cessation of hostilities and all parties involved in the war have signed the Peace Accord, but there are serious problems with implementation. It appears that both government and rebel groups are using this artificial lull to amass troops, weapons, and supplies. Allegations of wrongdoing and infraction are flying from all sides; no party seems sincerely to intend adhering to the cease-fire.

Jo Thompson is back home with her husband in Georgetown, Ohio, her bags permanently packed in anticipation of her return to the bonobos. In the Lukuru, many local people have been murdered and several of Thompson's workers have suffered beatings and threats against their lives. Some chose to flee the area and escape with their families. The rebel troops bring with them large numbers of migrant people, a taste for bush-meat, arms and ammunition, and a complete disregard for local traditions and laws about wildlife. No bonobo field researcher has been able to return to their site in over a year. When Thompson returns she will face a threat from the new migrant human population who do not know her or value her wildlife work. This thickly forested area of Central Africa is joint home to the continent's three great apes and to one of the bloodiest and widespread of human tribal wars. Both Liz Williamson and Jo Thompson intend to be back at work as soon as they can.

Disease
Neither the threat of war nor fatal disease is enough to permanently frighten women primatologists away.

It is predicted that, by the year 2000, the human AIDS pandemic will claim 110 million HIV-infected people. The majority of those people will be women. HIV is lethally concentrated in blood and sperm, and through sex the virus is more easily absorbed into a woman's body than into a man's. For many years scientists have suspected that AIDS originated in Africa and that chimpanzees were the reservoir of the disease. It has now been proved that some individual chimpanzees belonging to the sub-species of Central African chimpanzee *(Pan troglodytes troglodytes)* found living in Nigeria, Cameroon, Guinea, Gabon, Western Zaire and the Congo, carry the virus SIVcpz which in humans is known as HIV-1. When carried by a chimp the disease is not lethal to the animal, but if cross-species transmission occurs and SIVcpz skips directly to a human host, which apparently it can, HIV-1 can evolve into the deadly disease known as AIDS.

Virologist Beatrice Hahn from the University of Alabama believes the disease to be ancient and that there have been at least three separate waves of infection from chimpanzees to humans, the most recent and horrific cross-species transmission happening within the last fifty years. The bush-meat trade is inexorably linked to the spread of AIDS. Hahn is convinced this process of infection is still continuing through bush-meat hunters' exposure to chimpanzee blood. The crude dismemberment of the chimpanzees by the hunters could easily lead to the blood of an animal infected with SIVcpz finding its way into a human host, through a cut on the hunter's hand, for example. It has been noticed that villagers who live close to the forests and hunt regularly for chimp, bonobo or gorilla meat are often struck down with HIV; whole families have died after eating the meat of an ape.

AIDS may be thousands of years old but, due to the rapid growth in international travel, the latest wave of HIV keeps spreading. The AIDS virus is one of planet Earth's most successful species. The rampant spread of HIV through humans could be interpreted as the chimpanzee's last defence against man. Ultimately it seems the bush-meat trade is responsible for the deaths of millions of

HIV-infected people. The extermination of these animals is genocide, and in eating these animals it now seems we are also
committing suicide.

It was purely by chance that Hahn made the connection between
chimpanzees and AIDS. Marilyn was a twenty-six-year-old wild-
caught chimpanzee. She was taken when young from Africa to an
American laboratory (her exact geographical origins are unknown)
for use as a breeding animal. Although many of her infants were
intended for biomedical research, Marilyn was never artificially
infected with HIV. In 1985 she was discovered dead in her laboratory enclosure, having died while giving birth to still-born twins.
Her body was dissected and various organs were preserved. Some
time later Hahn was examining Marilyn's spleen and discovered
that she had been carrying the SIVcpz virus. The chimpanzee had
contracted the disease in the wild before being captured. This was
a vital clue to the origins of AIDS. Since then other wild-caught laboratory chimps have been discovered to carry SIVcpz.

Hahn believes it is a human imperative to conserve wild chimpanzee populations at all costs, allowing field primatologists to
study the animals in their natural habitat. For it is naturally infected
wild chimps that hold the clue to helping humans survive AIDS,
not laboratory chimps, as they do not fully develop HIV or AIDS
when artificially infected with the virus. If wild chimps are exterminated and their habitat destroyed, the chances of finding an HIV
vaccine for people will vanish too.

I asked some of the women primatologists I interviewed if this
discovery would dissuade them from working with chimpanzees
and I was emphatically told no, it would not. Stella Brewer, who
works in Gambia, says she has more chance of catching the AIDS
virus from another human than from a chimpanzee. Brewer
believes the chances of contracting rabies from the bite of a wild
chimp to be a greater threat to primatologists than the transmission
of the SIVcpz virus.

Birth

Panbanisha, the female bonobo that Sue Savage-Rumbaugh teaches language to, has now become a mother. Panbanisha's son, Nyota, was born in April 1998. Savage-Rumbaugh has decided that she wants to be Nyota's second, human mother and from his birth she has been in constant physical contact with the infant. While Panbanisha breastfeeds him, Savage-Rumbaugh talks verbally to the infant and uses a symbols board to expose him to English from the start. Savage-Rumbaugh no longer goes home at nights; she has moved into the research centre and shares a room with Panbanisha and Nyota, who are in acceptance of this extra full-time maternity care, or 'special enrichment', as Savage-Rumbaugh refers to it. Savage-Rumbaugh has high hopes for this second-generation language ape.

Wherever they are the work goes on

Stella Brewer received at an official ceremony in November 1999 the Golden Ark Award, presented by Prince Bernhard of Holland, for her chimpanzee rehabilitation and adoption scheme.

Janis Carter is now assisted in Africa by Esthel Raballand Ward, recently arrived from France. The loss of EU funding remains an enormous setback to the project..

Lisa Gadsby's Drill Rehabilitation and Breeding Centre (DRBC) in Nigeria now takes care of 72 drills. These animals were recovered from markets and bush-meat hunters. There have been 33 births within this captive group and Gadsby intends to release the drills that have formed natural groups once she can assure their safety. Since Gadsby's conservation project put a stop to hunting on the Afi Mountain there have been only two drills and one gorilla shot. The Afi Mountain, which is too steep for commercial logging, has become a sanctuary for drills, gorilla, chimpanzees and guenon monkeys. Gadsby's local education programme has been a major success and bush-meat hunting on the mountain is locally now taboo. The DRBC, which now employs 40 local people, has recently established a tree nursery.

Biruté Galdikas, in collaboration with the Indonesian Ministry of Forestry, has recently established a care clinic for orphaned orang-utans. The veterinary and general welfare clinic is located just outside the Tanjung Puting National Park and has a capacity to care for 20–30 orphans at any one time. Galdikas continues to divide her time between Borneo and Canada, where she lectures in anthropology at the Simon Fraser University, and Los Angeles, which is the location of the Orang-utan Foundation headquarters. Galdikas is President of the foundation which campaigns to save orang-utan habitat, and her primate lectures are in demand all over the world.

Jane Goodall is now a proud grandmother. Goodall's only son, Grub (Hugo Eric Louis Van Lawick), now in his early thirties, lives in Dar-es-Salaam. Grub decided not to follow in either of his parents' footsteps. Growing up on the shores of Lake Tanganyika gave Grub a love of fishing, not a love of chimps. Today he is an Indian Ocean deep-sea fisherman. Using cows' blood to lure them in Grub catches, on a single line, man-eating shark and also marlin. Grub says after his years in the cage at Gombe he grew to hate the displaying chimps that looked through the bars at him. He hasn't visited Gombe for thirteen years. His father, Baron Hugo Van Lawick, is now retired from wildlife filmmaking, he lives in his son's house with his son's Tanzanian ex-wife and his grandchildren, Angel and Merlin. Grub, meanwhile, lives in the house opposite with his second wife and new baby, in true patriarchal African style.

The Jane Goodall Institute, (JGI) continues to run four African chimpanzee sanctuaries that give protection to a total of 100

rescued chimpanzees. Goodall also supervises the Jane Goodall
Centre for Excellence in Environmental Studies. Based at Western
Connecticut State University the centre is dedicated to wildlife
research, education and conservation. Goodall also speaks to
schools about, Roots & Shoots, JGI's environmental education and
humanitarian program for young people. She also runs the
ChimpanZoo Program, which involves the study of 130 chimpanzees at twelve zoos in the United States. Zoo personnel, students
and JGI volunteers gather behavioural data for comparative studies on captive chimpanzees. Scientific papers are presented and
discussed at regular meetings. On top of all that Goodall continues
with her public awareness lecture tours and research at Gombe.
After forty years nothing could now deter her from this. On 14
July 2000 it will be forty years since Jane Goodall first stepped into
the Gombe Reserve, and she intends to celebrate this anniversary at
Gombe with old human and chimpanzee friends. Annually Goodall
embarks on months of globe-trotting to spread the chimpanzee
gospel.

Endnotes

Chapter 2 Leakey's Ladies

(1) page 33 line 1.

The Piltdown Man fossil was 'found' by Charles Dawson in 1912 at Piltdown, Sussex, England. The fossil was believed to be a skull from the earliest known ancestor of modern man. It was in fact a hoax. In 1953 it was proved that the lower jaw was that of an orang-utan and the upper cranium, although human, was neither prehistoric nor from Piltdown. Dawson's hoax remained unchallenged for many years because it was automatically assumed our ancestors should have a large cavity in their skull for a large brain. As a young man Louis Leakey had not been convinced by the Piltdown Man skull; he had written that it seemed the skullcap was human but the jawbone was from an ape. But Leakey had not guessed it was a forgery. We now know our brains grew rapidly in size from 1 million to as recently as 100,000 years ago.

(2) page 44 line 12.

The ant bear (also known as the aardvark) is a large, shy nocturnal rodent, found in Central and Southern Africa, where it lives on termites and other insects that it gathers up on its long, prehensile and sticky tongue.

(3) page 44 line 12.
Ichthyology is the biological study of fishes

Chapter 3 The Secret Life of the Female Baboon
(1) page 82 line 22.
An infant primate immediately inherits its mother's social rank; later, as it matures, its social rank will depend on its gender and whether the primate belongs to a matriarchal or patriarchal species. In matriarchal species a daughter will inherit her mother's rank and sometimes her mother's home range. In a patriarchal species the status of the infant's father will affect the individual's final social position. But the individual primate's own personality will also affect its social rank in adulthood.

Chapter 6 The Woman Who Loved Apes Too Much
(1) page 186 line 26.
Trying to coax an established gorilla group to adopt an infant is hard, but Dian Fossey did have some success at this. Bonne Année was nearly three years old when she was given to Fossey to rehabilitate. The young gorilla was recovered from poachers in January 1980 but her parentage was unknown. Fossey tried to introduce the infant to Group 5, and Beethoven, the silverback, was not aggressive to the youngster; in fact he stopped other members of his group from attacking her. But young adult males and females were hostile to Bonne Année, and fearing for her safety Fossey retrieved her from the group. Three weeks later she decided to introduce the tiny female to Group 4, which was at the time very unusual. It was an all-male group consisting of three silverbacks, three blackbacks and one immature male, none of which were closely related. When Fossey released Bonne Année, she confidently walked into the heart of the group and the males readily accepted her; there was no aggression. The young female remained an active member of the group until she died of pneumonia a year later. Male gorillas will kill

suckling infants that are not theirs to artificially bring the mother into oestrus so they can mate with her. But at three years old Bonne Année was becoming a juvenile, and without the presence of her fertile mother the male gorillas felt no hostility towards her. The adult males would have recognised Bonne Année as a prospective future mate and in this situation the adult males were protective of her.

(2) page 214 line 23.
American citizen Rosamond Carr runs an orphanage in Gisenyi, Rwanda. She is now in her 80s, and has almost 100 children in her care. She is an institution in Rwanda and is today kept informed on the progress of the gorillas by Liz Williamson.

Chapter 7 Aping For The Audience
(1) page 234 line 28.
On 9 April 1999 Mary Chipperfield and husband Robert Cawley were both convicted of cruelty to their three-year-old chimpanzee, Trudy, Lucy's daughter, and to a juvenile elephant. Trudy is now safely housed at the animal sanctuary, Monkey World in Dorset and the elephant now resides at Dudley Zoo. The Chipperfields were ordered to pay £8,500 in fines and nearly £12,000 in costs. The Animal Defenders welfare group run by Jan Creamer brought the conviction after a year-long under-cover investigation. The Animal Defenders' spokeswoman told me they are not sure of Lucy's whereabouts now. According to The Animal Defenders it is possible that the Chipperfields, who were not available for comment, have sold her to Cairo zoo.

Chapter 9 The Battle of the Sexes (and the uses of the female orgasm)
(1) page 295 line 9.
Lucy is often mistakenly thought of as the first human mother, but she should be considered more of a distant great aunt; immortality may have been achieved by one woman who possibly spawned all of us. Geneticists believe this archaic woman lived one hundred and thirty-five

thousand years ago in Africa and her descendants spread through Africa and northwards into Europe. She's known as Eve and her story is the *Mitochondrial Eve Hypothesis*. The function of the mitochondria organelles in a cell is to control energy absorption; the mitochondria contain many copies of DNA. This DNA mutates over time at a much faster rate than the DNA within the cell's nucleus. The nuclear DNA comes from both parents, the mitochondrial DNA comes only from the mother. Over the past few years more than three thousand people of all nationalities have been tested. It was found there was less than ten per cent difference between the mitochondria of modern people and that shared by chimpanzees, and within that ten per cent African people showed the greatest variation. This tells us that genetically skin colour is a superficial difference between people; it also suggests that the origins of modern humans are very recent and located in Africa. By tracing the evolution of the various mutations in the mitochondria's DNA a family tree one hundred and thirty-five thousand years old can be traced back to the existence of one woman.

She was not the first woman and she was not the only woman. But, out of ninety-three tribes consisting of approximately 10,000 people who left Africa to colonise the rest of the world, reaching Australia 50,000 years ago and Western Europe 40,000 years ago (the period that Neanderthals disappeared), it is only her DNA that has lasted.

Mitochondrial DNA is passed from mothers to daughters and so on. Women who only have sons or who have no children or no grandchildren do not pass the mitochondrial DNA on to the next generation. Of course other women lived 135,000 years ago but we cannot trace a lineage back to them because in one of the subsequent generations only sons were born. To put her into a linguistic context, Eve's ancestors had been talking, no doubt discussing a possible trip to Europe, for at least 250,000 years before she was born.

(2) page 295 line 23.

A German husband and wife team, Hohmann Goddfried and Barbara Fruth, have studied bonobos in the Lomako forests, a few hundred miles north of Jo Thompson's site. There they have seen female bonobos eating meat. Duikers often take naps during the day in well-secluded areas. The small antelope do not think of bonobos as predators and the bonobos know where the duikers sleep. The female bonobo creeps up on a duiker; Fruth and Goddfried have heard the duiker call as it is killed. Often it is the duiker's cry that alerts the scientists to the fact that a subtle hunt has just taken place. Because the deer is preyed upon in this way it does not pass on the fear of bonobos to other duikers. This understated manner of hunting is very different to male chimpanzee hunting. Male chimps may exhibit stealth at first but once the prey is in sight the organised hunt quickly becomes loud and frenzied as the group pursue their quarry and members of the prey species run for their lives.

The high-status female bonobos who sneak up on and kill duikers share the meat with their female friends. Male bonobos hang around begging for a share but the scientists have never seen a female give a piece of meat to a male. After a feast like this all that is left are the hooves. It is speculated that this sharing of meat, which makes up less than 10% of the female's diet, is a ritual to re-establish matriarchal bonds.

(3) page 297 line 16.
Kristen Hawkes, an anthropologist at the University of Utah, does not agree that the human heritage is a strictly patriarchal one. After studying the Hadza hunter-gatherer people of Africa's Rift Valley and non-human primates, Hawkes has found much variation and flexibility in the behaviour of males and females. In the Hadza after marriage sometimes the husband will move in with his wife's family. It is the Hadza women's foraging that keeps the children fed and alive (not the efforts of hunting men). In chimpanzees some high-ranking females do not migrate when reaching sexual maturity (like Fifi, who inherited Flo's range) but stay within their natal group. As more wild apes are studied, patterns of sex-biased dispersal are becoming more variable and thus Hawkes questions Foley and Lee's assumptions.

Chapter 10
(1) Page 347 line 1.

Sociobiology's Sexy Son hypothesis helps to explain why female primates often show an overriding sexual interest in a new male who has appeared on the horizon. Even if the female lives in a patriarchal society she will still seek out a stranger for mating opportunities. The theory states that if a female finds a new unknown male sexually attractive she can be sure other females will too. That being the case, if she has a son fathered by this sexy stranger, her son will grow up to become a sexy stranger himself. When her sexy son matures he will mate with a series of females as he travels from place to place, thus passing his mother's genes on to all the infants he fathers. If measured in numbers of progeny the sexy stranger is reproductively successful, but as he does not stay around long enough to care for the infants the long-term reproductive success is not so obvious from this reproductive strategy.

Select Bibliography

Internet Sites:

The Great Ape Project (GAP):-
http://www.enviroweb.org/gap/international/gapglobal.html
The Primate Society of Great Britain:-
http://www.ana.ed.ac.uk/PSGB
American Society of Primatologists – home page:-
http://www.asp.org/
Primate Info Net of the Wisconsin Primate Centre:-
http://www.primate.wisc.edu/
World Directory of Primatologists:-
http://www.primate.wisc.edu/pin/wdp.html
The International Primatological Society
http://www.primate.wisc.edu/pin/ips.html

Books, Journals and Academic Papers:

Baker, R., *Sperm Wars: Infidelity, Sexual Conflict and other Bedroom Battles*, London, Fourth Estate (1996).
Blum, D., *The Monkey Wars*, Oxford University Press (1994).

Brewer (Marsden), S., 'A Background to the Chimpanzee Rehabilitation Project in Gambia, West Africa' (May 1998).

Brewer, S., (Goodall, J., 'Preface'), *The Forest Dwellers*, Glasgow, William Collins Sons & Co. Ltd. (1978).

Carter, J., 'Freed from keepers and cages, chimps come of age on Baboon Island', *Smithsonian*, vol. 19, no. 3 (1988).

——, 'A Journey to Freedom', *Smithsonian*, vol. 12, no. 1 (1981).

Cheney, D. L., 'Function and intention in the calls of nonhuman primates', *Proceedings of the British Academy, Evolution of Behaviour Patterns in Primates and Man* (1995).

Cosmides, L. M., 'Evolutionary psychology, reasoning instincts and culture', *Proceedings of the British Academy, Evolution of Behaviour Patterns in Primates and Man* (1995).

Cosmides, L. M., Barkow, J.H., and Tooby, J., (eds.), *The Adapted Mind*, Oxford, Oxford University Press (1993).

Daly, M., and Wilson, M. I., 'Violence Against Stepchildren', *Current Directions in Psychological Science*, 5 (1996), 77–81.

——, 'Some Different Attributes of Lethal Assaults on Small Children by Stepfathers versus Genetic Fathers', *Ethology and Sociobiology*, 15 (1994), 207–17.

——, 'Evolutionary Social Psychology and Family Homicide', *Science*, vol. 242, pp. 519–24.

Darwin, C., *The Descent of Man, and Selection in Relation to Sex*, London, John Murray (1871).

——, *Origin of the Species by Means of Natural Selection*, London, John Murray (1859).

Dawkins, R., *The Selfish Gene*, Oxford University Press (1976).

Dunbar, R.I.M., *Grooming, Gossip and the Evolution of Language*, London, Faber and Faber (1996).

Dunbar, R.I.M., and Casperd, J.M., 'Modelling the Evolution of Reconciliation within the Primate Order', University of Liverpool, UK (1998).

De Waal, F., *Good Natured*, Cambridge MA, Harvard University Press (1996).

De Waal, F., and Lanting, *Bonobo: The Forgotten Ape*, University of California Press (1997).

Diamond, J. M., *Why Is Sex Fun?*, London, Weidenfeld and Nicolson (1997).

——, *The Rise and Fall of the Third Chimpanzee*, London, Vintage (1992).

Foley, R., 'An evolutionary and chronological framework for human social behaviour'. *Proceedings of the British Academy, Patterns of Social Evolution in Humans and Mammals* (April 1995).

——, 'Male kin bonded groups, expensive offspring and the paradox of hominid social evolution', *Proceedings of the British Academy, Evolution of Social Behaviour Patterns in Primates and Man* (1995).

Foley, R., and Lee, P.C., 'Finite Social Space, Evolutionary Pathways, and Reconstructing Hominid Behaviour', *Science*, vol. 243 (17 February 1998), pp. 901–6.

Fossey, D., *Gorillas in the Mist*, Boston, Houghton Mifflin (1983).

——, 'His Name Was Digit', *International Primate Protection League Commemorative Newsletter* 13 (1), pp. 10–15 (article originally published by IPPL, August 1978).

——, *Infanticide Comparative and Evolutionary Perspectives*, Glenn.

——, and Hrdy, S.B., (eds.), '11. Infanticide in mountain gorillas *(Gorilla gorilla beringei)* with comparative notes on chimpanzees', New York, Aldine (1984).

——, *The behaviour of the mountain gorilla*, University of Cambridge Ph.D. thesis (1976).

——, 'More Years with the Mountain Gorillas', *National Geographic*, 140 (1971), pp. 547–85.

——, 'Making Friends with Mountain Gorillas', *National Geographic*, 137 (1970), pp. 48–68.

Fouts, R., *Next of Kin: What my conversations with chimpanzees have taught me about intelligence, compassion and being human*, London, Penguin Putnam (1997).

Galdikas, B.M.F., *Reflections of Eden: My Life with the Orangutans of Borneo*, London, Victor Gollancz (1995).

——, 'Orangutans, Indonesia's "People of the Forest"', *National Geographic*, 148 (4) (October 1975) pp. 444–73.

Galdikas, B.M.F., and Crawford, C., *Infrahuman Rape in Sociobiological Perspectives*, International Society for Research on Aggression at Victoria B.C. (July 1983).

——, 'Rape in non-human animals: an evolutionary perspective', *Canadian Psychology*, vol. 27, no. 3 (1986), pp. 215–30.

Galdikas, B.M.F., and Shapiro, G., 'Attentiveness in Orangutans within the Sign Learning Context', International Orangutan Conference 5–7 March 1994, sponsored by California State University (1994).

Goodall, J., *Reason for Hope: A Spiritual Journey*, London, HarperCollins (1999).

——, *Through a Window: Thirty Years with the Chimpanzees of Gombe*, London, Penguin Books (1991).

——, *The Chimpanzees of Gombe: Patterns of Behaviour*, Cambridge MA, Harvard University Press (1986).

——, *In the Shadow of Man*, Glasgow, William Collins Sons & Co, Ltd. (1971).

Goodall, J., and Nichols, M., *Brutal Kinship*, London, Aperture (1999).

Goodall, J., Pusey, A., and Williams, J., 'The Influence of Dominance Rank on the Reproductive Success of Female Chimpanzees', *Science*, vol. 277 (8 August 1997), pp. 733–872.

Goodall, J., Zihlman, A.L., and Morbeck, M.E., 'Skeletal biology and individual life history of Gombe chimpanzees', *Journal of Zoology*, London 221 (1990), pp. 37–61.

Haraway, D., *Primate Visions: Gender, Race and Nature in the World of Modern Science*, New York, Routledge (1989).

Harrisson, B., *Orang-utan*, Oxford University Press (1987).

Hayes, H., *The Dark Romance of Dian Fossey*, London, Chatto and Windus (1990).

Hinde, R.A., Supplement to *Primate Eye*, no. 11 (January 1979).

Hrdy, S.B., *Mother Nature: A History of Mothers, Infants and Natural Selection*, London, Chatto & Windus (1999).

——, 'Raising Darwin's Consciousness, Female Sexuality and the

Prehominid Origins of Patriarchy', *Human Nature*, vol. 8, no. 1 (1997), pp. 1–49.

——, 'Mothers' Nature', *Natural History* (December 1995).

——, 'Infanticide: Let's Not Throw Out the Baby with the Bath Water', *Evolutionary Anthropology*, vol. 3, no. 5 (1994/5), pp. 149–86.

——, 'Sex Bias in Nature and in History: A late 1980s Reexamination of the "Biological Origins" Argument', *Yearbook of Physical Anthropology*, 33:25–37, Wiley-Liss, Inc. (1990).

——, no. 5 'The Primate Origins of Human Sexuality', and no. 7 'Raising Darwin's Consciousness: Females and Evolutionary Theory', *The Evolution of Sex, Nobel Conference XXIII*, (eds. Belling, R., and Stevens, G.), San Francisco, Harper and Row (1998).

——, *The Woman That Never Evolved*, Cambridge MA, Harvard University Press (1983).

——, Infanticide Among Animals', *Ethology and Sociobiology*, 1: 13–40.

——, *The langurs of Abu: female and male strategies of reproduction*, Cambridge MA, Harvard University Press (1977).

Hrdy, S.B., and Williams, G.C., *Social Behaviour of Female Vertebrates*, (ed. Wasser, S.K.), New York, Academy Press (1983).

International Primate Protection League Commemorative Newsletter, vol. 13, no. 1 (April 1986).

International Primate Protection League Newsletter, December 1988

Jolly, A., *Lucy's Legacy: Sex and Intelligence in Human Evolution*, Cambridge MA, Harvard University Press (1999).

Leakey, L.S.B., *By the Evidence: Memoirs*, New York, Harcourt Brace Jovanovitch (1974).

——, *Adam or Ape: A Sourcebook of Discoveries about Early Man*, Cambridge MA, Shenkman (1970).

Leakey, R.E., and Lewin, R., *The Sixth Extinction: Patterns of Life and the Future of Human Kind*, London, Anchor Books (1996).

Lee, P.C., *Modelling the Early Human Mind*, MacDonald Institute Monographs, (eds. Mellars, P., and Gibson, K.) Cambridge University Press (1996).

——, 'Biology and Behaviour in Human Evolution', *Cambridge Archaeological Journal*, 1:2 (1991), pp. 207–26.

Lenain, T., *Monkey Painting*, London, Reaktion Books (1997).

Lewin,R., *Human Evolution*, (3rd ed.), Cambridge, Blackwell Scientific Publications (1993).

McGuire, W., 'I didn't kill Dian Fossey', *Sunday Times*, 5 April 1987.

Morell, V., *Ancestral Passions: The Leakey Family and the Quest for Humankind's Beginnings*, London, Simon and Schuster (1995).

Mowat, F., *Woman in the Mists*, London, Macdonald (1989).

National Geographic Magazine, vol. 181, no. 3 (March 1992).

Parish, A.R., 'Female Relationships in Bonobos (*Pan paniscus*). Evidence for Bonding, Cooperation, and Female Dominance in a Male Philopatric Species', *Human Nature*, vol. 7, no. 1 (1996) pp. 61–96.

Redmond, I., 'The Death of Digit', International Primate Protection League Newsletter, 1988.

Ridley, M., *The Red Queen*, London, Penguin (1994).

Savage-Rumbaugh, S., 'Why Are We Afraid of Apes with Language?', Atlanta GA, Georgia State University (1996).

Savage-Rumbaugh, S., and Lewin, R., *Kanzi: The Ape at the Brink of the Human Mind*, London, Doubleday (1994).

Singer, P., *The Great Ape Project: Equality Beyond Humanity*, (Preface, Cavalierie, P., ed.), London, Fourth Estate (1993).

Smuts, B., 'Barbara Smuts', *Journeys of Women in Science and Engineering: No Universal Constants* (Ambrose, Dunkle, Lazarus, Nair, Harkus, eds.), Philadelphia PA, Temple University Press (1997), pp. 365–9.

——, 'Apes of Wrath', Commentary, *Discover Magazine* (August 1995).

——, 'American Scientist Interviews', *American Scientist*, vol. 76 (Sep.–Oct. 1988), pp. 494–9.

Smuts, B., Cheney, D., Seyfarth, R., Wrangham, R., and Struhsaker, T., (eds.), *Primate Societies*, Chicago, The University of Chicago Press (1987).

Strier, K. B., 'Behavioural Ecology and Conservation Biology of Primates and Other Animals', *Advances in the Study of Behaviour*, vol. 26, Academic Press (1997).

Temerlin, M. K., 'My daughter Lucy', *Psychology Today* (November 1975).

Tutin, C., et al, 'Protecting seeds from primates: examples from Diospyros spp. in the Lopé reserve, Gabon', *Journal of Tropical Ecology*, vol. 12, pt. 3 (1996).

Whiten, A., 'Imitation of the Sequential Structure of Actions by Chimpanzees *(Pan troglodytes)*', *Journal of Comparative Psychology* (1998).

——, *Machiavellian Intelligence II: Extensions and Evaluations*, Cambridge University Press (1998).

Williamson, S., and Nowak, R., 'The Truth About Women', *New Scientist*, no. 2145 (1 August 1998).

Wilson, E. O., *Consilience: The Unity of Knowledge*, New York, Alfred A. Knopf (1998).

Wrangham, R., and Peterson, D., *Demonic Males: Apes and the Origins of Human Violence*, London, Bloomsbury (1996).

Zihlman, A. L., 'The Paleolithic Glass Ceiling: Women in Human Evolution', *Women and Human Origins*, (ed. Lori Hager), Routledge (January 1995).

Zihlman, A. L., Morebeck, M. E., and Galloway, A., (eds.), *The Evolving Female: A Life-history Perspective,* Princeton University Press (1997).

Picture credits:
Reproduced by kind permission of:

1. Jo Thompson
2. The British Museum
3. Shirley McGrill & the IPPL
4. Bob Campbell
5. Jane Goodall Institute
6. Joan Travis
7. Kelly Stewart
8. Joan Travis
9. Rod Brindamour
10. Rod Brindamour
11. Anthrophoto File
12. Umeyo Mori
13. Umeyo Mori
14. Mariko Hiraiwa-Hasegawa
15. Stella Brewer-Marsden
16. Stella Brewer-Marsden
17. Janis Carter
18. Anthrophoto File

19. Ian Redmond
20. Joan Travis
21. Liz Williamson
22. Kobal Collection
23. Kobal Collection-
24. Kobal Collection
25. Kobal Collection
26. Peter Blake
27. Ailsa Berk
28. Sally Boysen
29. Amy Parish
30. Magnum Photos
31. Anthrophoto File
32. Stella Brewer-Marsden
33. Sarah Hrdy
34. Joan Travis
35. Debbie Martyr
36. Orangutan Foundation International
37. Peter Jenkins
38. Dan Westfall
39. Amy Parish
40. Amy Parish
41. Amy Parish
42. Adrienne Zihlman
43.
44. Bob Campbell
45. Jane Goodall Institute

Useful Addresses

The Animal Defenders internet site:-
http://www.cygnet.co.uk/navs/ad/news/index.html

Ape Alliance internet site:-
http://www.4apes.com

The Bonobo Protection Fund – USA
Georgia State University
Georgia State University Plaza
Atlanta, GA 30303
http://www.gsu.edu/~wwwbpf/bpf/

Chimpanzee Rehabilitation Project – Gambia
Janis Carter
Dept. of Parks and Wildlife Management
Ministry of Natural Resources
Banjul
The Gambia

Chimpanzee Rehabilitation Project – Guinea
c/o Stella Brewer-Marsden & Janis Carter
53 Gretton Road
Gotherington
Cheltenham
Gloucestershire
GL5 24QU
UK
Stella@jdmar.demon.co.uk
jcarter@qanet.gm

The Drill Rehabilitation and Breeding Centre
    and Afi Mountain Conservation Programme
Pandrillus Housing Estate
P.O. Box 826
Calabar, Nigeria
Tel. 234 (087) 234310
drill@infoweb.abs.net

Dian Fossey Gorilla Fund – Europe
110 Gloucester Avenue
London NW1 8JA
Tel. 0207 485 2681

Dian Fossey Gorilla Fund – International
800 Cherokee Avenue South East
Atlanta, GA 30315
USA
dfgf@mindspring.com

Fauna & Flora International
Great Eastern House
Tenison Road
Cambridge CB1 2DT
UK
Tel. 01223 571000,
Fax. 01223 461481
info@fauna-flora.org
http://www.wcmc.org.uk/ffi/members.html

The Jane Goodall Institute – UK
15 Clarendon Park
Lymington
Hants. SO14 8AX
Tel. 0159 067 1188

The Jane Goodall Institute – USA
Kim Stryker
Volunteer/Internship Coordinator
P.O. Box 14890
Silver Springs, MD 20911–4890
Fax: (301) 565–3188
kstryker@janegoodall.org
http://www.janegoodall.org/how/how_member_mem.html

The Great Ape Project internet site:-
http://www.enviroweb.org/gap/international/gapglobal.html

International Primate Protection League – UK
116 Judd Street
London WCIH 9NS
Tel. 0207 833 0661
clovenhoof@easynet.co.uk
http://www.ippl.org/

International Primate Protection League – USA
POB 766
Summerville, SC 29484
Tel. (843) 871–2280
ippl@awod.com

The Leakey Foundation
1002A O'Reilly Avenue
San Francisco, CA 94129
Tel. (415) 561–4646
Fax. (415) 561–4647
info@leakeyfoundation.org
http://www.leakeyfoundation.org/

Lukuru Wildlife Research Project
Dr Jo Thompson, Director
c/o P.O. Box 5064
Snowmass Village, CO 81615–5064
USA
http://members.aol.com/jat434/index.html

Monkey World – Ape Rescue Centre – UK
Longthorns
East Stoke
Wareham
Dorset BH20 6HH
Tel. 01929 462537

The Orangutan Foundation – International
822 S. Wellesley Avenue
Los Angeles CA 90049
USA
redape@ns.net

The Orangutan Foundation – UK
7 Kent Terrace
London NW1 4RP
Tel. 0207 724 2912

Index

Also from Virago

THE AGE OF ANXIETY
Sarah Dunant and Roy Porter (ed)

'Fighting talk' – *Observer*

As we come to the end of the 20th century, the idea that the future equals progress has sustained a philosophic body blow from which it will not easily recover. As technology expands our choices, it also serves to diminish our sense of control. Mix that with the insecurity of global economics, the threat of family breakdown, increased fear of crime and violence and you have a population which finds it difficult to handle today, let alone contemplate tomorrow. But how justifiable is our anxiety? And how do we stop ourselves being frightened of the millennium? Sarah Dunant and Roy Porter ask 10 prominent thinkers and writers to assess the general malaise that is settling over us. Not all the essayists are optimistic about life beyond the millennium, but they all think the world is a place worth fighting for.

Thought-provoking yet accessible, *The Age of Anxiety* is an important book for our times.

YOU JUST DON'T UNDERSTAND
Women and Men in Conversation
Deborah Tannen

'Tannen combines a novelist's ear for the way people speak with a rare power of original analysis . . . fascinating' – Oliver Sacks

Why do so many women feel that men don't tell them anything, but just lecture and criticise? Why do so many men feel that women nag them and never get to the point? In this pioneering book Deborah Tannen shows us how women and men talk in different ways, for profoundly different reasons. While women use language to make connections and reinforce intimacy, men use it to preserve their status and independence.

WHEN SHE WAS BAD

Patricia Pearson

'Effectively cremates the myth of innate female innocence by parading a female chain gang of gals gone wrong . . . compelling' – *Vanity Fair*

Our culture believes that women are not naturally aggressive. And yet, every day evidence proves otherwise: women kill their children, their husbands, their lovers, and their lovers' mistresses. Women join their lovers in torture and killings, women are psychopaths, women are terrorists and violent criminals. In this highly provocative book, Patricia Pearson demonstrates over and over again that the idea (ideal?) of female innocence is pure myth. She argues that the two main culprits of the tendency to overlook extreme behaviour in women are feminists who have claimed victimhood for women and male society which finds it impossible to see women as powerful.

Weaving the stories of violent women – from Myra Hindley and Rose West, from a mother who smothered eight of her children, from nurses who murdered their infant charges, to husband beaters – with the findings of criminologists, anthropologists and psychiatrists, Patricia Pearson makes a compelling case for redefining the debate about female violence and power.

WOMAN: AN INTIMATE GEOGRAPHY
Natalie Angier

Woman explores the essence of what it is to be female. In mapping the inner woman from organs to orgasms – Natalie Angier presents an extraordinary new vision of the female body as an evolutionary masterpiece.

'Women have long been regarded as slaves to biology and evolution, pioneers in a hormonal swamp. But now, some of the sacred tenets of evolutionary psychology that men are innately more aggressive, more promiscuous and more likely to fall for cute young things – have come under fresh challenge . . . *Woman* is a delicious cocktail of estrogen and amphetamine designed to pump up the ovaries as well as the cerebral cortex' –
Time Magazine

'How did we ever get by without it? This book explains your life' –
Fay Weldon

'Allows us to be delighted with ourselves, and gives us a sense of our innate strength and capacity for permanent revolution. The most radical work of feminism for a decade' –
Sunday Express

'Anyone living in or near a female body should read this book' –
Gloria Steinem

Now you can order superb titles directly from Virago

☐ The Age of Anxiety   Sarah Dunant and Roy Porter   £7.99
☐ You Just Don't Understand   Deborah Tannen   £7.99
☐ When She Was Bad   Patricia Pearson   £6.99
☐ Woman   Natalie Angier   £8.99

————————————— 🍎 —————————————

Please allow for postage and packing: **Free UK delivery.**
Europe: add 25% of retail price; Rest of World: 45% of retail price.

To order any of the above or any other Virago titles, please call our credit card orderline or fill in this coupon and send/fax it to:

Virago, 250 Western Avenue, London, W3 6XZ, UK.
Fax 020 8324 5678   Telephone 020 8324 5516

☐ I enclose a UK bank cheque made payable to Virago for £ . . . .
☐ Please charge £ . . . . . . to my Access, Visa, Delta, Switch Card No.

Expiry Date ☐☐☐☐   Switch Issue No. ☐☐

NAME (Block letters please) . . . .

ADDRESS . . . .

Postcode . . . . Telephone . . . .

Signature . . . .

Please allow 28 days for delivery within the UK. Offer subject to price and availability.

Please do not send any further mailings from companies carefully selected by Virago ☐